# FORMAL METHODS AND MODELS FOR SYSTEM DESIGN

# Formal Methods and Models for System Design

## A System Level Perspective

Edited by

**Rajesh Gupta**
*University of California at San Diego*

**Paul Le Guernic**
*INRIA-IRISA*

**Sandeep Kumar Shukla**
*Virginia Tech*

and

**Jean-Pierre Talpin**
*INRIA-IRISA*

SPRINGER - SCIENCE+BUSINESS MEDIA, B.V.

A C.I.P. Catalogue record for this book is available from the Library of Congress.

ISBN 978-1-4419-5464-0 ISBN 978-1-4020-8052-4 (eBook)
DOI 10.1007/978-1-4020-8052-4

*Printed on acid-free paper*

# Contents

# Preface

Perhaps nothing characterizes the inherent heterogeneity in embedded systems than the ability to choose between hardware and software implementations of a given system function. Indeed, most embedded systems at their core represent a careful division and design of hardware and software parts of the system functionality. To do this task effectively, models and methods are necessary to capture application behavior, needs and system implementation constraints. Formal modeling can be valuable in addressing these tasks.

As with most engineering domains, co-design practice defines the state of the art, though it seeks to add new capabilities in system conceptualization, modeling, optimization and implementation. These advances - particularly those related to synthesis and verification tasks - directly depend upon formal understanding of system behavior and performance measures. Current practice in system modeling relies upon exploiting high-level programming frameworks, such as SystemC, Esterel, to capture design at increasingly higher levels of abstraction and attempts to reduce the system implementation task. While raising the abstraction levels for design and verification tasks, to be really useful, these approaches must also provide for reuse, adaptation of the existing intellectual property (IP) blocks.

Current industrial trends in use of general-purpose language facilities - driven by the popularity of C/C++/Java and other high-level languages – provide an interesting and somewhat urgent backdrop to explore corresponding advances in formal models to reason with these high-level descriptions, to help understand the design, implement formal correctness proofs and predict performances and other metrics. It is important to bridge the gap between theoretical models of computations and the practical experience in general purpose programming based design. Doing so will benefit both the languages (from the rigor of models) and models (from the experience of programming practice).

This contrasted picture led to debates and discussions about bringing theory and practice together in an informal mailing list in 2002 which included various industry experts, and academic researchers. These discussions gave rise to the ACM/IEEE Conference on Formal Models and Methods for Co-Design (MEMOCODE), first organized at Mt. Saint Michel, France, in June 2003. This conference was organized to bring together practitioners of formal methods in the hardware and software industry, and academic researchers to create dialogue on the issues of convergence of formal methods with practice, as well as to create constructive exchange of ideas and views of people applying formal methods in software and those applying it to hardware design. Through extensive on-line discussions , five areas were identified. These areas were

(i) Conformance checking between designs at various abstraction levels,

(ii) Hierarchical verification to facilitate verification of designs expressed hierarchically without flattening the hierarchy,

(iii) Abstraction vs. optimization, as they lower level optimizations affect abstraction hierarchy,

(iv) Incremental verification to harness the existing verification results for effectively verify incrementally changed design, and

(v) Post-production patchability of designs and capturing such requirements in formal models.

These are by no means the only problems that confront us, but is a good sampling of areas that the members of the MEMOCODE community felt as crucial problems.

The idea of this edited volume arose through these discussions and propelled by two very interesting panel discussions at the conference that pointed out the need for documenting challenges and advances focused on high-level modeling and validation. Our goal in putting this volume is to provide a serious technical discussion of the advances in formal modeling and methodologies for embedded systems. The chapters of this book were invited from the attendees of the first MEMOCODE conference, complemented by articles written by the editors representing their own views and contributions to this problem area.

As mentioned already, the topics addressed in this book are chosen in order to encompass possible areas of formal methods that are believed to be useful in making the general purpose programming based methodology more successful and efficient. For organization purposes, we have divided the chapters into five major parts. The first part consists of papers on modeling methodologies, frameworks, and one case study on using formal modeling framework for system design. This part is meant to introduce the reader to the need for formalism and frameworks, the need for behavioral hierarchy, and heterogeneity support, and how certain frameworks such as Ptolemy II, BALBOA, or SystemC enhancements address these.

Part two of this volume focus on validation methodologies. The chapters in this part address the issue of abstraction and refinement, and automation of abstraction and refinement in detail. With the state-space explosion problem hampering the success of formal verification tools, we believe automated abstraction, refinement, and modular proof techniques would be crucial for successful use of formal verification in the industry. This part of the book reflects this view. Also abstract modeling itself is an arduous task, and hence a chapter on a methodology about incremental building of formal models has been added, to illustrate one possible way to tackling the model building problem.

Part three introduces how type theory, in particular behavioral type theory is gaining crucial importance in system integration, and composition of pre-verified IP components in a correct by construction manner. Type inference at the composition time could avoid a number of integration bugs, as is known from statically typed programming languages. However, unlike software procedures and functions, dynamic actor oriented concurrent models require that we extend type theory with behavioral types, and this part will provide the readers with some introduction to such theories.

Often times, system level models grow in size, and hence simulation or even implementation synthesis become unmanageable. Under such circumstances, model optimization may lead to efficient simulation or implementation. Part four of this volume introduces two chapters, one of optimizing Esterel programs for efficient execution, and another on SystemC model optimization. These represent a very small sample of this important research area.

Part five has a single paper on formal models to capture the requirements for post production patching of fabricated designs. To our knowledge there has not been much research done on this topic, and hence this paper stands by itself in part five.

Overall, we believe that this book will provide a reasonable sampling of the various topics relevant to formal methods in System Design, as understood and advocated by the MEMOCODE community. We invite our reader to participate and contribute to this growing community of researchers. More information on MEMOCODE is available at `http://www.irisa.fr/MEMOCODE`.

Rajesh Gupta

Paul Le Guernic

Sandeep Shukla

Jean-Pierre Talpin

I

# METHODS AND MODELS FOR SYSTEM LEVEL DESIGN

Chapter 1

# MODULAR HIERARCHIES OF MODELS FOR EMBEDDED SYSTEMS

Manfred Broy
Institut für Informatik, Technische Universität München, D-80290 München Germany, broy@in.tum.de http://wwwbroy.informatik.tu-muenchen.den

**Abstract**: Today, in general, software is embedded, distributed onto networks and structured into logical components that interact asynchronously. We study fundamental models of composed software systems and their properties, identify and describe various basic views, and show how they are related. We concentrate on models of composed systems that interact by message exchange. We consider, in particular, models of data, states, interfaces, functionality, hierarchical composed systems, and processes. We study relationships by abstraction and refinement as well as forms of composition and modularity. In particular, we introduce a comprehensive mathematical model for composed systems, its essential views and their relationships.

**Key words**: Software Engineering, Formal Models, Refinement, Views

## 1. MOTIVATION

Software development is today one of the most complex and powerful tasks in engineering. Modern software systems typically are embedded in technical or organizational processes, distributed, dynamic, and accessed concurrently by a variety of independent user interfaces. Just by formulating the right programs we obtain engineering artifacts that can calculate results, communicate messages, control systems, and illustrate and animate all kinds of information. Since programs are - implicitly or explicitly - defining models of system behavior and since well-chosen models are a successful way to understand software, modeling is an essential and crucial issue in software construction. In all scientific and engineering disciplines, models play a prominent role. For physics, mathematics has provided lots of models. The same holds for many engineering disciplines.



*R. Gupta et al. (eds.), Formal Methods and Models for System Design,* 3-32.

Economy works with models; biology works more and more with models, chemistry works with models. Constructing, analyzing, and arguing in terms of models is at the heart of science.

## 2. MODELS

In this section we define the concept of a model, the structure of models and define what a model view is.

In software engineering the word "model" is used in many different contexts with many quite different meanings. Examples are terms like "meta model", "process model", or "system model". In the following we are interested in two variations and meanings of the usage of the word model.

**Definition.** Thought Model (Gedankenmodell)
A thought model is a presentation of particular aspects of a (software) system and its application context. In essence, it provides an abstraction like any model. ❑

To find a good thought model is perhaps the most critical modeling task since the thought model determines the adequacy of the future development approach.

**Definition.** Mathematical Model
A mathematical model of a system or some of its aspects is a mathematical structure, in general an algebra, consisting of sets, functions, graphs, relations, and logical predicates. It represents a thought model in mathematical terms. ❑

A good mathematical model shows a number of properties such as modularity, flexibility as well as a number of essential logical and mathematical properties. We come back to this later. A mathematical model is an idealized abstraction that needs syntax to represent it directly. We need techniques to write down, to document, and to communicate mathematical and thought models.

**Definition.** Description Technique
A description technique is a set of syntactic concepts (text, formula, graphs, and tables) for the description of a thought model. Mathematical models provide the semantic theory for description techniques. ❑

In essence, we use description techniques (syntax) to represent a mathematical model (semantic) that formalizes the thought model

### 1.1.2 The Interface Model: Components

A (system) component is an active information processing unit that communicates asynchronously with its environment through its interface which is syntactically characterized by a set of input and output channels. This communication takes place within a global (discrete) time frame.

Let I be the set of input channels and O be the set of output channels of the component F. With every channel in the set $I \cup O$ we associate a data type indicating the type of messages sent along that channel. Then by $(I \blacktriangleright O)$ the *syntactic interface* of a system component is denoted. A graphical representation of a component with its syntactic interface and individual channel types is shown in Fig. 1. It has the syntactic interface

$(x_1 : S_1, ..., x_n : S_n \blacktriangleright y_1 : T_1, ..., y_m : T_m)$

By $M^\omega = M^* \cup M^\infty$ we denote the set of streams of elements from the set M that are finite or infinite sequences of elements from M. A stream represents the sequence of messages sent over a channel during the lifetime of a system.

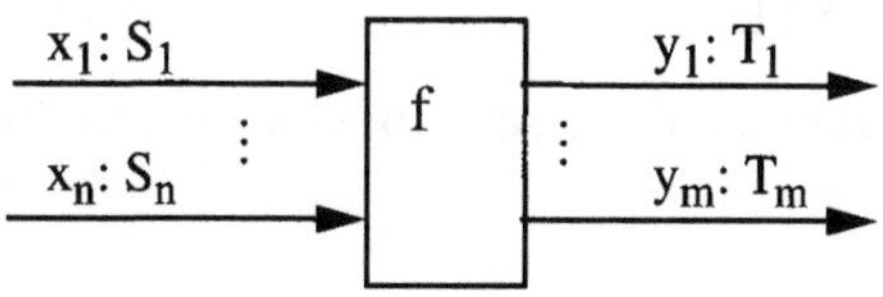


**Fig. 1** Graphical representation of a component as a data flow node with input channels $x_1$, ..., $x_n$ and output channels $y_1$, ... , $y_m$ and their respective types

In concrete systems this communication takes place in a time frame and it is often convenient to be able to refer to this. Therefore we work with *timed streams*.

Our model of time is simple. We assume that time is represented by an infinite sequence of time intervals of equal length. In each interval on each channel a finite, possibly empty sequence of messages is transmitted. By $(M^*)^\infty$ we denote the set of infinite streams of element sequences of set M. Mathematically, a timed stream in $(M^*)^\infty$ can also be understood as a function $\mathbb{N} \setminus \{0\} \rightarrow M^*$. We use the FOCUS model (see [6]). A system model consists of

- named components (with local state)
- named channels

driven by a global, discrete clock over which the components communicate via data streams.

The basic model is based on timed streams. This leads to a semantic model for interfaces and black-box-behaviors.

Throughout this paper we work with a few simple notations for streams. In particular, we use the following notations for a timed stream x:

| | |
|---|---|
| $z\hat{\ }x$ | concatenation of a sequence or stream z to a stream x, |
| x.k | k-th sequence in the stream x, |
| $x{\downarrow}k$ | prefix of the first k sequences in the timed stream x, |
| $\overline{x}$ | finite or infinite (untimed) stream that is the result of concatenating all sequences in x. |

Note that $\overline{x}$ defines a time abstraction for the timed stream x. Let C be a set of channels with types assigned by the function

$$\text{type}: C \to \text{TYPE}$$

Here TYPE is a set of types $\tau \in$ TYPE which are carrier sets of data elements. Let M be the universe of all messages. This means

$$M = \bigcup \{\tau: \tau \in \text{TYPE}\}$$

We define the valuations of the set C of channels by functions

$$x: C \to (M^*)^\infty$$

where for each channel $c \in C$ with type(c) = Stream $\tau$ the elements of the stream x.c are of the type $\tau$ (we denote the application of a function F to an argument b not only by F(b) but also by F.b to save the parenthesis):

$$x.c \in (\tau^*)^\infty$$

The set of valuations of the channels in C is denoted by $\vec{C}$. Let in the following I and O be sets of typed channels.

Let $r \in (M^*)^\infty$; $\overline{r}$ is called the *time abstraction* of the timed stream r. Similarly we denote for a channel valuation

$$x \in \vec{C}$$

by $\overline{x}$ its time abstraction, defined for each channel $c \in C$ by the equation

$$\overline{x}.c = \overline{x.c}$$

The operators easily generalize to sets of streams and valuations.

We describe the *interface* or *black box behavior* of a component by an *I/O-function* that defines a relation between the input streams and output streams of a component. An I/O-function is represented by a set-valued function on valuations of the input channels by timed streams. The function yields a set of valuations for the output channels for each valuation of the input channels.

$$F : \vec{I} \rightarrow \wp(\vec{O})$$

fulfills the following *timing property,* which axiomatises the time flow. It reads as follows (let $x, z \in \vec{I}, y \in \vec{O}, t \in \mathbb{N}$):

$$x{\downarrow}t = z{\downarrow}t \Rightarrow \{y{\downarrow}t{+}1 : y \in F(x)\} = \{y{\downarrow}t{+}1 : y \in F(z)\}$$

Here $x{\downarrow}t$ denotes the stream that is the prefix of the stream x and contains t finite sequences. In other words, $x{\downarrow}t$ denotes the communication histories in the channel valuation x until time interval t. The timing property expresses that the set of possible output histories for the first t+1 time intervals only depends on the input histories for the first t time intervals. In other words, the processing of messages in a component takes at least one time tick. We call functions with this property *time-guarded* or strictly *causal.*

By $\mathbb{F}[I \blacktriangleright O]$ we denote the set of all I/O-functions with input channels I and output channels O. By $\mathbb{F}$ we denote the set of all I/O-functions for arbitrary channel sets I and O. For any $F \in \mathbb{F}$ we denote by In(F) its set of input channels and by Out(F) its set of output channels.

### 1.1.3 Structuring Functionality

A system is called *multifunctional,* if it provides a number of self-contained logical functions. Typically systems of today like mobile phones or automobiles provide hundreds or even thousands of different functions.

The interface view may be quite complex for a multifunctional system. To understand such a huge functionality we need structuring mechanisms. Use cases are a step in that direction. They help to isolate functions. A further step is to define relationships between the functions that express various dependencies or non-dependencies between the functions. This helps in the reduction of the complexity and thus supports the understanding of multifunctional systems.

## 1.2 System States

One common way to model a system and its behavior is to describe it in terms of a *state space* and its *state transitions.* Each state of a system consists of the states of its internal and external channels and the states of its components. This leads to a local and global *state view* of the system.

### 1.2.1 States of State Transition Systems: State Attributes

We describe the data state of a transition machine by a set of typed attributes V that can be seen as *program variables.* A data state is modeled by the mapping

$$\eta: V \rightarrow \bigcup_{v \in V} \text{type}(v)$$

It is a valuation of the attributes in the set V by values of the corresponding type. By $\vec{V}$ we denote the set of valuations of the attributes in V. By Σ we denote the set of all valuations.

Often it is helpful to use a special attribute that refers to the control state. In general, it has an enumeration type Control (a finite set of control state identifiers) as its type. With each control state we associate in the set of reachable states of a state machine the set of states that can be assumed in that control state. This set can be characterized by an assertion, which is a predicate on the data state, called the *control invariant.*

A system that is modeled by a state machine has a *local* data state. Each local data state is an element in the set $\vec{V}$, which is the set of valuations for its attributes. In addition to the channel attributes (see below), we use the attributes of the local states of the systems to refer to the data and control states of a system.

### 1.2.2 States of Channels

In this section we deal with the fact that a channel has a state, too. It is represented by the stream of messages sent over it (its communication history) and the stream of messages that have been sent but not received so far and therefore are still stored in its buffer. The content of the buffer of a channel is a stream, which is a postfix of the communication history of the channel. To take care of pending (buffered) messages in state assertions, we associate with each channel c the following two derived variables in logical formulas:

$c^+$ denotes the sequence of messages sent but not received yet on the channel c, called the *channel buffer*,

c denotes the sequence of messages sent on the channel c, called the *channel history.*

$c^+$ denotes the content of the buffer. The identifiers c and $c^+$ are called the *channel attributes.*

Given a state space Σ and a set C of typed channels we denote by Σ(C) the state space complemented by the states of the set of channels defined by

$$\Sigma(C) = \Sigma \times (C \rightarrow M^*)$$

$\Sigma(C)$ is the state space of a component with the state of its channels C and internal states from $\Sigma$.

### 1.2.3 State Machine Model: State Transitions

Often a component is described in a well-understandable way by a state transition machine with input and output. A state transition is one step of a system execution leading from a given state to a new state.

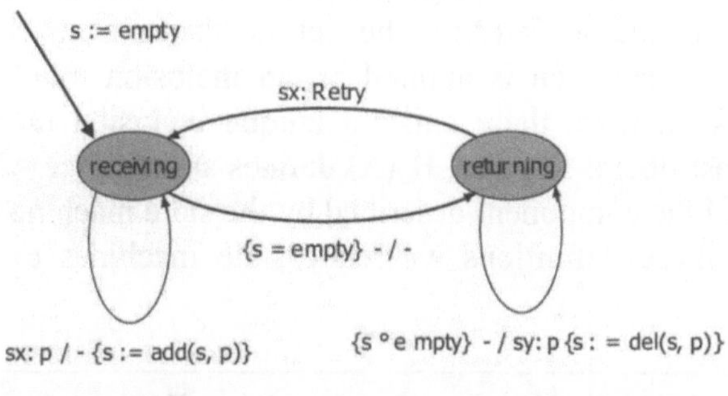


**Fig. 2 Graphical representation of a state machine**

A state machine $(\Delta, \Lambda)$ with input and output according to the syntactic interface $(I \blacktriangleright O)$ is given by a set $\Lambda \subseteq \Sigma(O)$ of pairs $(\sigma_0, y_0)$ of an initial state $\sigma_0$ and an initial output sequence $y_0 \in (O \rightarrow M^*)$ as well as a state transition function

$$\Delta: (\Sigma \times (I \rightarrow M^*)) \rightarrow \wp(\Sigma \times (O \rightarrow M^*))$$

For each state $\sigma \in \Sigma$ and each valuation $u: I \rightarrow M^*$ of the input channels in I by sequences we obtain by every pair $(\sigma', s) \in \Delta(\sigma, u)$ a successor state $\sigma'$ and a valuation $s: O \rightarrow M^*$ of the output channels consisting of the sequences produced by the state transition. By $SM[I \blacktriangleright O]$ we denote the set of all state machines with input channels I and output channels O. By SM we denote the set of all state machines.

### 1.2.4 From State Machines to Interfaces

In this section we define the interface abstraction for state machines. Each state transition function

$$\Delta: (\Sigma \times (I \rightarrow M^*)) \rightarrow \wp(\Sigma \times (O \rightarrow M^*))$$

induces a function

$B_\Delta: \Sigma(O) \to (\vec{I} \to \wp(\vec{O}))$

$B_\Delta$ provides the black box or interface view onto the state transition function $\Delta$. For each state set $\Lambda \subseteq \Sigma(O)$, each input pattern $z \in (I \to M^*)$, and each input channel valuation $x \in \vec{I}$, we specify the black box function $B_\Delta$ for the given state $\sigma$ by the inclusion maximal solution of the equation

$$B_\Delta(\Lambda)(\langle z\rangle\hat{\ }x) = \{\langle s\rangle\hat{\ }y: y \in B_\Delta(\{(\sigma',s') \in \Sigma(O): (\gamma',s) \in \Delta(\gamma,z) \wedge (\sigma,s) \in \Lambda\}).x\}$$

The right-hand side of the equation above is inclusion monotonic in $B_\Delta$. If we add elements to $B_\Delta(\Lambda).x$ the set is also increased. Since in the definition of $B_\Delta$ recursion is applied on an inclusion monotonic function, which even is guarded, there exists a unique inclusion maximal solution. This is the most liberal fixpoint. $B_\Delta(\Lambda)$ defines an interface which represents the behavior of the component described by the state machine $(\Delta, \Lambda)$.

Based on these definitions we relate state machines to their interface abstractions.

> Given a state transition function $\Delta$ and $\Lambda \subseteq \Sigma(O)$, $B_\Delta(\Lambda)$ provides the interface abstraction of the behavior of the state transition machine $\Delta$ for the initial states from $\Lambda$.

This way we define a function

$$abs_{SM/IF}: \mathbb{SM}[I \blacktriangleright O] \to \mathbb{F}[I \blacktriangleright O]$$

by

$$abs_{SM/IF}((\Delta, \Lambda)) = B_\Delta(\Lambda)$$

Note that we get a refinement mapping between a state machine $(\Delta, \Lambda)$ and $B_\Delta(\Lambda)$.

### 1.2.5 From Interfaces to State Machines

In this section we show that an interface abstraction defines itself an abstract state machine. Given an interface

$F: \vec{I} \to \wp(\vec{O})$

we define the state space by

$\Sigma = \mathbb{F}[I \blacktriangleright O]$

We get a state machine

$\Delta_F: (\Sigma \times (I \rightarrow M^*)) \rightarrow \wp(\Sigma \times (O \rightarrow M^*))$

by the following definition (let $G \in \mathbb{F}[I \blacktriangleright O]$, $z \in (I \rightarrow M^*)$)

$\Delta_F(G, z) =$

$$\{(H, s): \exists\ s': \forall\ x \in \vec{I}: G(\langle z\rangle\hat{\ }x) = \{\langle s'\rangle\hat{\ }\langle s\rangle\hat{\ }y: \langle s\rangle\hat{\ }y \in H(x)\}\}$$

Note that for $\langle s\rangle\hat{\ }y \in H(x)$ the value s does not depend on x according to the causality assumption. The function H in the formula above is called a *resumption*. It represents the new state of the machine after the transition represented by an I/O-function. The initial states of the state machine are defined by the set

$$\{(F, s): \exists\ x \in \vec{I}, y \in \vec{O}: \langle s\rangle\hat{\ }y \in F(x)\}$$

The value s in $\langle s\rangle\hat{\ }y \in F(x)$ does not depend on the input history x according to the causality assumption. On this state machine we can apply all the classical notions and concepts of state machines.

We define the function

$emb_{SM/IF}: \mathbb{F}[I \blacktriangleright O] \rightarrow \mathbb{SM}[I \blacktriangleright O]$

by

$emb_{SM/IF}(F) =$

$$(\Delta_F, \{(G, s): \exists\ x \in \vec{I}, y \in \vec{O}: \langle s\rangle\hat{\ }y \in F(x) \wedge \forall\ x \in \vec{I}: G(x) = \{\langle s\rangle\hat{\ }y: \langle s\rangle\hat{\ }y \in F(x)\}\})$$

We get an abstract implementation (for the formal definition of the relation $\approx\!\!>\!\!>_{IF/SM}$ see later)

$$F \approx\!\!>\!\!>_{IF/SM} emb_{SM/IF}(F)$$

Furthermore, we easily define a function

$abs_{SM/SM}: \mathbb{SM}[I \blacktriangleright O] \rightarrow \mathbb{SM}[I \blacktriangleright O]$

by

$abs_{SM/SM}(\Delta, \Lambda) =$

$$((\Delta_F, \{(F, s): \exists\ x \in \vec{I}, y \in \vec{O}: \langle s\rangle\hat{\ }y \in F(x)\})$$

where $F = B_{\Delta}(\Lambda)$

In fact, this way we get the most abstract refinement of a state machine. We come back to this in the section on property refinement.

## 1.3 The Distributed System Model: Composed Systems

An interactive composed system consists of a family of interacting components (in some approaches also called *agents* or *objects*). Components interact by exchanging messages via the channels that connect them. A network of communicating components gives a structural system view, also called *system architecture*. Its nodes represent components and its arcs communication lines (channels) on which streams of messages are sent.

We model composed systems by data flow nets. Let $\mathbb{K}$ be a set of identifiers for components and I and O be sets of input and output channels, respectively. A composed system (ν, O) with syntactic interface (I ► O) is represented by the mapping

$$\nu: \mathbb{K} \rightarrow \mathbb{F}$$

that associates with every node a component behavior in the form of a black box view, formally, an interface behavior given by an I/O-function. O denotes the output channels of the system.

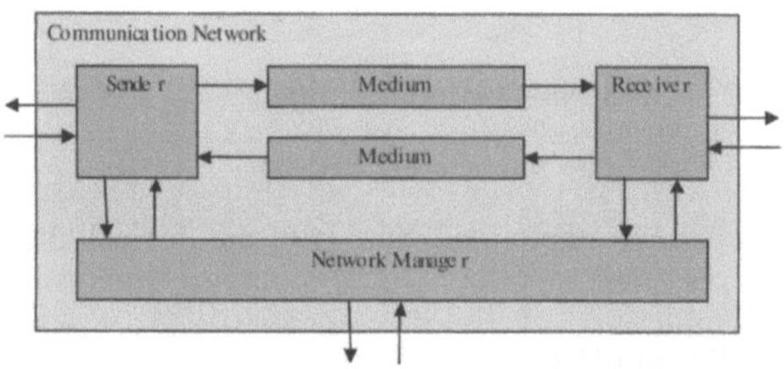


**Fig. 3** Graphical representation of an architecture

As a well-formedness condition for forming a net from a set of component identifiers $\mathbb{K}$, we require that for all component identifiers k, j ∈ $\mathbb{K}$ (with k ≠ j) the sets of output channels of the components ν(k) and ν(j) are disjoint. This is formally guaranteed by the condition

$$k \neq j \Rightarrow \text{Out}(\nu(k)) \cap \text{Out}(\nu(j)) = \emptyset$$

In other words, each channel has a uniquely specified component as its source[1]. We denote the set of all (internal and external) channels of the net by the equation

$$\text{Chan}((\nu, O)) = O \cup \{c \in \text{In}(\nu(k)): k \in \mathbb{K}\} \cup \{c \in \text{Out}(\nu(k)): k \in \mathbb{K}\}$$

*The set*

$$I = \text{Chan}((\nu, O)) \backslash \{c \in \text{Out}(\nu(k)): k \in \mathbb{K}\}$$

denotes the set of input channels of the net. $CS_{IF}$ denotes the set of all composed systems with component behaviors represented by interfaces.

## 1.4 The Process Model

In this section we introduce several examples of basic process models. To keep the approach intelligible we consider only more or less sequential processes. More sophisticated notions of concurrent processes such as studied in Petri-net theory or general event structures are not considered. A process is a description of a system behavior in terms of the runs of the system.

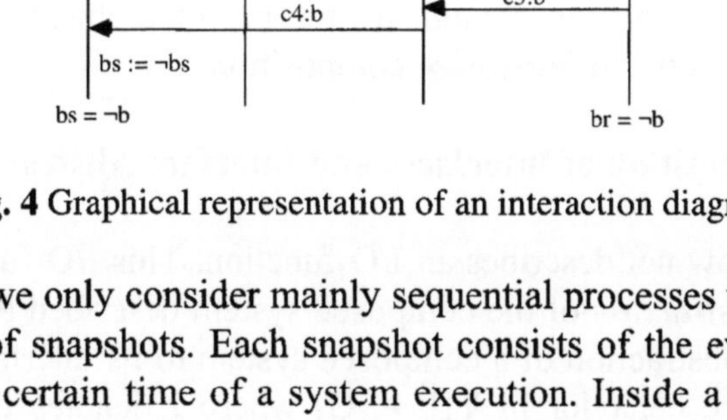


**Fig. 4** Graphical representation of an interaction diagram

For simplicity we only consider mainly sequential processes where a process is a sequence of snapshots. Each snapshot consists of the events and states that occur at a certain time of a system execution. Inside a snapshot, some

[1] Channels that occur as input channels but not as output channels have the environment as their source.

concurrency may be modeled. We work with a number of different concepts of snapshots.

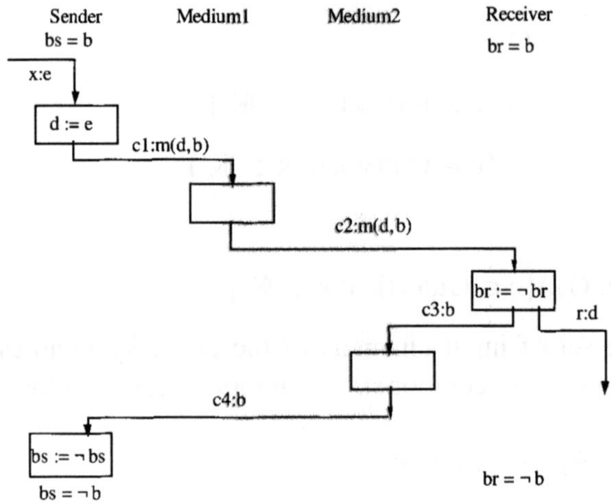


**Fig. 5** Graphical representation of a process

Given a snapshot set H a process over H is a finite or infinite sequence of snapshots, with other words, a stream

$H^{\omega}$

Dependent on the choice of the snapshot set, we get processes for different system views.

## 1.5 Composition

In this section we study forms of composition for systems. To cope with large systems composition should always be hierarchical. We distinguish *vertical* composition and *horizontal* composition.

### 1.5.1 Composition of Interfaces and Interface Abstraction

Each data flow net describes an I/O-function. This I/O-function is called the *interface abstraction* of the composed system described by the data flow net. We get an abstraction of a composed system to its interface by mapping it to an interface behavior in $\mathbb{F}[I \blacktriangleright O]$ where I denotes the set of input channels and O denotes the set of output channels of the data flow net. This interface view is represented by the component behavior $F \in \mathbb{F}[I \blacktriangleright O]$ specified by the following formula (note that $y \in \vec{C}$ where $C \equiv \text{Chan}((v, O))$ as defined above):

$$F_{(\nu, O)}(x) = \{y|_O: y|_I = x \wedge \forall k \in \mathbb{K}: y|_{Out(\nu(k))} \in \nu(k)(y|_{In(\nu(k))})\}$$

The formula essentially expresses that the output history of a data flow net is the restriction of a fixpoint for all the net-equations for the output channels.

We define a function

$$abs_{\mathbb{CS}/\mathbb{IF}}: \mathbb{CS}_{\mathbb{IF}} \to \mathbb{F}$$

by

$$abs_{\mathbb{CS}/\mathbb{IF}}((\nu, O)) = F_{(\nu, O)}$$

It defines the interface abstraction for the composed system.

### 1.5.2 Composed System States

Given a set $\mathbb{K}$ of component identifiers a composed system associates a behavior with each component. So far we have considered composed systems where each component was represented by its interface abstraction. Of course, we may also represent the components by state machines. Then we work with a mapping:

$$\nu: \mathbb{K} \to \mathbb{SM}$$

Again we assume all the conditions of a composed system and the channels of its components are introduced for systems composed of interface abstractions. The variables and derived variables for the channels and the attributes of the systems form the alphabet of a system. Each system state is represented by a valuation of the system alphabet. Thus a system state provides values for all the buffers and histories of its channels as well as for all the local data and control states of its components.

Let V be the set of all attributes of the state machines $\nu.k$ (for simplicity let us assume that all are distinct). Let $\Sigma_k$ be the state space of state machine $\nu.k$ for each $k \in \mathbb{K}$. The composed state $\Sigma$ of the composed system is defined by:

$$\Sigma = \{\sigma: \mathbb{K} \to \cup \{\Sigma_k: k \in \mathbb{K}\}: \forall k \in \mathbb{K}: \sigma(k) \in \Sigma_k\}$$

It is isomorphic to the set $\vec{V}$.

Let C be the set of typed internal and output channels the system state space is defined by the direct product of their sets of valuations:

$$\Sigma(C) = \Sigma \times (C \to M^*)$$

For states $\sigma \in \Sigma(C)$ and state or channel attributes x, we write $\sigma.x$ to denote the value associated with x in the state $\sigma$.

### 1.5.3 State Machine Composition: Distributed State Machines

Given a set $\mathbb{K}$ of components and a state machine $\nu(k) = (\Delta_k, \Lambda_k)$ for each component $k \in \mathbb{K}$ where $\Lambda_k$ is a set $\Lambda_k$ of pairs $(\sigma_0, y_0)$ and $\Delta_k$ a state transition function

$$\Delta_k: (\Sigma_k \times (I_k \rightarrow M^*)) \rightarrow \wp(\Sigma_k \times (O_k \rightarrow M^*))$$

we define a state machine $(\Delta, \Lambda)$ for the composed system by the state machine with transition function

$$\Delta: (\Sigma \times (I \rightarrow M^*)) \rightarrow \wp(\Sigma \times (O \rightarrow M^*))$$

where its set of input channels is defined as follows

$$I = \{c \in I_k: k \in \mathbb{K}\} \backslash \{c \in O_k: k \in \mathbb{K}\}$$

and

$$O = \{c \in O_k: k \in \mathbb{K}\}$$

$$\Sigma = \{\sigma: \mathbb{K} \rightarrow \cup \{\Sigma_k: k \in \mathbb{K}\}: \forall k \in \mathbb{K}: \sigma(k) \in \Sigma_k\}$$

We specify $\Delta$ by

$$\Delta(\sigma, x) = \{(\sigma', y): \forall k \in \mathbb{K}: (\sigma'(k), y|O_k) \in \Delta_k(\sigma(k), x|I_k)\}$$

and $\Lambda$ by:

$$\Lambda = \{(\sigma, y): \forall k \in \mathbb{K}: (\sigma'(k), y|O_k) \in \Lambda_k\}$$

This yields the composed state machine. $(\Delta, \Lambda)$ is a state machine like the state machines $(\Delta_k, \Lambda_k)$ but with a distributed state. In each of its state transitions all state machines execute concurrently.

### 1.5.4 System Hierarchies

Let HCS denote this set of all composed hierarchical systems.

We define hierarchical (and by a more sophisticated construction - using ideas from domain theory - even recursive hierarchical) systems. In most cases and also for our purpose, however, finite hierarchies are sufficient. We define in full generality that a hierarchical composed system is a pair $(\nu, O)$ were $\nu$ is a mapping:

$$\nu: \mathbb{K} \rightarrow \text{HCS}$$

By the notion of component identifiers and hierarchical composed systems we introduce a new concept into our system model namely that of an *instance*. In particular, we include composed systems where one interface or one state machine is used several times by assigning it to different component identifiers.

Since we have an interface abstraction function for state machines and for processes we get an interface abstraction for hierarchical composed systems:

$abs_{\mathbb{IF}}: \mathbb{HCS} \rightarrow \mathbb{F}$

It maps every element $s \in \mathbb{HCS}$ onto its interface abstraction. If s is an interface, the $abs_{\mathbb{IF}}(s) = s$. Of course we may also introduce an abstraction function

$abs_{\mathbb{CIF}}: \mathbb{HCS} \rightarrow \mathbb{CS}_{\mathbb{IF}}$

that maps a composed system $(\nu, O)$ with

$\nu: \mathbb{K} \rightarrow \mathbb{HCS}$

onto a system $(\nu', O) = abs_{\mathbb{CIF}}(\nu)$ composed of interfaces. It is easily defined by

$\nu'(k) = abs_{\mathbb{IF}}(\nu(k))$

for all $k \in \mathbb{K}$.

## 2. REFINEMENT

In requirements engineering, in the design and implementation phase of system development many issues have to be addressed such as requirements elicitation, conflict identification and resolution, information management as well as the selection of a favorable software architecture. These activities are connected with development steps. Refinement relations are the medium to formalize development steps and in this way the development process.

We formalize the following basic ideas of refinement:

- *property refinement* - enhancing requirements - allows us to add properties to a specification,
- *glass box refinement, implementation refinement* - designing implementations - allows us to decompose a component into a composed system or to give a state transition description for a component specification,
- *interaction refinement* - relating *levels of abstraction* - allows us to

change the representation of the communication histories, in particular, the granularity of the interaction as well as the number and types of the channels of a component.

In fact, these notions of refinement describe the steps needed in an idealistic view of a strictly hierarchical top down system development. The three refinement concepts mentioned above are formally defined and explained in detail in the following.

## 2.1 Property refinement

Property refinement is a classical concept in program development. It is based - just as deduction in logics - on logical implication. In our model this relates to set inclusion.

### 2.1.1 Property Refinement of Interfaces

Property refinement allows us to replace an interface behavior or an interface with one having additional properties. This way a behavior is replaced by a more restricted one. An interface

$$F: \vec{I} \rightarrow \wp(\vec{O})$$

is refined by a behavior

$$\hat{F} : \vec{I} \rightarrow \wp(\vec{O})$$

if

$$F \approx>_{\mathbb{IF}} \hat{F}$$

This relation stands for the proposition

$$\forall x \in \vec{I}: \hat{F}.x \subseteq F.x$$

Obviously, property refinement is a partial order and therefore reflexive, asymmetric, and transitive. Moreover, the inconsistent specification logically described by false refines everything.

A property refinement is a basic refinement step adding requirements as it is done step by step in requirements engineering. In the process of requirements engineering, typically the overall services of a system are specified. Requiring more and more sophisticated properties for components until a desired behavior is specified, in general, does this.

### 2.1.2 Property Refinement of Composed Systems

Given two composed systems $S_1$ and $S_2$, we write

$S_1 \approx> _{CS} S_2$

if

$abs_{CS/IF}(S_1) \approx>_{IF} abs_{CS/IF}(S_2)$

Here we require only that the interface abstractions of both systems are in the property refinement relation. The two systems may have completely different composition structures, however.

### 2.1.3 General Property Refinement of Systems

It is the idea of property refinement that we do not take care of the internal structure of systems, but only compare their interface abstractions. Given two hierarchical composed systems $S_1$ and $S_2$ in HCS, we write

$S_1 \approx> S_2$

if

$abs_{IF}(S_1) \approx>_{IF} abs_{IF}(S_2)$

Property refinement is a straightforward concept. Basically we can see property refinement always as combined with interface abstraction.

## 2.2 Compositionality of Property Refinement

In our case, the proof of the compositionality of property refinement is straightforward. This is a consequence of the simple definition of composition. Let (v, O) and (v', O) be composed systems with the set $\mathbb{K}$ of components given by their interfaces. The rule of compositional property refinement reads as follows:

$$\frac{\forall\, k \in \mathbb{K}:\ v(k) \approx> v'(k)}{F_{(v,\,O)} \approx> F_{(v',\,O)}}$$

Compositionality is often called *modularity* in system development. Modularity allows for a separate development of components.

Modularity guarantees that separate refinements of the components of a

of the communication.

An *interaction refinement* is described by a pair of two functions

$$A: \vec{C}' \to \wp(\vec{C}) \qquad R: \vec{C} \to \wp(\vec{C}')$$

that relate the interaction on an abstract level with corresponding interaction on the more concrete level. This pair specifies a development step that is leading from one level of abstraction to the other one as illustrated by Fig. 6. Given an abstract history $x \in \vec{C}$ each $y \in R.x$ denotes a concrete history representing x. Calculating a representation for a given abstract history and then its abstraction yields the old abstract history again. Using sequential composition, this is expressed by the requirement:

$$R \circ A = Id$$

Let Id denote the identity relation and "∘" the sequential composition defined as follows:

$$(R \circ A).x = \{y \in A.z: z \in R.x\}$$

A is called the *abstraction* and R is called the *representation*. R and A are called a *refinement pair*. For untimed components we weaken this requirement by requiring $R \circ A$ to be a property refinement of the untimed identity, formally expressed by the following equation:

$$\overline{(R \circ A).x} = \{\bar{x}\}$$

This defines an identity under time abstraction.

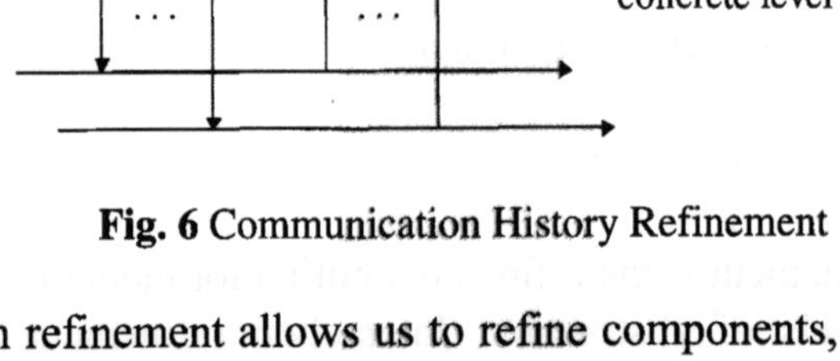


**Fig. 6** Communication History Refinement

Interaction refinement allows us to refine components, given appropriate refinement pairs for their input and output channels. The idea of an interaction refinement is visualized in Fig. 7 for the so-called $U^{-1}$-simulation. Note that here the components (boxes) $A_I$ and $A_O$ are no longer definitional in the sense of specifications, but rather methodological, since they relate two levels of abstraction. Nevertheless, we specify them as well by the specification techniques introduced so far.

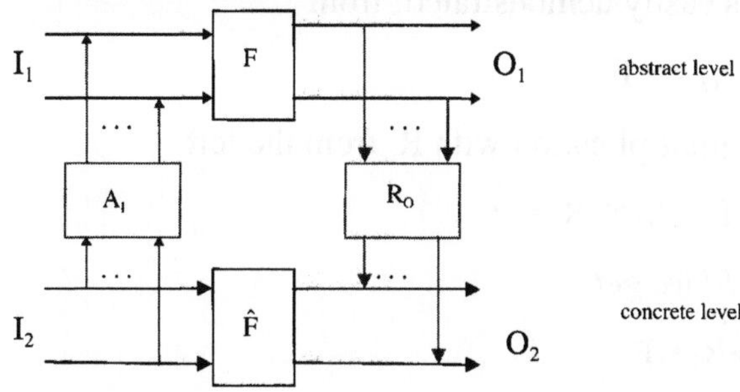


**Fig. 7** Interaction Refinement ($U^{-1}$-simulation)

Given refinement pairs

$A_I: \vec{I}_2 \rightarrow \wp(\vec{I}_1)$ $R_I: \vec{I}_1 \rightarrow \wp(\vec{I}_2)$

$A_O: \vec{O}_2 \rightarrow \wp(\vec{O}_1)$ $R_O: \vec{O}_1 \rightarrow \wp(\vec{O}_2)$

for the input and output channels we are able to relate abstract to concrete channels for the input and for the output. We call the interface

$\hat{F}: \vec{I}_2 \rightarrow \wp(\vec{O}_2)$

an *interaction refinement* of the I/O-behavior

$F: \vec{I}_1 \rightarrow \wp(\vec{O}_1)$

if the following proposition holds: $\approx>$

$A_I \circ F \circ R_O \approx> \hat{F}$ *$U^{-1}$-simulation*

This formula essentially expresses that $\hat{F}$ is a property refinement of the component $A_I \circ F \circ R_O$. Thus for every "concrete" input history $\hat{x} \in \vec{I}_2$ every concrete output $\hat{y} \in \vec{O}_2$ can be also obtained by translating $\hat{x}$ onto an abstract input history $x \in A_I \cdot \hat{x}$ such that we can choose an abstract output history $y \in F.x$ such that $\hat{y} \in R_o.y$.

There are three further versions of interaction refinement obtained by replacing in Fig. 7 the upward function $A_I$ by the downward function $R_I$ or the upward function $A_O$ by the downward function $R_O$ or both. We get the following characterizations:

$F \circ R_O \approx> R_I \circ \hat{F}$ *Downward Simulation*

$A_I \circ F \approx> \hat{F} \circ A_O$ *Upward Simulation*

$F \approx> R_I \circ \hat{F} \circ A_O$ *U-simulation*

These are different useful relations between levels of abstractions. $U^{-1}$-simulation is the most restrictive, "strongest" notion which implies the other three. This fact is easily demonstrated: from

$$A_I \circ F \circ R_O \approx> \hat{F}$$

we derive by multiplication with $R_I$ from the left

$$R_I \circ A_I \circ F \circ R_O \approx> R_I \circ \hat{F}$$

*and by $R_I \circ A_I = Id$ we get*

$$F \circ R_O \approx> R_I \circ \hat{F}$$

Which is the property of downward simulation. By similar arguments we prove that an $U^{-1}$-simulation $\hat{F}$ is also an upward simulation and an U-simulation.

We choose an $U^{-1}$-simulation as the most general form of interaction refinement and define:

$$F \rightsquigarrow_{IF} \hat{F}$$

if there exist refinement pairs such that

$$A_I \circ F \circ R_O \approx> \hat{F} \qquad U^{-1}\text{-simulation}$$

This way we establish the relation of interaction refinement of interfaces. It is again a partial order.

### 2.4.2 Granularity Refinement of State Machines

Also for state machines we can define a refinement both of the input and output channels and of the states. Given two state machine $(\Delta_k, \Lambda_k)$ for k: = 1, 2 where $\Lambda_k$ is a set of pairs $(\sigma_0, y_0)$ and $\Delta_k$ is a state transition function

$$\Delta_k: (\Sigma_k \times (I_k \rightarrow M^*)) \rightarrow \wp(\Sigma_k \times (O_k \rightarrow M^*))$$

we call $\Delta_2$ a granularity (or vertical) refinement of $\Delta_1$ if there exists a mapping

$$abs: \Sigma_2 \rightarrow \Sigma_1$$

such that

$$\{(abs.\sigma, A_O.y_0) : (\sigma_0, y_0) \in \Lambda_2\} \subseteq \Lambda_1$$

and we have for all $\Theta \subseteq \Sigma_2(O_2)$

$$A_I \circ B_{\Delta_1}(\{(abs.\sigma, A_O.y) : (\sigma, y) \in \Theta\}) \circ R_O \approx> B_{\Delta_2}(\Theta)$$

This formula expresses that for every state σ the interface abstraction is a refinement of the interface abstraction of abs.σ. By a state machine refinement both the state space, the types of messages and the granularity of interaction can be refined.

If this holds we write:

$(\Delta_k, \Lambda_k) \sim>_{IF} (\Delta_k, \Lambda_k)$

This way we establish the relation of vertical refinement in terms of interfaces of the state machines.

### 2.4.3 Granularity Refinement of Processes

Also for processes we can define a refinement of the level of abstraction. Since processes are canonically obtained from interfaces or state machines, we get an induced notion of refinement. We do not define this notion explicitly, since it is rather straightforward.

### 2.4.4 Granularity Refinement of Composed Systems

Composed systems are refined in levels of abstraction by refining their components. Of course, we have to make sure, that all channels are refined consistently. Since a composed system can be seen either as one big interface or one big state machine the refinement notions introduced above carry over in a straightforward manner.

Given two composed systems $S_1$ and $S_2$, we write

$S_1 \sim>_{CS} S_2$

if

$abs_{CS/IF}(S_1) \sim>_{IF} abs_{CS/IF}(S_2)$

Again this is a straightforward concept.

### 2.4.5 General Refinement of Systems through Levels of Abstractions

It is the idea of granularity refinement that we do not take care of the internal structure of systems, but only compare their interface abstractions in terms of refinement pairs. Given two systems $S_1$ and $S_2$ in HCS, we write

$S_1 \sim> S_2$

if

$abs_{IF}(S_1) \sim>_{IF} abs_{IF}(S_2)$

Again this is a straightforward concept. Basically we can see granularity refinement always in terms of interface abstraction.

# 3. MODELS IN THE DEVELOPMENT PROCESS

Building models is a complex and laborious task. To get the best return on investment of modeling the task of modeling should be integrated in the development process such that it supports all steps of software development.

In the following we sketch a rather straightforward development process and identify for each phase the role of modeling techniques.

## 3.1 Requirements Engineering

Requirements engineering starts, in general, by informal documents and activities fixing the overall goals of a software development task. Only if for a particular domain models happen to be given we can expect model-related activities from the very beginning.

When entering into the detailed requirements modeling issues start to play a role - even if requirements are formulated informally. In fact, even in informal descriptions of requirements often implicitly a kind of thought model is used that corresponds to a system model. Often it is useful to make that model explicit, even if requirements are expressed informally.

Starting from informal use cases the functional requirements and the user interface requirements can be captured in models. In general, it is useful to begin with informal requirements and then formalizing them in a model. This leads to a functional feature model formalizing use cases and also indicates the dependencies between features. This helps to identify inconsistencies and spots of incompleteness.

If the models are expressive enough, then indeed they can capture some of the nonfunctional requirements, too.

The models constructed during requirements engineering prove to be very helpful in the validation of the requirements specification. Especially criticality requirements then can be verified on the basis of the models at this stage.

The model includes in general a data model in the form of an extended E/R-model that defines the state space of the system by defining all data attributes, data sets, and their encapsulation in components.

In requirements engineering it may also be useful to provide – especially in the case of embedded systems – an environment and domain model that captures essential properties of the environment or the application domain.

Such a model requires a comprehensive description of data, structures, rules, and behaviors of the system environment or domain.

## 3.2 Design Engineering: Architecture

In the design phase the architecture of a software system is worked out. This includes a logical component architecture, its component structure and connection as well as the interaction between the components in the architecture. For each component an interface description has to be derived.

For that we need a design data model that is an extension and distribution of the extended E/R-model that defines the state space of the system architecture by defining all data attributes, data sets, and their encapsulation in components. Furthermore from the interaction model the component interface specifications can be derived and represented by an appropriate interface model. A very concrete way to describe the interface is a state transition model.

This way we get a formal model of the architecture of the software system. On this basis the architecture can be validated and verified.

Especially for embedded systems in addition to the software architecture a model of the hardware architecture is to be given as well as a software/hardware mapping modeling the deployment.

A performance model can be used to capture, validate and verify some of the nonfunctional properties for the architecture.

The design phase does go on with a hierarchical component decomposition and a detailed component design.

## 3.3 Implementation Engineering

An implementation is given by code written in some programming language targeted for a certain run time environment and a certain hardware platform. In principle such code can be generated from high-level design models at least for prototyping purposes.

Many interesting properties of large code junks can be illustrated with the help of models. We can, for instance, describe the code structure by a code architecture. In principle we may also use modeling techniques the other way around to extract models out of given code by model extraction.

## 3.4 Quality Assurance, Verification

Quality engineering aims at the verification of the development steps, in particular, to establish confidence that the implementation realizes the requirements. In practice, this is usually done by inspection, review, and testing. Here models can contribute to guide the review and inspection work.

Logical verification and model checking used to be more theoretical approaches to verification. However, step-by-step this technology gets more practical and is applied in some areas such as hardware design in industry.

In particular, practically interesting seems the derivation of test cases from models. Here models do not only help to provide coverage criteria. They also help to generate the complete test case such as input to the system to drive the test as well as the expected output to verify the results. This leads to enormous savings (up to 50%) in the testing task, which may consume up to 50% of the overall development costs. A result of only this is a cost reduction of the development up to 25%.

## 4. SUMMARY AND OUTLOOK

Why did we present this setting of mathematical models and relations between them? First of all, we want to show how rich and flexible the tool kit of mathematical models is and how far we can go in integrating and relating them. Perhaps, it should be emphasized that we get first of all an integrated system model very close to practical approaches by SDL or UML where a system is a set or hierarchy of components. In this tree of components the leaves are state machines. In our case the usage of streams and stream processing function is the reason for the remarkable flexibility of our model toolkit and the simplicity of the integration.

This work demonstrates a number of significant properties that a modeling technique should show to support a methodology. The properties lead to a comprehensive theory of modeling.

Software development is a difficult and complex engineering task. It would be very surprising if such a task could be carried out properly without a proper theoretical framework. It would at the same time be quite surprising if a purely scientifically theoretical framework would be the right approach for the practical engineer. The result has to be a compromise as we have argued between formal techniques and theory on one hand and intuitive notations based on diagrams on the other hand. Work is needed along those lines including experiments and feedback from practical applications. But as our example and experiment already show a lot is to be gained that way.

## Acknowledgements

It is a pleasure to thank Markus Pizka, Leonid Kof, Ingolf Krüger, and Bernhard Schätz for stimulating discussions and helpful remarks on draft versions of the manuscript.

## REFERENCES

[1] J.C.M. Baeten, J. Bergstra: Process Algebras with Signals and Conditions. In: M. Broy (ed.): Programming and Mathematical Method. Springer NATO ASI Series, Series F: Computer and System Sciences, Vol. 88, 1992, 273-324

[2] G. Booch: Object Oriented Design with Applications. Benjamin Cummings, Redwood City, CA, 1991

[3] M. Broy: Towards a formal foundation of the specification and description language SDL. Formal Aspects of Computing 3, 1991, 21-57

[4] M. Broy: Refinement of Time. M. Bertran, Th. Rus (eds.): Transformation-Based Reactive System Development. ARTS'97, Mallorca 1997. Lecture Notes in Computer Science 1231, 1997, 44-63, To appear in TCS

[5] M. Broy, C. Hofmann, I. Krüger, M Schmidt: A Graphical Description Technique for Communication in Software Architectures. Joint 1997 Asia Pacific Software Engineering Conference and International Computer Science Conference (APSEC'97/ICSC'97)

[6] M. Broy, K. Stølen: Specification and Development of Interactive Systems: FOCUS on Streams, Interfaces, and Refinement. Springer 2001

[7] G. Berry, G. Gonthier: The ESTEREL Synchronous Programming Language: Design, Semantics, Implementation. INRIA, Research Report 842, 1988

[8] D. Harel: Statecharts: A Visual Formalism for Complex Systems. Science of Computer Programming 8, 1987, 231 - 274

[9] R. Hettler: Zur Übersetzung von E/R-Schemata nach SPECTRUM. Technischer Bericht TUM-I9409, TU München, 1994

[10] C.A.R. Hoare: Communicating Sequential Processes. Prentice Hall, 1985

[11] I. Jacobsen: Object-Oriented Software Engineering. Addison-Wesley, ACM Press 1992

[12] G. Kahn: The Semantics of a Simple Language for Parallel Processing. In: J.L. Rosenfeld(ed.): Information Processing 74. Proc. of the IFIP Congress 74, Amsterdam: North Holland 1974, 471-475.

[13] R. Milner: A Calculus of Communicating Systems. Lecture Notes in Computer Science 92, Springer 1980

[14] ITU-T (previously CCITT) (March 1993) Criteria for the Use and Applicability of Formal Description Techniques. Recommendation Z. 120, Message Sequence Chart (MSC), 35pgs.

[15] ITU-T. Recommendation Z.120, Annex B: Algebraic Semantics of Message Sequence Charts. ITU-Telecommunication Standardization Sector, Geneva, Switzerland, 1995.

[16] B. Selic, G. Gullekson. P.T. Ward: Real-time Objectoriented Modeling. Wiley, New York 1994

[17] J. Rumbaugh: Object-Oriented Modelling and Design. Prentice Hall, Englewood Cliffs: New Jersey 1991

[18] B. Rumpe: Formale Methodik des Entwurfs verteilter objektorientierter Systeme. Ph.D. Thesis Technische Universität München, Fakultät für Informatik 1996. Published by Herbert Utz Verlag

[19] Specification and Description Language (SDL), Recommendation Z.100. Technical Report, CCITT, 1988

[20] M. Broy, C. Facchi, R. Hettler, H. Hußmann, D. Nazareth, F. Regensburger, O. Slotosch, K. Stølen: The Requirement and Design Specification Language SPECTRUM. An Informal Introduction. Version 1.0. Part I/II Technische Universität München, Institut für Informatik, TUM-I9311 / TUM-I9312, May 1993

[21] G. Booch, J. Rumbaugh, I. Jacobson: The Unified Modeling Language for Object-Oriented Development, Version 1.0, RATIONAL Software Cooperation

[22] P. Zave, M. Jackson: Four dark corners of requirements engineering. ACM Transactions on Software Engineering and Methodology, January 1997

Chapter 2

# ACTOR-ORIENTED MODELS FOR CODESIGN

## *Balancing Re-Use and Performance*

Edward A. Lee, and Stephen Neuendorffer
*University of California at Berkeley*
{eal,neuendor} @eecs.berkeley.edu

**Abstract** Most current hardware engineering practice is deeply rooted in discrete-event modeling and synchronous design. Most current software engineering is deeply rooted in procedural abstractions. The latter says little about concurrency and temporal properties, whereas the former lacks many of modularity capabilities of modern programming languages. Actor-oriented design emphasizes concurrency and communication between components while maintaining modularity. Components called actors execute and communicate with other actors. In contrast to the interfaces in object-oriented design (methods, principally, which mediate transfer of the locus of control), interfaces in actor-oriented design (which we call ports) mediate communication. But the communication is not assumed to involve a transfer of control. This paper discusses the structure of actor-oriented models and shows how data and behavioral type systems enhance modularity and re-use potential while enabling designs that embrace concurrency and time. This paper shows how components can be designed for re-use through parameterization and behavioral polymorphism, and how component specialization can offset the performance costs of doing so.

**Keywords:** Actor-oriented design, behavioral types, behavioral polymorphism, component specialization, code generation, parameterization, Ptolemy.

# Introduction

Current engineering practice for the codesign of embedded hardware and software systems draws heavily on the existing technology in these fields. For hardware design, register-transfer level (RTL) descriptions combined with a variety of optimization and transformation techniques have been effectively



*R. Gupta et al. (eds.), Formal Methods and Models for System Design,* 33-56.

leveraged to design ASIC and FPGA systems. For software systems, high-level procedural languages, combined with sophisticated architecture-specific compilers have been leveraged to program a wide variety of microprocessors. However, while these approaches have flourished separately, there has been relatively little bridging the gap between the hardware and software design worlds.

One difficulty that both hardware designers and software designers share is the problem of constructing large designs with high-level abstractions. Both hardware and software languages have developed abstraction mechanisms based on the principles of data hiding and encapsulation, enabling design reuse. In software, significant progress has been made extending procedural languages with *objects* that associate data with a set of *methods* that access that data. Languages for building *models* of object relationships, such as the Unified Modeling Language and *design patterns* of common relationship structures, have further organized object-oriented design techniques. On the hardware side, *modules* are often used to hide the details of complex logic blocks. The internal logic of a module can only be accessed through externally visible *signals. Communication protocols* allow modules to declare high-level communication patterns and for those interfaces to be automatically replaced with the correct signals and control logic.

We claim that one of the reasons that technologies for hardware design and software design have remained separate is the vast difference between methods and signals. In particular, a method represents a *transfer of control*, forcing sequential execution of code. This sequential transfer of control manifests itself as the possibility of deadlock when concurrent threads execute the same piece of code. In contrast, signals represent pure *dataflow* between modules, allowing modules to operate concurrently. Because of the inherently sequential nature of object interaction, the concurrency of circuits, real-time software, and distributed systems-of-systems are difficult to represent in an object-oriented system.

Another significant difference between technology for hardware design and software design concerns *static structure*. In object-oriented languages, objects and object structures are dynamically created during execution. In most hardware design languages, however, the relationships between modules are fixed and do not change at run-time. Static relationships between modules are a significant aid for circuit synthesis, but stand in the way of using hardware design techniques for software and even complicate the design of systems with reconfigurable logic.

In this chapter, we describe a style of system modeling that is capable of expressing systems that are both concurrent (such as logic circuits and distributed systems), and also dynamically modifiable at run-time (as in software or reconfigurable logic). We call this style *actor-oriented design*. Actor-oriented models concentrate on the dataflow through a system and are thus able to elegantly describe concurrent systems. The interaction between actor-oriented components is described using well-understood patterns of interaction, called *models of computation*. Models of computation play the same role in actor-oriented models as design patterns in object-oriented languages and communication protocols in hardware design languages. Actor-oriented models can also be reconfigured at different levels of granularity. In order to soundly compose components and to provide for synthesis of hardware and software for these models, we describe several techniques for analyzing the static structure of these models.

## 2 ACTOR-ORIENTED DESIGN

A recent thrust in embedded systems research has been to develop domain-specific languages and synthesis tools for those languages. For example, Simulink, from The MathWorks, was originally created for control system modeling and design, and has recently come into significant use in embedded software development (using Real-Time Workshop and related products). Simulink is one of the most successful instances of *model-based design* [28]. It provides an appropriate and useful abstraction of control systems for control engineers.

Simulink also represents an instance of what we call *actor-oriented design*. Actors are concurrent components that communicate through *ports* and interact according to a common pattern of interaction, or *model of computation*. Primarily, actor-oriented design allows designers to consider the interaction between components distinctly from the specification of component behavior. Actor-oriented design allows us to compose objects with simple behavior to create more complex behaviors and to perform such compositions repeatably and robustly.

Our notion of actor-oriented design is related to the work of Gul Agha and others. The term "actor" was introduced in the 1970's by Carl Hewitt of MIT to describe the concept of autonomous reasoning agents [8]. The term evolved through the work of Agha and others to describe a formalized model of concurrency [1]. Agha's actors each have an independent thread of control and communicate via asynchronous message passing. We additionally embrace a larger family of models of concurrency that are often more constrained than general message passing. Our actors are still conceptually concurrent, but unlike Agha's actors, they need not have their own thread of control. Moreover, although communication is still through some form of message passing, it need not be strictly asynchronous.

## 2.1 Actor-oriented models

In actor-oriented design, components called actors execute and communicate with other actors in a model. Actors have a well-defined component interface. This interface abstracts the internal state and behavior of an actor, and restricts how an actor interacts with its environment. The interface includes ports that represent points of communication for an actor, and parameters that are used to configure the operation of an actor. Often, parameter values are part of the *a priori* configuration of an actor and do not change when a model is executed. The configuration of a model also contains explicit communication channels that pass data from one port to another. The use of channels to mediate communication implies that actors interact only with the channels that they are connected to and not directly with other actors.

Like actors, which have a well-defined external interface, *models* (which are compositions of interconnected actors) may also define an external interface. External interfaces allow for *hierarchical abstraction*, illustrated in Fig.1. This interface consists of external ports and external parameters, which are distinct from the ports and parameters of the individual actors in the model. The external ports of a model can be connected by channels to other external ports of the model or to the ports of actors that comprise the model.

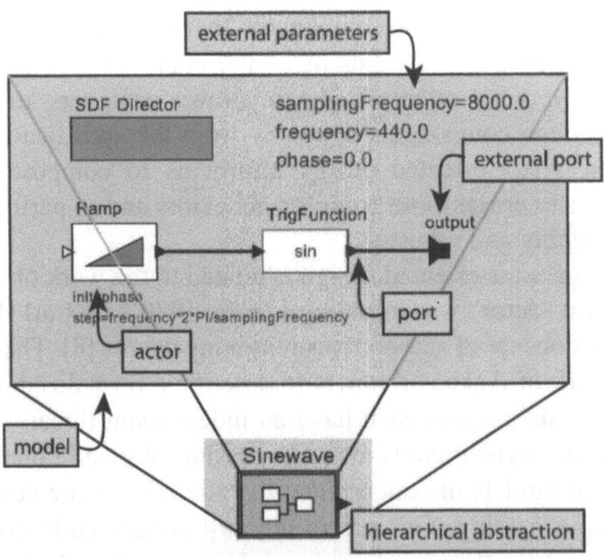


Figure 1: Hierarchical abstraction in actor-oriented design.

External parameters of a model can be used to determine the values of the parameters of actors inside the model.

Taken together, the concepts of models, actors, ports, parameters and channels describe the abstract syntax of actor-oriented design. This syntax can be represented concretely in several ways, such as graphically, as in Fig.1, in XML, or in a program designed to a specific API. Ptolemy II [18] offers all three alternatives.

In addition to Simulink, there are many examples of actor-oriented languages, frameworks, and software techniques, including Labview (National Instruments), Modelica (Linkoping), GME: generic modeling environment (Vanderbilt) [12], Easy5 (Boeing), SPW: signal processing worksystem (Cadence), System studio (Synopsys), ROOM: real-time object-oriented modeling (Rational) [26], VHDL, Verilog, SystemC (Various), Polis & Metropolis (UC Berkeley) [6], and Ptolemy & Ptolemy II (UC Berkeley) [18].

Many of these, like Simulink, use a visual syntax to represent actor-oriented designs. Some are used for designing hardware, some for software, and some for both. An example of a model from Ptolemy II is shown in Fig.2. This model uses the actor defined in Fig.1. Of course, many different syntaxes are compatible with actor-oriented modeling, and for some applications, visual syntaxes like that in Fig.2 are entirely inappropriate.

## 2.2 Models of Computation

It is important to realize that the syntactic structure of an actor-oriented design says little about the semantics. The semantics is largely orthogonal to

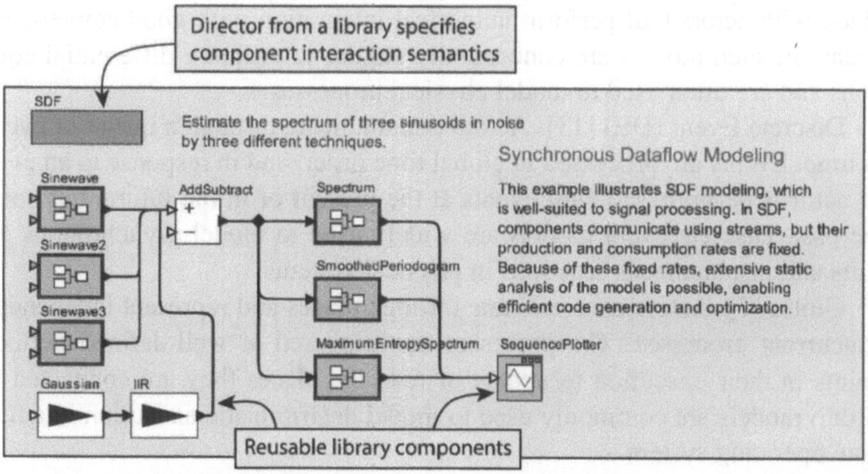


Figure 2: Ptolemy II model using the "synchronous dataflow" (SDF) model of computation.

the syntax, and is determined by a *model of computation*. The model of computation defines the nature of communication between components, while the components themselves encapsulate computation. The model of computation consists of operational rules for executing a model. These rules determine when actors perform internal computation, update their internal state, and perform external communication.

There is a wide variety of useful models of computation with different semantics for representing behavior. Additionally, different models of computation can be translated down to lower level platforms in different ways, with different design tradeoffs. A few interesting models of computation are summarized below:

Synchronous Dataflow (SDF) [14]: Actors communicate through queues and send or receive a fixed number of tokens each time they are fired. Actors are fired according to a predetermined static schedule. Synchronous dataflow models are highly analyzable and have been used to describe hardware and software systems.

Synchronous Reactive (SR) [3]: Actors communicate through places (single variables), which may additionally take on the special value *Unknown*. At each execution point, or *tick*, actors update the values of their output, until the system stabilizes. Synchronous reactive models are semantically similar to synchronous languages, which are widely used to describe hardware and software systems.

Continuous Time (CT) [20]: Actors communicate through places, which represent the value of a continuous time signal at a particular point in time. At each time point, actors compute their output based on their previous input and the tentative input at the current time, until the system stabilizes. When combined with actors that perform numerical integration with good convergence behavior, such models are conceptually similar to ordinary differential equations and are often used to model physical processes.

Discrete Event (DE) [13]: Actors communicate through a queue of events in time. Events are processed in global time order, and in response to an event an actor is permitted to emit events at the present or in the future, but not in the past. Discrete event models are widely used to model asynchronous circuits and instantaneous reactions in physical systems.

Giotto [7]: Actors communicate through places and represent independent concurrent processes. The processes are triggered at well-defined periodic points in their execution to update or read the places they are connected to. Giotto models are commonly used to model deterministic execution in a real-time operating system.

In Ptolemy II, the model of computation is indicated by a *director*, represented by the boxes at the upper left of each of Fig.1 and Fig.2. Ptolemy II is unique among the frameworks listed above in that it has no built-in preferred model of computation, but rather supports a variety of models of computation via components called directors. This capability enables hierarchically heterogeneous design [4], where the modeling properties engendered by different models of computation can be combined.

Fig.3 shows a hierarchically heterogeneous Ptolemy II model that uses a continuous time (CT) director at the top level of the hierarchy and a Giotto director at the next level down. The CT director includes an ODE solver and has semantics somewhat similar to Simulink. The Giotto director implements the semantics of the Giotto language [7], which supports periodic hard-real-time tasks and mode switching. The key here is that the Giotto modeling framework (the Giotto director) is not designed specifically to interact with a continuous-time director. Instead, it exports a well-defined behavioral interface for hierarchical composition.

## 2.3 Relation to other design techniques

In this section, we relate actor-oriented design to other design techniques, using the notion of a *platform*. Sangiovanni-Vincentelli has articulated clearly

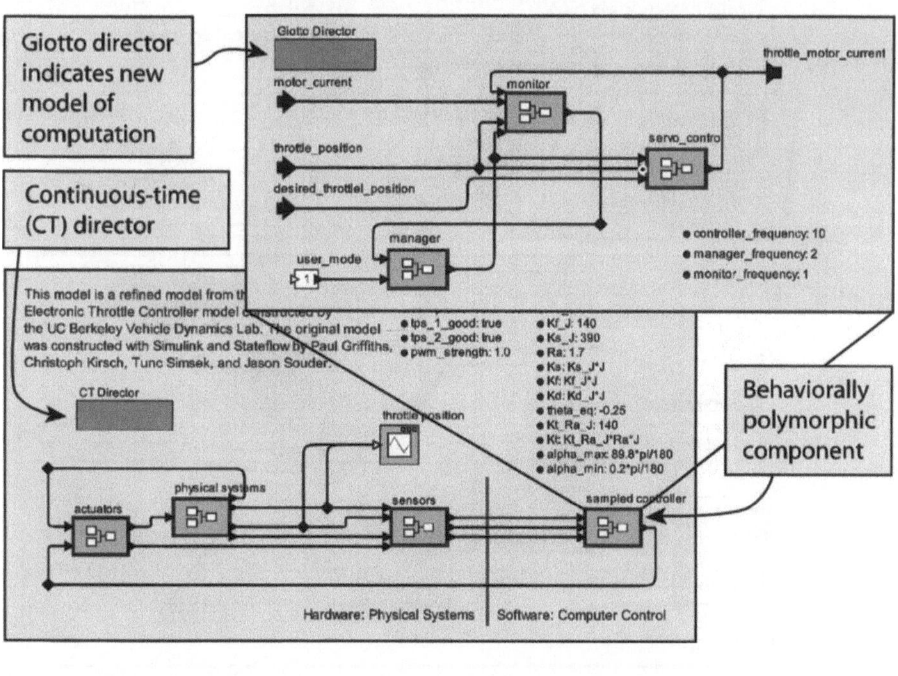


Figure 3: Hierarchical heterogeneity enables an actor refinement to use a different model of computation than that overall framework.

the benefits of *platform-based design* [25]. We have defined platforms to be sets of designs [15]. Examples of such sets are:

- The set of all boolean functions.
- The set of all x86 binaries.
- The set of syntactically correct Java programs.
- The set of all Java byte-code programs.
- The set of all Wintel PCs.
- The set of all ANSI C programs.

Fig.4 illustrates some platforms and their interrelationships. Each box is a platform. For example, the set of all Java programs has a downward arrow labeled "javac" which represents a function whose domain is the set of all syntactically correct Java programs, and whose range is the set of Java byte

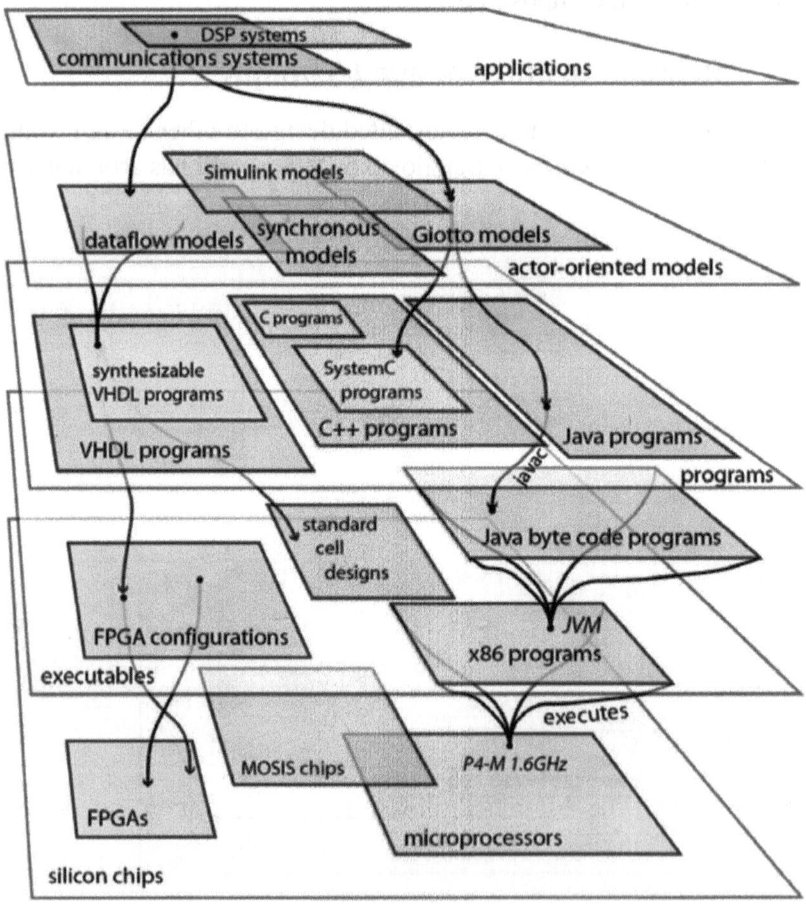


Figure 4: An illustration of some *platforms* (sets of designs) and their interrelationships.

code programs. Similarly, the set of x86 programs contains a program, an implementation the Java virtual machine, that is related to the set of all Java byte code programs by its ability to execute members of the set. For more details about this view of platforms, see [15].

Model-based design [28] concentrates on the specification of designs in platforms with useful modeling properties. For example, Simulink block diagrams represent control systems as visual representations of systems of ordinary differential equations. The set of Simulink designs, therefore, inherits the formal properties and analytical properties of systems of ODEs. These properties are useful to control systems engineers for analyzing, for example, transient responses and stability. Model-based design is also beginning to focus on the modeling of properties themselves via so-called "metamodeling".

Fig.4 shows a level labeled "actor-oriented models" at a level above programs. The platforms at this level are relatively immature, but play a central role in the future of codesign. In particular, actor-oriented models focus on the separation of concerns between communication and computation [10]. We place actor-oriented models at a level between programs and applications for several reasons. Firstly, actor-oriented design provides a platform which abstracts both hardware and software. Without this level, the gap between "applications" and "programs" is large enough to force very early decisions about hardware vs. software implementations. Secondly, the implementation technologies for early successes in this domain, such as Simulink, take the form of generators (such as Real-Time Workshop) that transform actor-oriented models into program-level models.

### 2.3.1 Comparison With Object-Oriented Programs

The level in Fig.4 labeled "programs" represents conventional object-oriented software design practice today. Much of the action here is about attempting to give useful modeling properties to designs at this level. For example, UML (the unified modeling language) and the MDA (model-driven architecture) from OMG (the object management group) are about giving modeling structure to designs at this level. Much of the work in design patterns that was kicked off by Gamma *et al.* [5] is about identifying modeling structure in programs at this level.

Fig.5 illustrates the difference between object orientation and actor orientation. In current practice, as defined by languages such as C++ and Java, object-oriented components interact with one another principally by method calls, which represent a transfer of control. The result is that the coordination structure of object-oriented programs is often poorly expressed and described. Managing concurrency, for instance, is notoriously difficult using threads, mutexes and semaphores. Conventions that ensure deadlock avoidance, such as acquisition of locks in a fixed order (see for example [11]), are not supported by the languages (nothing about a method signature declares what locks it will acquire, for instance). As a consequence, these conventions are difficult to apply in practice, and seemingly innocent changes to code can create disastrous failures such as deadlock.

As a result, it is difficult to treat objects in object-oriented languages as *components*. Despite the success of software component technologies such as CORBA and KOALA [21] for representing concurrent software, they fundamentally suffer from fragile composition. The interaction between two components can be broken by simply adding more components to the system. Higher-level patterns are codified only through the APIs of services, and usage patterns for these APIs are expressed only informally in documentation. The communication mechanism for these components is an integral part of a component design, making them difficult to reuse.

In actor-oriented abstractions, low-level implementation mechanisms of threads and semaphores do not even rise to consciousness, forming instead the "assembly-level" mechanisms used to deliver much more sophisticated models of computation. Moreover, actor-oriented abstractions can embrace time and concurrency, and therefore match much better the modeling of hardware systems, which are intrinsically concurrent.

### 2.3.2 Comparison with Hardware Description Languages

In comparison with object-oriented languages, hardware description languages, such as Verilog and VHDL, are more actor-oriented. A signal in these languages represents the flow of data between concurrent components. For the most part, hardware modules can be composed freely without concern for introducing primitive semantic errors, such as deadlock.

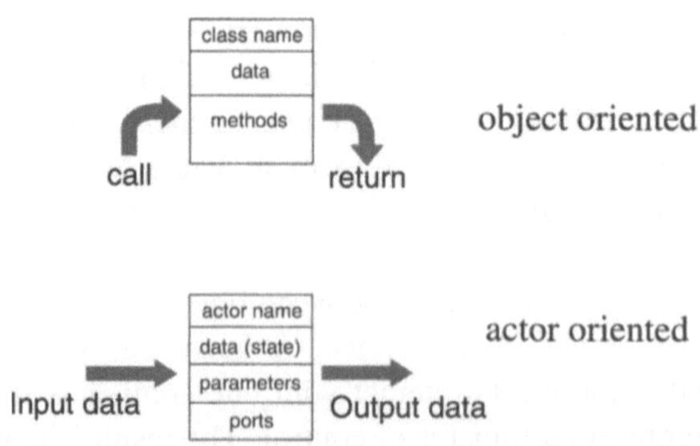


Figure 5: An illustration of the relationship between actor-oriented models and object-oriented components.

However, patterns of communication and interaction in these languages are difficult to specify directly. In order to robustly compose different modules, they must conform to rather arcane interface rules [9]. In other words, the set of signals that a hardware component exposes provides insufficient information for composition.

A key difficulty in hardware description languages arises from the emphasis on clocked synchronous models for synthesis. This emphasis forces designers to describe complex communication behaviors over many clock cycles, complicating module interfaces and the design of modules with truly asynchronous interfaces. Although the description of high-level communication protocols and the use of *communication refinement* can hide some of the complexity, the complexity is still present in the design. In actor-oriented models, high-level models of computation are often sufficient to replace communication protocols over low-level signals. Signals are recovered when the actor oriented model is transformed onto a synchronous platform for synthesis in a process similar to communication refinement. Additionally, the use of high-level models of computation often implies properties of a model that are expensive to check otherwise, such as deadlock-freedom of a set of communication protocols.

A second difficulty in hardware design languages arises from the enforcement of static structure. Although static structure is natural for many hardware applications, such as ASICs, it is inadequate in other cases. For instance, coarse- and fine-grained reconfiguration is often present software and reconfigurable logic designs. Although some kinds of parameterization can be represented manually in hardware design languages, it is left to designers to construct these idioms using registers and control logic. Actor-oriented models not only directly express these kinds of reconfiguration, but also provide for them in a semantically robust way that can be analyzed and constrained when necessary.

## 2.4 Actors are Components

We consider actors to be *abstract system components* in the following senses:

- They have a well-defined interface, given by ports and parameters, through which they can interact with the rest of the model. Other interactions are not allowed.
- Actor interfaces provide necessary information for checking compatibility with other interfaces, such as data type and behavioral constraints.
- They are generalizable across different uses of the same component. For instance, a component may be given new parameter values or may be connected to a different set of channels.

- They may have vastly different levels of granularity. For instance, an actor may represent algorithms ranging from the boolean OR of two bits to the boolean satisfiability solution of an arbitrary logic formula.
- Larger components may be assembled from smaller components. Component composition is the primary mechanism by which complex behavior is constructed.
- They are executable specifications of behavior. (i.e. Actor models have an operational semantics)

Of these, the generality of actors is something that often varies from one actor to the next. Many actors are designed specifically to fill a particular role in a single model, whereas other actors are designed to be more widely applicable. Although general actors are often more difficult to design, they can be reused in a wider variety of contexts. By maximizing the possibilities for *component reuse*, systems can be designed more quickly and design effort can be leveraged more easily.

Component generality also allows for the possibility of *dynamic reconfiguration* during execution. For example, an actor that is not specialized to a connection with another actor has the potential to be dynamically reconnected to a different actor with no modification. Similarly, an actor that is not specialized to particular parameter values has the potential to be dynamically reconfigured with new values. The use of generalized components increases the available design possibilities when using components.

Object-oriented languages primarily offer generality through variable function arguments. Any set of actual arguments that are compatible with the formal arguments of a function may be passed to that function. Furthermore, a function may be called from many places in a program using different arguments in each case. We call such generality *data polymorphism*. Many object-oriented languages add the notion of *type polymorphism*, where a program can operate uniformly on different types of objects, as long as those types expose similar operations. For the most part, hardware description languages lack rich forms of generality and primarily provide data polymorphism. We identify several types of generality in actor-oriented models, all of which are exhibited in Ptolemy II:

- data polymorphism: An actor may be given data with different values.
- type polymorphism: An actor may be given data of different types.
- parameter generality: An actor may be given a different set of parameter values.
- connection generality: An actor may be connected to a different set of actors.
- behavioral polymorphism: An actor may interact with other actors according to different models of computation.

## 2.5 Leveraging Static Structure

As mentioned previously, many object-oriented modeling techniques are concerned with expressing the static object-oriented structures. Actor-oriented models also emphasize static structure in a model. We often build models whose structure does not change during the execution of a model, and when we do, we can exploit the fact that the structure is static. Such models have a fixed set of actors, interacting according to a specific model of computation. The ports of an actor form the declared interface of the actor, and this interface does not change. The connections between actors in a model are specified during the construction of the model and do not change during execution of a system. In fact, the Ptolemy II system is specifically designed to allow the construction of models that dynamically evolve during their execution, and yet this capability has been rarely leveraged.

From a software architect's point of view, an emphasis on static structure is a convenient design abstraction. A UML entity-relationship diagram can be considered as the permanent structure of the system, and the fact that this structure is dynamically created in an object-oriented system can often be ignored. Hardware architects, on the other hand, must carefully choose the static computational structure that will be placed in an ASIC since this structure cannot, in most cases, be modified after synthesis. Traditionally, hardware design languages have forced designers to manually provide for circuit reconfiguration.

We have leveraged the static structure of actor-oriented models in two key ways. Firstly, static model structure combined with appropriate type analysis enables detection of unsafe actor compositions. Analysis of both traditional data types and modern behavioral types support generic, reusable components. It is well known how to make such a component polymorphic in a data type sense (*type polymorphism*). It can be designed to be able to add numbers (int, float, double, Complex), add strings (concatenation), add composite types (arrays, records, matrices), and add user-defined types. The formal structure of such type systems is much better known than the formal structure of parameterization and behavioral polymorphism, both of which have re-use benefits at least as great. For this reason, the following sections describe type systems for static analysis of reconfiguration and heterogeneous compositions.

Secondly, we have developed the notion of *actor-oriented specialization* which makes use of the static context of an actor to improve performance in software. Specialization enables practical application of type polymorphism, parameterization, and behavioral polymorphism, since providing these abstractions at run-time incurs significant performance cost. Specialization leverages the static structure in models, as determined through static model analysis, to enable component-based design at all levels of hierarchy. Specialization also enables ASIC and FPGA implementation synthesis without explicit management of reconfiguration.

# 3 PARAMETERIZATION

Actor parameters are a simple way to configure the behavior of a hierarchical model. At design time, parameters help keep the size of actor libraries manageable and allow models to be quickly modified or tuned for performance. At run time, actor parameters allow for dynamic reconfiguration of actors (and models) while a model is running. Dependencies between parameter values allow simple reconfiguration options to be exposed while hiding the reconfiguration of individual parameters from a designer. However, unconstrained reconfiguration can be problematic during design, since semantic modeling constraints can be violated.

## 3.1 Representing Reconfiguration

There are many common modeling syntaxes for generically representing reconfiguration in actor-oriented models. For instance, reconfiguration can be represented using a *modal model* controlled by a finite state machine. Each state of the finite state machine corresponds to a hierarchical model that gives the behavior of the modal model in that state. Finite state machine transitions represent switching to a new configuration of a model. Transitions can also set the value of actor parameters. In a hierarchical model, modal models appear as just another level of abstraction. An example of a modal model is shown in Fig.6.

Reconfiguration can also be tied directly to dataflow in a model, using *reconfiguration ports*. A reconfiguration port is associated with an actor parameter and data received by the port reconfigures the value of that parameter. Reconfiguration ports exist in many commercial dataflow environments, such as AVS/Express (Advanced Visual Systems, Inc.). An example of a reconfiguration port is shown in Fig.7.

Reconfiguration can also be represented using a special actor with a single input port. This `SetParameter` actor is bound to a single parameter of a model in a similar fashion to the reconfiguration port. However, since the `SetParameter` actor does not require the presence of hierarchy in the model, it is often more convenient. It can also be more difficult to use, since the point in execution of a model that the `SetParameter` actor is executed is more dependent on its interaction with other actors in the model. One approach to this difficulty is to treat the execution of the `SetParameter` actor as a *request* for reconfiguration at the next possible time, resulting in delayed reconfiguration at a well-defined point in the execution of the model.

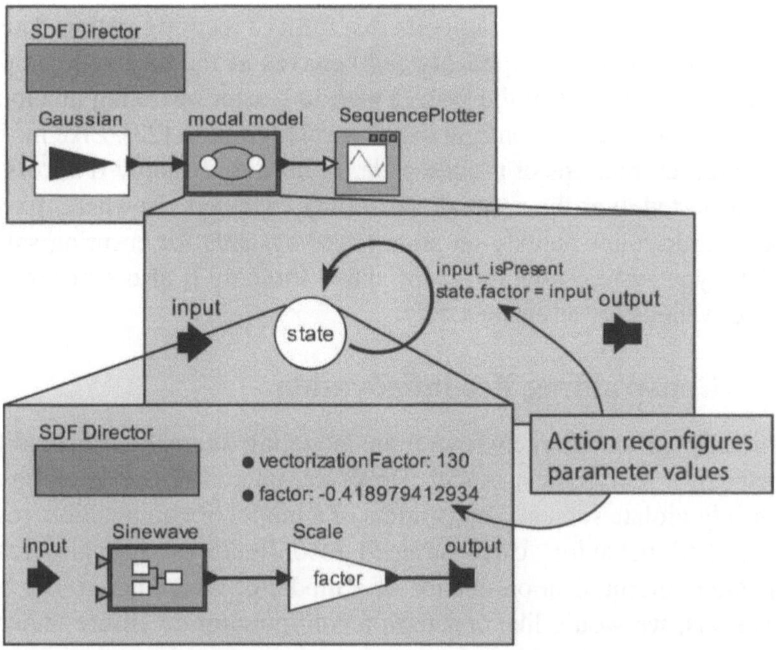


Figure 6: Ptolemy II model using a modal model for reconfiguration.

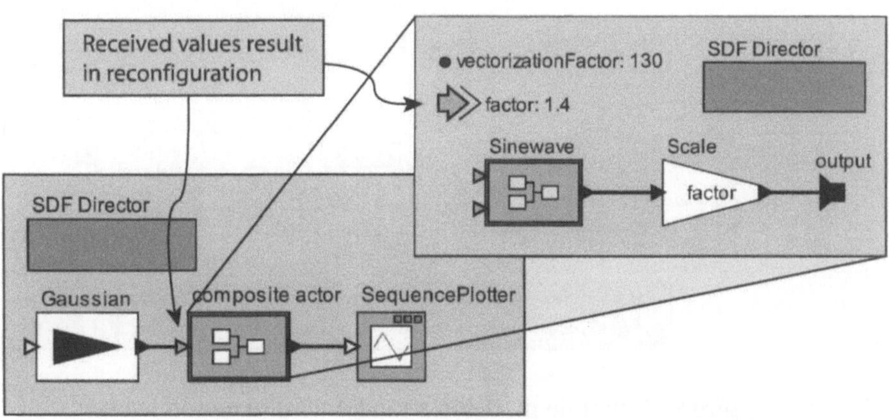


Figure 7: Ptolemy II model illustrating reconfiguration ports.

## 3.2 Mobile Models

Although most reconfiguration replaces simple data values in a model, we note that reconfiguration in actor-oriented models can also replace portions of an actor-oriented model. The `MobileModel` actor[1], which acts as a placeholder for other components, supports this form of reconfiguration. This actor receives actor-oriented components and behaves as the received component. Fig.8 shows an example of the `MobileModel` actor operating in a model of a control system with a reconfigurable control algorithm [23]. Like more conventional representations of mobile code, as in Java, Ptolemy II allows actors to be transported over the network and safely executed elsewhere. Execution of mobile code relies heavily on strong type systems for ensuring safety. In Java, data type systems are sufficient, while Ptolemy II also relies on behavioral type systems for ensuring safety.

## 3.3 Constraining Reconfiguration

Generally speaking, reconfiguration modeling syntaxes can modify any parameter in a hierarchical model. However, unconstrained reconfiguration can possibly violate semantic constraints of a model of computation, resulting in unexpected behavior. Unconstrained reconfiguration can also result in unstructured communication outside of a model of computation. For both of these reasons, we would like constrain reconfiguration to ensure understandable models with predictable behavior.

We have focused on *hierarchical reconfiguration* as the key interesting modeling constraint. Informally, hierarchical reconfiguration implies that a

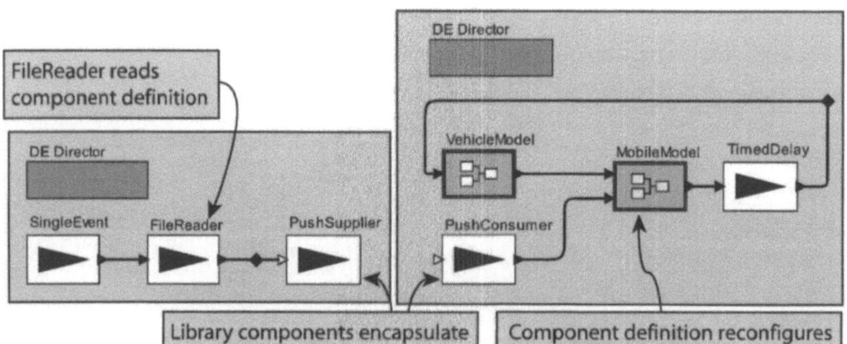


Figure 8: A mobile model is a model that, like mobile code, can be transported over the network and safely executed elsewhere.

1. Designed by Yang Zhao.

parameter associated with a particular actor in a hierarchical model is not reconfigured while that actor is executing. Hierarchical reconfiguration solidifies our intuition that the behavior of an actor is a function of its parameters and the interesting fine-grained interaction of an actor with its environment is strictly expressed through ports.

In order to ensure that reconfiguration in a model is hierarchical, we have developed a *reconfiguration type system* that approximately analyzes how each parameter in a model is reconfigured [24]. This theory determines a *least change context* for each parameter that summarizes the effect of reconfiguration and parameter dependency on that parameter. Reconfiguration constraints, such as hierarchical reconfiguration, can be easily expressed as a constraint on the least change context and efficiently checked. This type system is built-in to Ptolemy II and used to guarantee safe uses of reconfiguration.

# 4 BEHAVIORAL POLYMORPHISM

The ability to construct hierarchical models using multiple models of computation is a key capability of Ptolemy II, but it creates an interesting challenge. Primarily, a hierarchical component like that in Fig.1 must be able to operate within a foreign model of computation. That is, the semantics of component interaction inside a hierarchical component must be able to differ from the semantics of component interaction outside the hierarchical component. To achieve this, Ptolemy II has pioneered the use of *behavioral polymorphism*. A component is behaviorally polymorphic in that its behavior will depend on the external context in which it is placed. Behavioral polymorphism enables hierarchical heterogeneity.

## 4.1 Motivating Example

Consider the AddSubtract actor in the center of Fig.2, which adds or subtracts signals that are provided at the input ports. We notice that the function performed by this actor is entirely independent of how data is transported to the actor. For example, the actor can be used in a dataflow framework, where it will add when all connected inputs have data. Or it can be used in a time-triggered framework, where it will add when the clock ticks. Or it can be used in a discrete-event framework, where it will add when any connected input has data, and add in zero time. Or it can be used in a process network framework, where it will execute an infinite loop in a thread that blocks when reading empty inputs and adds when it has read data from all connected inputs. Or it can be used in a CSP-based framework, where it will execute an infinite loop that performs rendezvous on input or output. Etc.

By not choosing among these when defining the component, we get a huge increment in component reusability. Essentially, the interaction pattern between actors is an aspect that is independent from the actors' functionality. Behavioral polymorphism is essential to managing the complexity of hierarchically heterogeneous models. Without behavioral polymorphism, each hierarchical component would have to be designed for a specific combination of the inside and outside models.

A key problem when describing behavioral polymorphism is defining exactly how polymorphic a component actually is. How do we describe the behavioral requirements of a component, in order to ensure that a component will work in a particular circumstance? In a data type system, a type checker ensures that the data type requirements of a component are satisfied in a given context. A corresponding *behavioral type system* is needed to robustly support behavioral polymorphism.

## 4.2 Object-Oriented Approach to Achieving Behavioral Polymorphism

In Ptolemy II, the director in a model defines the model of computation, which includes the communication mechanism between components. As illustrated in Fig.9, the director instantiates an object that implements a Java interface called Receiver, shown in UML form at the upper right of Fig.9. The six methods of the Receiver interface have different implementations depending on the model of computation. These implementations can, for example,

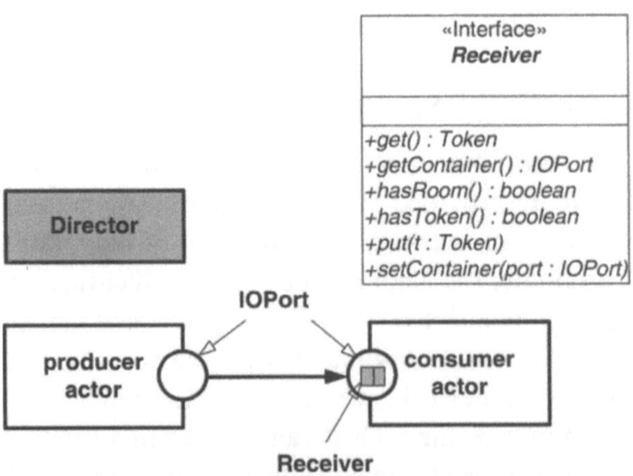


Figure 9: The Ptolemy II Receiver interface.

queue messages, or implement rendezvous, or post events to an event queue, for example. Since the receiver instance used in communication is supplied by the director, not by the component, the component becomes behaviorally polymorphic. Whether a call to the get() method (which reads an input) reads from a queue, blocks, or waits for a rendezvous is not up to the component designer. It is up to the director designer.

## 4.3 Behavioral Types

The object-oriented approach of the previous section has its limitations. What if:

- The component requires data at all connected input ports?
- The component can only perform meaningful operations on two successive inputs?
- The component can produce meaningful output before the input is known (enabling it to break potential deadlocks)?
- The component has a mutex monitor with another component (e.g. to access a common hardware resource)?

None of these is expressed in the object-oriented interface definition, yet each can interfere with behavioral polymorphism.

The problem is that the Receiver interface only describes object-oriented structure between method calls. It defines abstract data types, but not the dynamic behavior that is central to these communication protocols. Instead of relying on inadequate object-oriented data types, we use a *behavioral type system* (which we previously called a "system-level type system" [16]) to obtain benefits analogous to data typing.

An example of a behavioral type signature for communication protocols is shown in Fig.10. This signature describes a component's interaction with its environment using an extension of *interface automata* [2][17]. Type checking is done through automata composition, which will detect component incompatibilities. Subtyping order is given by the alternating simulation relation [2], supporting behavioral polymorphism. An alternative representation of behavioral types would be pre/post conditions [19], which may be essentially equivalent, but lacks the intuitive visualization of Fig.10. For details on how to specify behavioral types, see [16][17].

The use of a behavioral type system enables key quality control techniques.

**Checking behavioral compatibility of components that are composed.** Composition of components that cannot effectively work together because of conflicts in their dynamics (e.g. differing assumptions about communication protocols) is identified as an error. This is analogous to type conflicts in conventional type systems.

**Checking behavioral compatibility of components and frameworks.** Use of a component in a framework that cannot effectively meet the assumptions of the component is identified as an error.

**Behavioral subclassing enables interface/implementation separation.** Behavioral type signatures define interfaces that can be implemented by a number of components. For example, components that are simple memoryless stream transformers (which react to input data, transform it in some way, and produce output results) all share the same behavioral interface definition.

**Helps with the definition of behaviorally-polymorphic components.** Behavioral type signatures declare the minimal assumptions that are required for the component to work. This maximizes the number of contexts in which the component can be used.

## 5 ACTOR-ORIENTED SPECIALIZATION

Although high-level models can be useful simply for behavioral design and conceptualization, these models become truly powerful when they become a designer's primary programming paradigm. Design tools are more than capable of compiling designs from a modeling platform into a executable platform. Object-oriented modeling tools emphasize round-trip engineering for maintaining consistency between models and code. Many actor-oriented design tools provide code generation capability for automatically translating a model into executable software and hardware. In particular, Simulink's real-

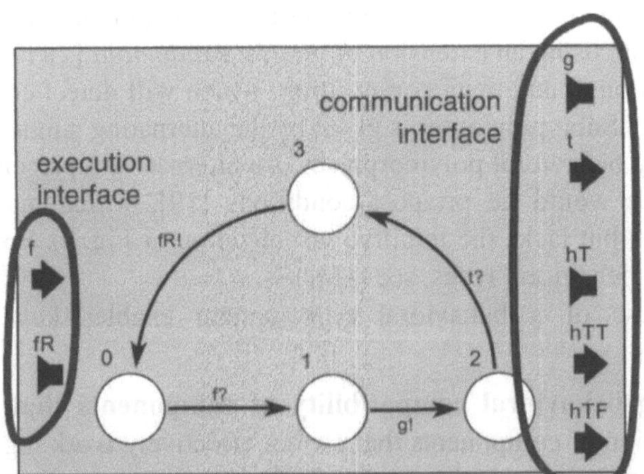


Figure 10: Behavioral type signature using interface automata.

time workshop has become commonly used in the control-systems community.

Specialization of object-oriented programs has grown out of the long history of partial evaluation for functional languages and method inlining optimization in procedural languages. Object-oriented specialization relies on program analysis and extra program annotations to indicate opportunities for specialization [27], and 'optimal' specialization is known to be an undecidable problem. Whereas object-oriented specialization concentrates on transforming the method calls between objects, actor-oriented specialization concentrates on transforming interaction patterns between actors. Because the interaction patterns between actors are explicitly described, actor-oriented specialization is straightforward to apply and relatively simple to understand.

In the context of actor-oriented design, we claim that specialization of a model to a particular context is the primary operation in transforming actor-oriented models into program-level models. Executability alone is not sufficient for resource constrained systems. If it were, then we would simply embed a complete simulation environment in every system and declare success. In reality, optimization to reduce resource usage is crucial for reducing system cost and satisfying timing requirements. Actor-oriented specialization results in optimized behavior while still allowing generic design-time constructs [22]. Specialization implements a transformation from a modeling platform to an efficient implementation platform.

We consider several forms of actor-oriented specialization according to the various generic aspects of Ptolemy II components:

- Parameter specialization
- Type specialization
- Connection specialization
- Domain specialization

Parameter specialization replaces actor parameters that are not reconfigured with a constant value. This specialization relies on analysis of an actor-oriented model to determine constraints on reconfiguration, as described in Section 3. Efficient computation of the values of parameters is possible through partial evaluation of parameter expressions dependent on constant parameters. Additionally, reconfiguration constraints allow for further optimization, since parameter reconfiguration occurs only at well determined points in the execution of a model.

Type specialization replaces generic data types with specific types inferred from the model context. For software implementation, specific types allow the removal of abstract data type indirection through *token unboxing*. In Java implementations synthesized from Ptolemy II, token unboxing greatly reduces load on the garbage collector, resulting in faster and more predictable execution. For hardware implementations, specific types are crucial to selecting efficient representation of data values and operations.

Connection and domain specialization deal directly with the implementation of a model of computation on a particular implementation platform. These specializations are possible when the structure of models and the modeling semantics are fixed at design time. We expect that a provider of a synthesis system will be largely concerned with the efficient implementation of component interaction in a particular system. Given that a designer wishes to target a particular implementation platform, he will select a model of computation that has been provided for that platform and begin designing. Retargeting the model to another implementation platform that provides the same high-level semantics is easy, while targeting an implementation platform without direct support of a model of computation, or migrating a model from one semantics to another is likely to be much more difficult.

## 6 CONCLUSION

We have outlined a class of design techniques that we call actor-oriented design, and have related it to model-based design, platform-based design, and object-oriented design. We have described mechanisms for generalizing actors, promoting re-use in various contexts. We have focused on parameterization and behavioral polymorphism, and presented described type systems for statically determining valid modeling compositions. These type systems bring to actor-oriented design benefits similar to what abstract data types and their corresponding type systems have brought to object-oriented design. Lastly, we have argued that component specialization can offset the performance costs associated with component-based design.

## 7 ACKNOWLEDGMENTS

The work described here is part of the Ptolemy project, which is supported by the National Science Foundation (NSF award number CCR-00225610), the Defense Advanced Research Projects Agency (DARPA), and Chess (the Center for Hybrid and Embedded Software Systems), which receives support from the State of California MICRO program, and the following companies: Daimler-Chrysler, Hitachi, Honeywell, Toyota and Wind River Systems.

## References


[1] G. Agha, "Concurrent Object-Oriented Programming," *Communications of the ACM*, 33(9): 125–140, September 1990.

[2] L. de Alfaro and T. A. Henzinger, "Interface Automata," *Proceedings of the Ninth Annual Symposium on Foundations of Software Engineering* (FSE), ACM Press, 2001.

[3] Stephen A. Edwards and Edward A. Lee, "The Semantics and Execution of a Synchronous Block-Diagram Language," *Science of Computer Programming*, Vol. 48, no. 1, July 2003.

[4] J. Eker, J. Janneck, E. Lee, J. Liu, X. Liu, J. Ludvig, S. Neuendorffer, S. Sachs, Y. Xiong, "Taming Heterogeneity---the Ptolemy Approach," *Proceedings of the IEEE*, v.91, No. 2, January 2003.

[5] E. Gamma, R. Helm, R. Johnson and J. Vlissides, *Design Patterns: Elements of Reusable Object-Oriented Software*, Addison Wesley, 1994.

[6] G. Goessler and A. Sangiovanni-Vincentelli, "Compositional Modeling in Metropolis," Second International Workshop on Embedded Software (EMSOFT), Grenoble, France, LNCS 2491, Springer-Verlag, October 7-9, 2002.

[7] T. A. Henzinger, B. Horowitz and C. M. Kirsch, "Giotto: A Time-Triggered Language for Embedded Programming," First International Workshop on Embedded Software (EMSOFT), Lake Tahoe, CA, LNCS 2211, Springer-Verlag, October 8-10, 2001.

[8] C. Hewitt, "Viewing Control Structures as Patterns of Passing Messages," *Journal of Artificial Intelligence*, 8(3): 323–363, June 1977.

[9] M. Keating and P. Bricaud, *Reuse Methodology Manual for System-On-A-Chip Designs*, Kluwer, 1998.

[10] K. Keutzer, S. Malik, R. Newton, J. Rabaey, and A. Sangiovanni-Vincentelli. "System Level Design: Orthogonalization of Concerns and Platform-Based Design", *IEEE Transactions on Computer-Aided Design*, 19(12), December 2000.

[11] D. Lea, Concurrent Programming in Java$^{TM:}$ Design Principles and Patterns, Addison-Wesley, Reading MA, 1997.

[12] A. Ledeczi, M. Maroti, A. Bakay, G. Karsai, J. Garrett, C. Thomason IV, G. Nordstrom, J. Sprinkle and P. Volgyesi., "The Generic Modeling Environment," Workshop on Intelligent Signal Processing, May 2001.

[13] E. A. Lee, "Modeling Concurrent Real-time Processes Using Discrete Events," *Annals of Software Engineering, 7:*25–45, 1999.

[14] E. A. Lee and D. G. Messerschmitt, "Synchronous Data Flow," *Proceedings of the IEEE*, 75(9):55-64, September 1987.

[15] E. A. Lee, S. Neuendorffer and M. J. Wirthlin, "Actor-Oriented Design

of Embedded Hardware and Software Systems," *Journal of Circuits, Systems, and Computers*, 12(3): 231-260, 2003.

[16] E. A. Lee and Y. Xiong, "System-Level Types for Component-Based Design," First International Workshop on Embedded Software (EMSOFT), Lake Tahoe, CA, LNCS 2211, Springer-Verlag, October 8-10, 2001.

[17] E. A. Lee and Y. Xiong, "A Behavioral Type System and Its Application in Ptolemy II," *Formal Aspects of Computing Journal*, special issue on Semantic Foundations of Engineering Design Languages, to appear.

[18] E. A. Lee, "Overview of the Ptolemy Project," *Technical Memorandum UCB/ERL M03/25*, University of California, Berkeley, July 2, 2003.

[19] B. H. Liskov and J. M. Wing, "A Behavioral Notion of Subtyping," *ACM Transactions on Programming Languages and Systems*, 16(6): 1811-1841, November 1994.

[20] Jie Liu and Edward A. Lee, "A Component-Based Approach to Modeling and Simulating Mixed-Signal and Hybrid Systems," *ACM Transactions on Modeling and Computer Simulation*, 12(4):343-368, October 2002.

[21] R. van Ommering, "The Koala Component Model for Consumer Electronics Software, *IEEE Computer*, 33(3):78-85, March 2000.

[22] S. Neuendorffer, "Automatic Specialization of Actor-Oriented Models in Ptolemy II," Master's Report, *Technical Memorandum UCB/ERL M02/41*, University of California, Berkeley, CA 94720, December 25, 2002.

[23] S. Neuendorffer, "Implementation Issues in Hybrid Embedded Systems," *Technical Memorandum UCB/ERL M03/22*, University of California, Berkeley, CA 94720, June 2003.

[24] S. Neuendorffer and E. A. Lee, "Hierarchical Reconfiguration of Dataflow Models," Conference on Methods and Models for Codesign (MEMOCODE), June 2004, to appear.

[25] A. Sangiovanni-Vincentelli, "Defining Platform-Based Design," *EEDesign of EETimes*, February 2002.

[26] B. Selic, G. Gullekson and P. Ward, *Real-Time Object-Oriented Modeling*, New York, NY, John Wiley & Sons, 1994.

[27] U. P. Schultz, J. L. Lawall and C. Consel, "Automatic Program Specialization for Java," *ACM Transactions on Programming Languages and Systems*, 25(4): 452 - 499, July 2003.

[28] J. Sztipanovits and G. Karsai, "Model-Integrated Computing," IEEE Computer: 110–112, April 1997.

Chapter 3

# STRUCTURAL COMPONENT COMPOSITION FOR SYSTEM-LEVEL MODELS

Frederic Doucet[1], Sandeep Shukla[2], Rajesh Gupta,[1]

[1]*Computer Science and Engineering, University of California, San Diego*

[2]*Electrical and Computer Engineering Department, Virginia Tech*

**Abstract** There are many frameworks for high-level modeling of system-on-chip architectures. Most offer a virtual-to-concrete architecture paradigm and a trade-off in efficiency vs. convenience. We present a framework that implements a split-programming model to provide a composition environment with a type abstraction, close to a "box-and-arrow" specification style. From an abstract specification of component interconnections, a structural type inference is used to produce architectures that can be simulated. We highlight the importance of reflection and introspection to provide component type abstraction. We show the structural type resolution problem to be NP-complete, present a heuristic to solve it and conclude with a packet switch example.

**Keywords:** System-level design, component composition framework, interfaces, split-level programming, type system, type inference, SystemC

## Introduction

Due to advances in microelectronic integration process technologies, it is now possible to have a complete system, from an end application point of view, residing on a single chip. Designing a system-on-chip (SOC) consists of composing a range of building blocks from processors, memory, software to communication and networking elements to reside on a single piece of silicon. While the amount of functionalities implemented such devices have steadily been growing, it has outstripped manual design capabilities and the capacity of electronic design automation tools.



*R. Gupta et al. (eds.), Formal Methods and Models for System Design,* 57-81.

Consequently, much of the recent effort in this CAD research area has been focused on the specification methodologies and languages for modeling designs at the system level. A number of strategies are being explored by the microelectronic designers in an attempt to improve the design productivity and the quality of designs through advances in modeling and validation techniques. Raising the abstraction level at which designs are entered and validated has a direct impact on the design quality and design time. Application functionality is structured into various hardware and software components that as a whole meet the functional and performance goals. Many component-based design frameworks have been propose as tools for design, validation and verification of system-on-chips. However, the level of details for describing such designs is very large, and the problem of defining the right abstraction to ensure design efficiency is difficult. For instance, when communication protocols needs to be redesigned, a repartitioning of hardware and software functionality might be necessary and tedious. The fundamental CAD problem we discuss in this paper is to find a tradeoff between convenience and efficiency in a component composition environment.

The focus of this paper is on SOC component composition frameworks which provide support for correct composition of existing components and automatic or semi-automatic means for validation and verification checks. We are looking for a solution that has both high abstraction and extended composition mechanisms. We show the techniques that are used to improve component reuse and design productivity using an example framework currently under development. We will first review the fundamentals of component frameworks. Then, using the BALBOA component composition framework, we will present an approach to structural composition, abstraction and type inference. We conclude with an example.

## 1. Component Composition Frameworks

A component is a block of functionality implemented in software or as a dedicated piece of silicon hardware or a combination of the two. A component composition framework (CCF) is used integrate and execute components as a system. The CCF often provides a virtual substrate on which components are connected and executed and is an underlying part of the design. A good composition framework provides reasoning capabilities and tools to (i) help design a model of the application and its evolution (ii) help deploy the application. SOC component composition frameworks are difficult to build because of the various ways integrated circuit, electronic and software components are designed, assembled and how system performance and correctness are analyzed.

## 1.1 Fundamentals of Component Frameworks

First, let us review software component frameworks that are used to deploy component application. These frameworks are commonly used in distributed computing and transaction processing over networks. In software component framework, we find recurring themes that we can identify as fundamentals of component frameworks. These are the notions of components, interfaces and connectors.

A component implements a number of interfaces. An interface is a collection of methods and events that provide a way to communicate between components. A connector is a component with the specific purpose of establishing connections between components. Connectors have being identified as essential for structured systems design by many researchers [1]. In order to capture and isolate component interactions, connectors are considered as first class constructs, separately from "computation components". Connector encapsulate communication protocols that can be described in a variety of ways such as in C, or formally such as message sequence charts [2] or interface automata [3]. There have been many papers about methodologies and benefits for interface-based design [4], the separation of communication from computation.

Another fundamental aspect of CCFs is the importance of structural design. Most models are interconnections or netlists [5] [6] of components that are specified statically. This is because functional hierarchy is structured through message and channels. Such a composition presupposes a structural representation for components. Even if a component is not structural, but behavioral, it can often be composed using special components (e.g., protocol modules, directors, controllers) interconnecting the components.

While component assembly can be done in the component implementation language, it is often better to use a component assembly language. The role of a composition language is to instantiate and connect components. In software engineering research, architecture description languages (ADL) [7] have addressed parts of the problems of orthogonolizing component definition from component assembly to solve typing mismatches [8] by validating component-connector composition through some formalisms. However, even if many successful solutions have been implemented, the problem of component composition in software is not considered solved, especially in the verification and performance analysis context.

## 1.2 Model Interoperability

In the SOC design context, CCFs are used to build SOC models, thus adding a modeling or abstract dimension to the composability problem. One often is faced with composing components described at different levels of abstraction, one component at behavioral level (e.g., as an algorithm) and another in a register transfer level (RTL) description. Model composability can be defined along a number of modeling dimensions [9]. For this paper we focus on the structural and behavioral composition dimensions.

#### 1.2.1 Structural Composition.

Structural composition is about the level of detail in the structure of a model. For instance, think about it like a block diagram with sub-block diagrams and so on. The level of depth of a model is the level of structural details. This also includes the precision of data types which expresses the representation or format of data values in a model. Data values can be active messages, objects, abstract data types, enumerated values, word-level values, integers, bit-true representations, etc.

#### 1.2.2 Behavioral Composition.

Behavioral composition is about the level of functional detail ranging from mathematical formulas to detailed intermediate operations (gate-level or instruction level). This also includes the temporal detail which expresses the precision of the event orderings in the model. Tradeoffs in this space are partial-ordered event accurate models, token-cycle accurate models, instruction-cycle accurate models, clock-cycle accurate models, clock-phase accurate models and so on.

#### 1.2.3 Tradeoffs in composability.

The abstraction implemented by a component framework will exhibit tradeoffs in the modeling dimensions. For instance, a VHDL simulator will have very fine data value details, with discrete event time dimension, but then the design will be functionally very simple or temporally very precise. When designers use a programming language to build models, they also decide how to combine the modeling dimensions with design dimensions, thus creating more composability challenges that need to be carefully managed. With multiple abstractions in a system-level design environment, the correctness of the composition has many aspects that may not be straight forward for models built with general purposed programming language. We next consider several component-based design approaches for software or SOCs.

## 1.3 Virtual Component and Wrappers

A virtual component has a well-defined functionality without an associated hardware or software implementation. Therefore, a virtual component is a part of a model, with some abstractions. A virtual architecture can usually be simulated and analyzed in a high-level models of computation such as a data-flow network. A virtual component is turned into a concrete implementation by either instantiating its implementation model, by combining with other components, or by it having the capability to execute.

Platform-based design (PBD) is often defined as the creation of a stable core-based architecture that can be rapidly extended, customized for a range of applications, and delivered to the customer for rapid deployment. Platform-based design requires a "standard" architecture, to which components are interfaced, and to which wrapper can be generated and PBD often uses "read() and write()" interfaces. Generally, a PBD provides structure to pure component-based design by providing architectural constraints on SOC implementations. Coral [10] and Colif [11] use standard architectures and effectively implement platform-based design in the architectural domain using the interface unit as the CoreConnect bus or some other industry standard bus like an AMBA bus. These approaches use virtual channels (as connectors) between processes or components and a notion of substitution to produce an implementation model. A virtual channel is refined to a real channel by selecting the appropriate bus or operating system connector with same interface from a library. Furthermore, connectors can also implements conversions between levels of abstractions [12]. In many co-simulation environments bus-functional models are often used for interfacing low level register-transfer models with high-level process models [13].

Bus interfacing, or wrapper generation is more complicated in general component frameworks than in platform frameworks due to the absence of standards or diversity of interfaces. The Ptolemy framework [14] takes a very different approach to composition. Ptolemy views components composability as the problem of interoperating between their models of computations. While other frameworks adopt clear models and refinement rules, Ptolemy merges the dimensions of interoperability and composability into the notion of models of computation. This requires that the components be designed in the style of actors that conform to one of the Ptolemy domains. However, it is not clear how to compose multiple different languages, styles, and levels of abstractions of IPs implemented in general purpose programming language- unless their models of computation are clearly and correctly identified. This kind of integration is often done by creating wrappers around the existing library components to enable communication of data values between different modules and coordination between them. For instance, wrappers can be used to transform and adapt components to different domains of computation- they are pieces of code that enable reuse of existing component models.

There are several ways in which such wrappers can be built. A common strategy is by using inheritance available in most object-oriented programming languages. In this approach, the wrapper is programmed by manually inserting code to align various design concerns inside the inherited class. The component and the wrapper have a common self in this implementation; i.e. the wrapper and the component are the same object. As a result, the interoperability issues related to typing are resolved at compile time, and the wrapper and component have strong dependencies. An alternative is to use a wrapper that, if needed, delegates to the design component. In this case, the component is not modified, and the wrapper and the component are two distinct objects. Modules from different libraries can be imported as is, and dynamically placed in wrappers at runtime. While inheritance can quickly achieve interoperability in some cases, it is not a recommended approach for many reasons. For one, it actually hinders reusability in the long term [15]. In this article, we are going to describe a framework that uses automatic and semi-automatic wrapper generation to adapt IP components to a component framework. Furthermore, we are going to show how these wrappers can be used to abstract component types and composition.

## 2. Composability in the Structural Domain

The BALBOA [16] component composition framework is used to build system models with an architectural perspective. The BALBOA component composition environment is a layered environment that provides a component model and partial typing capabilities.The BALBOA framework implements a split programming model, where a high-level language is used to assemble a design topology and create the validation environment.

### 2.1 Split Programming Model

Split programming refers to system model generation and component programming in two different levels that are strongly connected by a matching class hierarchy and methods [17]. The key idea behind split-level programming is to relieve system engineers of programming artifacts and software engineering concerns and lets them focus on system architecture. The BALBOA framework uses the same split programming foundation as the NS-2 environment [18], but we customized it for system-level design. Our split programming model a separation between an abstract architecture and a concrete architecture. We defined a domain-specific component integration language (CIL) to manipulate a set of C++ components representing SOC blocks.

The CIL has a type abstraction feature that is used to abstract underlying C++ types. An abstract CIL component can be only partly specified and not necessarily executable in this form. Think of that as an abstract UML model, where an architecture is a set of boxes and arrows that are not yet fully defined, but do carry a design semantic sufficient for the engineers to understand what they will do. Using a procedure called type inference, the BALBOA environment will produce a "concrete" architecture that will be executable- by automatically selecting the underlying C++ component types.

## 2.2 Layers in the Environment

In BALBOA, the key in providing the component abstraction is the environment layering because it adds a level of indirection, implementing a difference between an abstract and a concrete architecture. The abstract layer has incomplete type information while the concrete layer has very detailed types assigned. Figure 3.1 shows the layers in BALBOA. Languages are on the left side and the run-time structure on the right side. The description of the layers of Figure 3.1 is as follows.

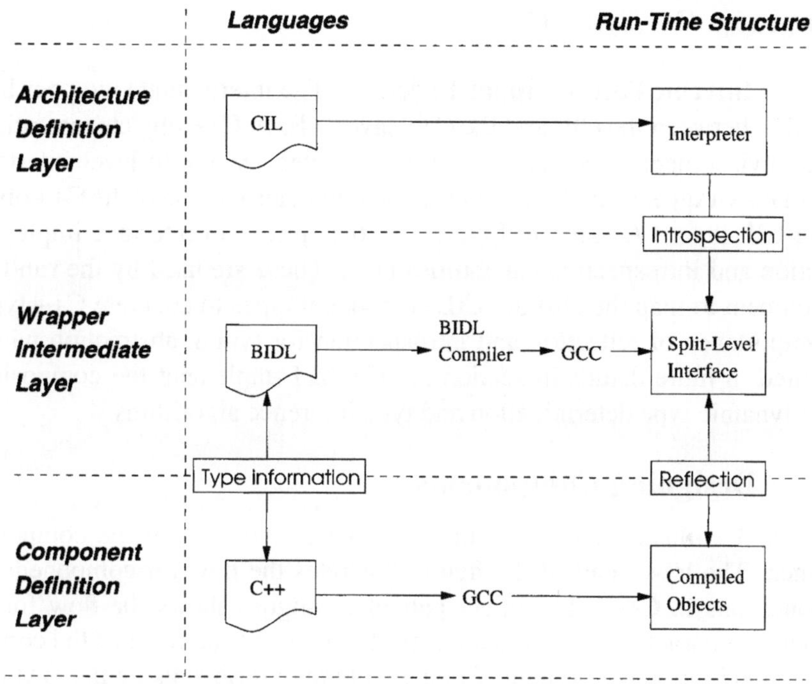


*Figure 3.1.* Layering in the BALBOA environment: the languages are on the left side and the run-time structures on the right side of the figure

**2.2.1 The Architecture Definition Layer.** The design of an architecture is done by a system architect, where the focus is on module instantiation and interconnection by using the architectural support in the component integration language (CIL). A system architect builds the overall system architecture by instantiating, connecting, configuring components, and establishing relationships among components using the CIL. The CIL is an interpreted scripting language used to build connections, and to build new components or compose attributes or behaviors to existing ones. The partial typing abstraction is implemented through the CIL. Our aim is for the CIL to be a language for architects to specify a design structure of boxes and arrows with fewer details as possible.

**2.2.2 Component Definition Layer.** The component definition layer is a set of C++ IP components stored in libraries. Any C++ class or object can be placed in this layer without it needing to derive from a specific C++ class. Ideally, this layer can accommodate C++ IP models in a range of libraries without affecting the implementation of the two other upper layers. A component library designer implements C++ classes to populate the IP library. The design of a library component has to be done by a designer who understands well the language (e.g. C++/SystemC).

**2.2.3 Intermediate Wrapper Layer.** The intermediate wrapper layer is the link between the CIL and the C++ layers. Each C++ objects instantiated in the environment is contained and manipulated by a split-level interfaces (SLI). This wrapper provides the mechanism for manipulation of the C++ object by the CIL layer. As seen in Figure 3.1, the split-level interface implement reflection and introspection capabilities [19]. These are used by the run-time environment to map the abstract CIL component types to concrete C++ types. The importance of reflection and introspection for typing abstraction will be presented in more details in section 3. The SLI implement the composition rules, dynamic type determination and type inference algorithms.

## 2.3 Designing Components

Figure 3.2 illustrates the tool flow and design process for the component designer. The lower part of the figure illustrates the flow for component implementations in C++. The upper part of the figure shows the flow for the component characterization and the exportation of the interface of the components to the interpreted domain. Component are described using the BALBOA Interface Description Language (BIDL). The component designer write a class declaration header in the BIDL description and add keywords for component families, versioning and template classes handling and specification of available types. The BIDL compiler translates and expands the description of the type of the component to a format that the CIL interpreter can understand. The BIDL compiler generates C++ code to create and configure the split-level interface of a component and to generate C++ code necessary for the type abstraction. The BIDL has a role similar to the CPP preprocessor- however it does not expand macros but generates a custom split-level interface specific for every component.

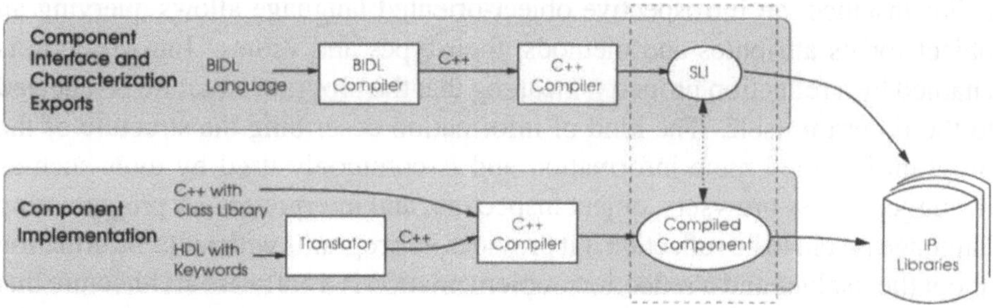


*Figure 3.2.* **Component design tool flow: write the component in C++, use the BIDL to characterize/export it and generate a split-level interface (SLI)**

## 2.4 Simulation and Validation

As shown in Figure 3.1, the split-level interfaces are the links between the CIL components and the C++ objects. Composition requests from the CIL script language are interpreted through the SLIs and the simulation commands are delegated to the C++ objects. The simulation control flow is usually only in the C++ layer because the interpretation of CIL commands can be slow. However, the CIL layer can also have a part in simulation. For instance we have a number of stimuli generators and monitors, assertion checkers in CIL libraries. This is like validation languages such as 'e' where automatic inference of structural details is possible- and often implemented.

Our motivation in providing the hardware designer to design in a "loose" typing environment is from the observation that strict typing in C++ often results in excessive programming effort on the part of the system designer. This effort is particularly notable in case of using predefined IP libraries, where a component port type may be restricted to a subset of possible interface types. Our goal is to be able to provide an environment that makes it easy to instantiate components and connect ports without specifying C++ types completely. This is what we call the abstract typing capability of the CIL layer of our design environment. However, actual instantiation and running of a simulation cannot work without instantiating the correct underlying C++ types. In our case we rely on split programming with reflection and introspection capabilities to implement type abstraction for partially specified structures in the BALBOA component composition environment. The next section will describe these capabilities.

# 3. Importance of Reflection and Introspection for Type Abstraction

The reflection and introspection capabilities are used to implement the type abstraction in our split programming model. Introspection is the capability of a program to explicitly see, understand and modify its own structure [19] [20]

. For instance, an introspective object-oriented language allows querying an object for its attributes and methods, their types and values. Introspection is enabled by a reflection property, meaning that the program structure is reflected to the program itself. The kind of information describing the structure of the program is called meta-information and is commonly used by tools such as debuggers, class browsers, object inspectors, and interpreters. A programming language is said to be reflective if it provides its programs with meta-information about themselves and a reflection implementation is a software architecture that separates the meta-information from the base program. Among the popular programming languages, C++ carries only very little type information through the virtual function mechanism for polymorphic objects and the supporting run-time type identification (RTTI) mechanisms. In order to have introspection in the following two capabilities are required:

**Reification** : a data structure to capture the reflected meta-information about the structure and the properties of the program.

**Introspection** : the capability to query and modify the reified structures.

These concepts are illustrated in the reflection capabilities of the Java programming language where it is possible to query the Java virtual machine (JVM) to know the structure of an object. To such a query, the JVM returns meta-information that fully describes the class of the object, what base class, attributes and methods it has. Because the Java virtual machine understands the structure of the program, it can be queried by a program to know information about itself. In other words, a Java program can understand its own structure at run-time because the run-time infrastructure understands and reifies the structure. Understanding what are the structural attributes of objects is knowing their types- and check them through type systems. Introspection and reflection are often used in component-based design frameworks. In the BALBOA context, we use the introspection and reflection to provide the component type abstractions.

## 3.1 Run-time Reflection Structures

Our implementation of the reflection and introspection framework uses a declarative meta-language, and replication of composition information through the split-level interfaces [21]. The SLI are implementations of part of the type system. It is the meta-layer that contains the reified information for the component types. This separation of concerns through between the layers reduces coupling among the reflection mechanisms and the rest of the environment and enables automatic wrapper generation. Figure 3.3 shows the introspection architecture implemented in the BALBOA component framework environment. A design database is added into the run-time environment to reflect the reified

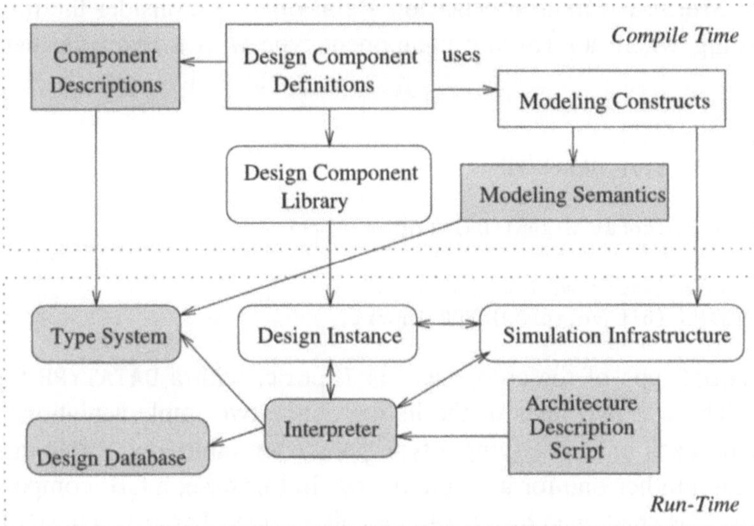


*Figure 3.3.* Reflection and introspection implementation in BALBOA framework

structure to the introspection. Secondly, in the model source, statements are added to store component instantiations and connection bindings into the design database. This information is a reification of the structural design information that is implicit in the design database. It uses composition replication that requires that the component definition code be separated from the architecture composition code. The design database captures and manipulates the implicit structure of a design. In the BALBOA implementation, the reification is done through a design database and a type system, but does not require the usage of sub-types. The type system is used to reify the full C++ type information of a component, and its modeling semantics. It captures structural information, some behavioral information and non-functional properties of components.

## 3.2 Type Parameterization and Abstraction

The simplest version of the CIL abstraction implement C++ data type and parameter value abstraction. We call this a type parameterization abstraction. The type information is declared through the BIDL and the BIDL compiler will build a custom SLI for every C++ component class and fill the necessary reification tables. In the simplest case, the BIDL compiler parses a component description and searches for type parameterization relations (TPR), where type parameters are shared between two components. There are two kinds of TPRs: (a) through types of method arguments (b) through types of the component attributes compositions.

**3.2.1 Methods Parameterization Relation.** Consider the following BIDL listing where an `Inport` component type with a `bind_to` method is declared:

```
Component Inport {
  PARAMETERS {type DATATYPE};

  void bind_to(const Signal<DATATYPE>& sig);
};
DEFINE_PARAMS {unsigned int} FOR Inport;
DEFINE_PARAMS {bit_vector32} FOR Inport;
```

The partial type of the component is `Inport`, with a `DATATYPE` type parameter. The last two lines of the listing define two implementations of the component, each one specifying a type parameter value, one for an unsigned integer and another one for a 32 bit vector. In this case, a CIL component of type `Inport` is shadowed by a C++ class also named `Inport`, whose instance is used for the internal object. When an `Inport` CIL component is instantiated, the parameter `DATATYPE` is undefined- and the C++ object not allocated until there it is set. The BIDL compiler will find in this listing one type parameterization relationship between the argument `sig` of method `bind_to` and the `Inport` component, sharing parameter `DATATYPE`. When the `bind_to` method is called on an inport component, the run-time environment can inspect the type of the `sig` argument for the value of the shared `DATATYPE` parameter. If defined, it can be propagated from the `sig` parameter to the inport component (or vice-versa).

**3.2.2 Composition Parameterization Relation.** Consider the following BIDL listing where a `Buffer` parameterized component type is declared with two ports `in` and `out` and two implementations, one for unsigned integer and one for 32 bit vector :

```
Component Buffer {
  PARAMETERS {type DATATYPE};

  Inport<DATATYPE>  in;
  Outport<DATATYPE> out;
};
DEFINE_PARAMS {unsigned int} FOR Buffer;
DEFINE_PARAMS {bit_vector32} FOR Buffer;
```

The BIDL compiler will find two type parameterization relationships in the listing. The first one is between the attribute (and sub-component) `in` and the `Buffer` component where they share the parameter `DATATYPE`, and the second one is between the attribute `out` and the component also with `DATATYPE`. Type parameter definitions can be propagated to the `Buffer` component along the composition relation when the type parameter of a sub-component in or out is defined (or vice-versa).

## 3.3 Type Parameterization Example: FFT

Consider the following BIDL interface declaration code for a FFT class interface, with the real and imaginary inputs and outputs ports, and a type parameter for the data widths:

```
template<class DATATYPE>
class FFT {
  sc_in_clk          CLK;
  sc_in<bool>        data_valid;
  sc_in<bool>        data_ack;
  sc_out<bool>       data_req;
  sc_out<bool>       data_ready;
  sc_in<DATATYPE>    in_real;
  sc_in<DATATYPE>    in_imag;
  sc_out<DATATYPE>   out_real;
  sc_out<DATATYPE>   out_imag;
};
DEFINE_PARAMS {unsigned int} FOR FFT;
DEFINE_PARAMS {        bit16} FOR FFT;
DEFINE_PARAMS {        bit32} FOR FFT;
DEFINE_PARAMS {   sc_lv<32>} FOR FFT;
```

This declaration is mapped to an appropriate CIL component type same signature through the BIDL. Consider a sampling system using a Fast Fourier Transform (FFT) illustrated on Figure 3.4. Components connections are abstract and the architect do not have to specifying their types.

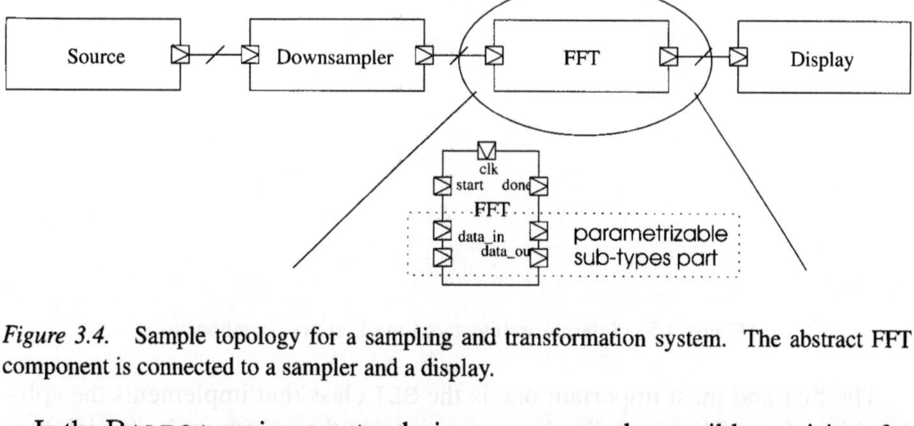


*Figure 3.4.* Sample topology for a sampling and transformation system. The abstract FFT component is connected to a sampler and a display.

In the BALBOA environment, a designer can query the possible precisions for the FFT CIL component and select one. Using the connections in the topology, the CIL run-time type inference system will propagate the selected precision to the the other components connected to the FFT. Other components involved in the connections need to have implementations for the chosen type parameter. If not, the type propagation will back track and request another different starting parameter. For instance, the downsampler may not have a C++ implementation available in the library for a specific data width used as input to the FFT. In this example, back tracking is not difficult, but as the design grows in size its gets more complicated.

## 3.4 Implementation and Code Generation

Figure 3.5 illustrates the internal architecture of a BALBOA component consisting of four blocks: the internal object (C++/SystemC), the type system information, the interpreted attributes/methods (can be a reflection of the compiled attributes/methods), and the split-level interface routines. These structure are to implement the reification of the C++ type into the SLI and the CIL. The BIDL compiler generates a set of C++ classes that are specialization of classes in the BALBOA framework. Figure 3.6 shows the simplified class diagram for the type system module run-time environment. There are 6 classes in the figure and three of them are used as base classes for extensions generated by the BIDLs. Those classes are in gray boxes on the figure.

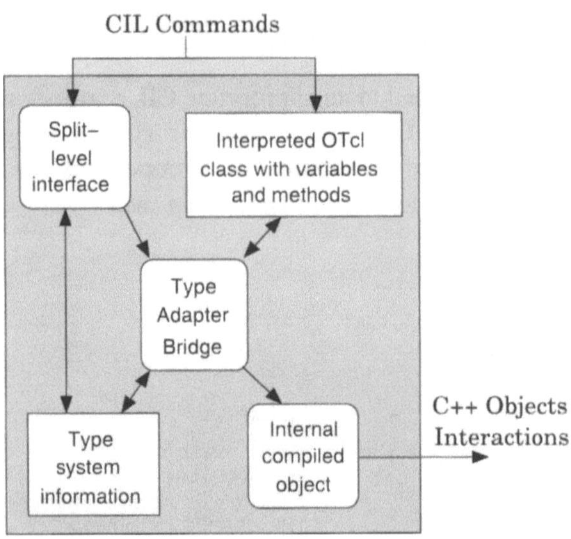


*Figure 3.5.* Internal architecture for a BALBOA component

The first and most important one is the `SLI` class that implements the split-level interface. The specialized version contains the values saying which type parameters are declared and defined in the `params_decl` and `params_defs` vectors. It also contains a vector of attribute type parameter relations `atprs`

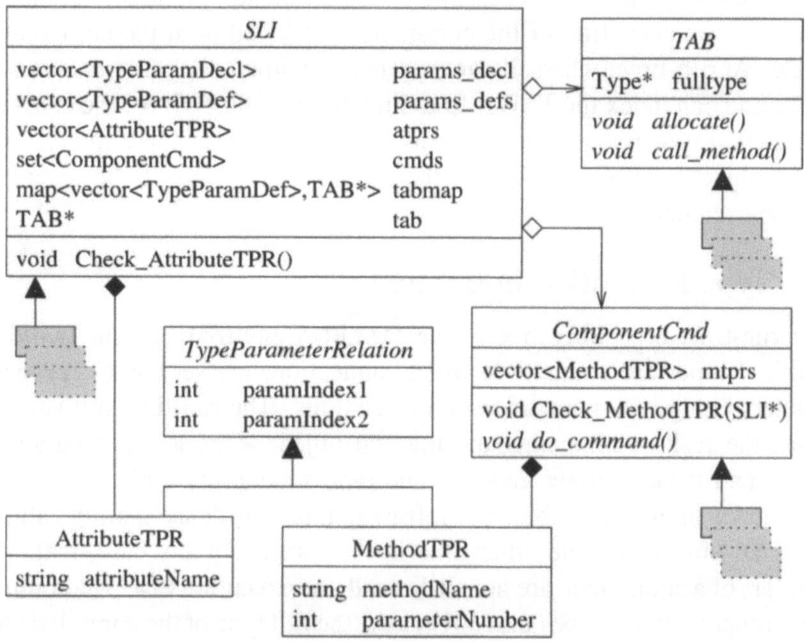


*Figure 3.6.* Type parameterization framework: the split-level interface (SLI) contains the type parameterization relationship by composition (through attributes) and the commands contain the type parameterization relations through methods. Classes in gray are extension generated by the BIDL compiler when adding new component types.

for parameters shared in composition. Type parameter relations are pairs of vector indexes to the shared parameters in the type declaration vectors of the two components. The second important class is `ComponentCmd` that executes a command on a component. This class is also specialized by the BIDL compiler to implement method calls to the internal object (such as a call to the `bind_to` method of the Inport class). The `ComponentCmd` class contains the type parameterization relations between command parameters and the component. The third class that gets extended by the BIDL compiler is the Type Adapter Bridge (`TAB`) class. This class is the proxy the C++ object. Only the TAB knows the specific type and it is the only access to the compiled internal object. The type availability table is implemented through a table of TABs, and it is indexed by the SLI type parameter definition vector `params_defs`. A specific TABs is derived by the BIDL compiler for every available C++ class usable as the internal object. Note that the C++ class does not need to inherit form any base class. Thus, a CIL component type can be used as an umbrella types for many different C++ components that have the same port kinds.

We say a CIL component is abstract when its internal C++ object is not allocated. In this state the C++ object cannot be executed, and only the interpreted parts of the component are accessible. If a command is issued to an abstract component, the execution of the command is delayed until the component is concrete. At run-time, when all parameters are defined, the `params_defs` vector is set and can index the TAB map to find the specific TAB for the exact C++ type match, a process analogous to row matching in a database query. When the C++ object is allocated, we say the component is concrete because it can totally be executed.

## 4. Type Resolution in BALBOA

The run-time type inference in the BALBOA environment has two tasks: (i) verify if types are compatible when connections are set and (ii) propagate parameters along type parameterization relations. The run-time infrastructure also has the task of allocating the internal object when all type parameters are valid and have a single match in the type availability table. If the types are unknown or incompatible, the infrastructure can delay the allocation of the internal object to some other time in the future. In any case, if the type parameters of a component are not all defined, the execution of a C++ command on that component has to be delayed because the full type of the compiled object is unknown; hence we cannot allocate the object and invoke its behavior.

There are two ways to do the type inference: do it once when the architectural topology is fully specified (on an elaboration phase), or do it progressively as the topology is specified (like in ML [22] by progressively applying type reduction rules). In this case, parameters can be defined multiple times for a component; therefore commands have to be delayed until the sub-types definitions vector is committed. There are two policies for the allocation of the internal object of the component. The run-time inference engine can be opportunistic and do the allocation as soon as it is possible– when the type parameter definition vector is defined and there is a type allocation table match. The environment can also be lazy and wait until the user of the system issues a command to allocate the objects. The system will then do the allocation, but it will not run the delayed commands until the user also instructs the system to do so. We now describe the type resolution problem and how we implemented the resolution in the BALBOA run-time environment.

### 4.1 Type Parameter Matching Problem

In C++, type parameterization is often done using templates. In hardware design, this is especially convenient to express designs with various data widths. On compilation, a template type is turned into a concrete C++ type by parameter passing. This is useful in instantiating components whose interface and internal data types can be changed depending on the design requirements. For instance, we may have an adder with various implementations in the libraries. It can have implementations for integers, float, double, 4, 8 and 16 bit vectors and so on. When building architectural models, specific port types is a function of the architecture which may change as the design evolves. Ideally, a system architect would like to instantiate the component in an architecture without specifying all the type parameters for each ports, and let the run-time environment resolve the detailed implementations automatically.

**4.1.1 Split-typing.** In the BALBOA environment, a partial or abstract type is also viewed as "split type" because it is managed through the split-level interface. A component has one type at the CIL layer and can have multiple possible implementation types at the C++ level. In the adder example, at the CIL level all the ports of the adder are of partial "port" type. Based on the containment hierarchy and the connections to other components, type inference will map specific C++ data types to each of the ports. The following notation is used to illustrate the algorithms for incremental type inference. Let $T$ be a set of all concrete C++ data types, let $P$ be a set of ports and $S$ be a set of signals in the design created in the CIL. In the interpreted layer, ports and signals are abstract types, but in the compiled layer they must be mapped to a concrete type in $T$. Each port or signal is associated with a data type via mappings, $dt_p$ and $dt_s$, such that $dt_p : P \rightarrow T \bigcup \{\bot\}$, and $dt_s : S \rightarrow T \cup \{\bot\}$. A port is untyped if $p \in P : dt_p(p) = \bot$, and a signal is untyped if $s \in S : dt_s(S) = \bot$, where $\bot$ denotes the fact that no concrete type has been specified or inferred yet. When the type inference is done, $dt_p$ and $dt_s$ must not map any element to $\bot$. If that cannot be achieved, the type-inference algorithm must detect and report this condition, and if possible back track to search for a new solution.

**4.1.2 Components.** In the CIL a component $c$ has a set of ports denoted by $ports(c) = p_{c_1}, p_{c_2}, \ldots, p_{c_n} \in P$ where $n$ is the number of ports. Component $c$ is said to be "polymorphic" if there are many compiled versions of $c$ with ports of different types. The function $dt_p$ has a limited choice in assigning $p_{c_1}, p_{c_2}, \ldots, p_{c_n}$ to the available compiled types. One can view this as a choice of assignment to a vector of ports $\triangleleft p_{c_1}, p_{c_2}, \ldots, p_{c_n} \triangleright$, from one of the possible rows from a type availability table of ordered rows $T_c \subset T \times T \times T \times \cdots \times T$, each row corresponding to a compiled version of $c$. The rows of the table are filled through the BIDL description. Signals are used to link ports and are parameterized by the data type they carry. A signal must be assigned the same data type as the ports it is connected to. Hence, signals are constraints in the port assignment problem.

## 4.2 Type Resolution as a Table Row Selection

Let us consider the simple adder example with 4 ports, $in_1, in_2$, as input ports and $out_1, out_2$ as output ports. The availability of library implementation of the adder is illustrated in the type availability table of table 4.2. The table list all available combination of type parameter values, each row corresponding to a specific C++ implementation type. Let us consider the following scenario. An adder is declared in the CIL layer without any port types, then $dt_p(p)$ is undefined for all ports $p$. As types are propagated through the design, it turns out that $in_4$ gets instantiated to *bool*. Now, the type management system will immediately recognize that all implementation in the table have this parameter value, so no conclusion is available. Let us assume next that $in_1$ gets typed to $int16$. The SLI can now match the row: $(in1, in2, out1, out2) = (int16, int16, int16, bool)$. Then it will be inferred that $in2$ and $out1$ should also be of type $int16$. At this point, any signal that is connected to $in2$ and $out1$ will have the type propagated to them as $int16$. This means that $dt_p(in1) = int16$, $dt_p(in2) = int16$, and $dt_p(out1) = int16$).

| $In_1$ | $In_2$ | $Out_1$ | $Out_2$ |
|---|---|---|---|
| float | float | float | bool |
| int8 | int8 | int8 | bool |
| int16 | int16 | int16 | bool |
| int32 | int32 | int32 | bool |
| int64 | int64 | int64 | bool |
| bool | bool | bool | bool |

*Table 3.1.* A type availability table lists all valid combinations of parameter values for the C++ implementations of a CIL interface. This table is an example for a component with two input ports and two output ports.

## 4.3 Formulation of the Type Inference Problem

Given a design with a set $P$ of ports, a set $S$ of signals, and the partition of $P$ into $k$ disjoint sets where $k$ is the number of components and the partition of $P$ is disjoint because, components do not share ports, the type inference problem is as follows. For each component $c$, with its port vector $\lhd p_{c_1}, p_{c_2}, \ldots, p_{c_n} \rhd$, assign a row from its type availability table $T_c$, such that if there is a signal $s \in S$, which connects a port $p_{c_i}$ in component $c$ to another port $p_{d_j}$ of a component $d$, then the type assigned to $p_{c_i}$ and to $p_{d_j}$ must be the same. This restriction makes the problem complex, and we show here that the problem is NP-complete.

THEOREM 3.1 *The* BALBOA *type inference problem is NP-Complete.*

**Proof:** Given a type assignment for all the ports, it is easy to check in polynomial time that the assignments in the type tables are correct, and the ports connected via signals have the same type. Hence the problem is clearly in NP.

For the NP-hardness proof, we reduce the problem of one-in-three monotone 3SAT [23], to the BALBOA type-inference problem. 3SAT is the following problem: given a set $U$ of boolean variables and a collection of disjunctive clauses over $U$, such that each clause is a disjunction of exactly 3 literals, find if there is a truth assignment of the variables in $U$, such that all the clauses are satisfied. One-in-three 3SAT, is a special case of 3SAT problem, where the truth assignment of $U$ has the restriction that if $x_i$ and $x_j$ are two literals appearing in the same clause, then both cannot be assigned the truth value of 1. Monotone one-in-three 3SAT, has the further restriction that no negated literal appears in any of the clauses. Given an instance of monotone One-in-three 3SAT, for each clause $c$, which has three literals $x_{c_1}, x_{c_2}, x_{c_3}$, one can create a table $T_c$, with exactly 3 rows $\{<1,0,0>, <0,1,0>, <0,0,1>\}$. Given the set of vector of variables in $c$, $\lhd x_{c_1}, x_{c_2}, x_{c_3} \rhd$, one has to assign one of the rows in table $T_c$ with the restriction that if the variable that appears as $x_{c_1}$ in $c$ also appears in another clause $d$ as $x_{d_i}$, then the choice of the rows from $T_c$ and $T_d$ must be such that the assigned truth values are the same for $x_{c_1}$ and $x_{d_i}$. This is a version of the BALBOA type inference problem and hence it is at least as hard as the monotone one-in-three 3SAT, which is known to be NP-complete [23]. Given that the problem is NP-complete, we use a heuristics to solve the problem. Moreover, the problem is solved incrementally because components are added or deleted in the CIL- the type assignments keep changing. The heuristic implemented in the BALBOA type inference uses a delaying mechanism which delays the type instantiation and allocation of components until all type parameters associated with a component are resolved. The resolution is done on-the-fly, as components are added type parameters get propagated to partial types.

## 4.4 Type Inference Heuristics

There are multiple ways of solving these problems, including dynamic programming, local search [24] etc. A sequence of natural join operations may take a long time since the different $T_c$ have different sizes. The order in which the joins are computed can be optimized using dynamic programming, hence we will follow that approach. However, since we build models incrementally, when a new component is added and connected, the new required join may not be in the order that the dynamic programming would have yielded. In the current implementation of the BALBOA environment, the run-time type inference mechanism may be summarized as follows. We first verify compatibility of types when connections are set between two components. We join the type availability tables of both components, renaming the connected ports to the same name if necessary. If the join returns empty, the components are not compatible, if it is not empty, the system will remember the result. Since the split-level interface allocates the component, it can decides to delay the allocation until the type parameterization is resolved. Now considering this join result as the type availability table for this pair of components, we then select a new component that is connected to one of these, and apply the procedure. If the result of the final join is empty, the architecture described in the CIL cannot be instantiated with the available implementations. If it is singleton, then we have found a unique combination of type parameters that can be instantiated. If the final join yields multiple rows, we have to choose a type parameter combination from the final table.

```
/* Type Inference Heuristic: Propagates Type Parameters to Components in an Architecture */
CheckTyping(A,C,Cnx): A_t
    Input:          Untyped architecture A, of component C, and connections Cnx
    Output:         Typed architecture A_t
    Intermediate:   Open:Set, Done:Set
  for all (c_i, c_j) ∈ C
      join_result = RunNaturalJoin (c_i, c_j, Cnx)
      if (join_result == empty)
          Backtrack(c_i, c_j)
      else
          Put (c_i, c_j) in Done;
          Put all components that are connected to c_i and c_j in Open;
          for all c_k in Open {
              Add all components connected to c_k that are not already in Done to Open;
              join_result = RunNaturalJoin(join_result, c_k, Cnx);
              if (join_result == empty)
                  Backtrack(join_result, c_k)
              Move c_k from Open to Done;
          } /* end for */
      } /* end else */
  } /* end for all */
  if join_result has a singleton row,
      Compiled model can be allocated;
  if join_result has multiple rows
      Designer needs to chose one row.
```

*Figure 3.7.* Simplified type inference heuristics in the BALBOA run-time environment

Figure 3.7 lists a simplified version of the type inference heuristic. The *CheckTyping* heuristic chooses two components that are connected, both with small type availability table sizes. It does a join locally between the two components to assess that the types are compatible. While the designer is using the CIL, this join is used to propagate type parameters as the architecture is built. All connection information is passed to the join procedure as *Cnx*, which is used to decide which columns of the tables to run the join on. In the current implementation, the run-time environment does not build the join tables explicitly, but walks the architectural graph and builds the join tables implicitly in the SLIs. In this case, the run-time environment does an incremental version of this procedure by remembering all the intermediate join results through the

SLIs. In future works, we will study optimizations of this procedure, including caching joins, incremental joins etc.

## 5. Packet Switch Example

Let us now illustrate the use of the type inference and IP selection features of BALBOA through a simple packet switch specification. System models built using programming languages can be parameterized in various ways [25]. However, it is often ad hoc and different from design to design. The CIL leverages the parameterization capabilities of programming languages by clearly defining three specific ways to do it. To illustrate this, let us consider the example of a very simple packet switch system shown on Figure 3.8. It consists of packet senders `s`, receivers `r` and a switch `pkt_switch`. The configurable parameters of the switch are the number of ports and the type of packets processed. A 4x4 configuration is shown in the figure and there is no sender connected to the first port of the switch because we assume another component will be connected there.

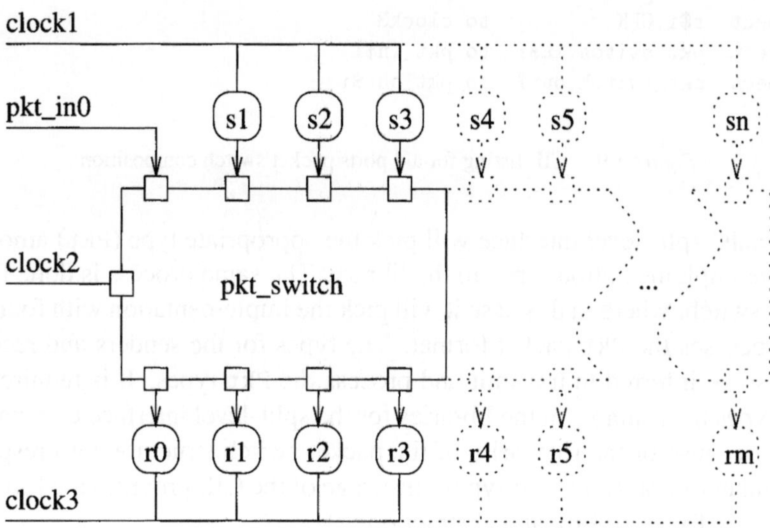


*Figure 3.8.* Packet Switch CIL Example

Figure 3.9 shows the CIL listing for a switch topology composition. We first start by setting a variable for the number of ports to `4x4`, and then instantiates the switch component with that parameter. The third line sets a variable to `Pkt` for the type of packets processed and then instantiates a signal with that sub-type. The next three lines instantiate clocks for the senders, the switch and the receivers. The "for" loop is parameterized to instantiate packet senders and receivers, and connect them to the input and output ports of the switch. No signals in the description have their data type specified. When connected,

```
/* CIL listing to setup a switch with environment */
set NUMBER_OF_PORTS 4X4
Pkt_Switch pkt_switch -number_of_ports \
        $NUMBER_OF_PORTS
set PACKET_TYPE Pkt
Signal pkt_in0 -subtypes {$PACKET_TYPE}
Clock clock1 -period 75 -duty_cycle 0.5
Clock clock2 -period 30 -duty_cycle 0.5
Clock clock3 -period 15 -duty_cycle 0.5
connect pkt_switch.CLK to clock2
for {set i 0} {$i<$NUMBER_OF_PORTS} {incr i} {
  if {$i>0} {
    Sender   s$i -id $i
    Signal   pkt_in$i
    connect  s$i.CLK          to clock1
  }
  Receiver r$i -id $i
  Signal   pkt_out$i
  connect  s$i.pkt_out       to pkt_in$i
  connect  r$i.pkt_in        to pkt_out$i
  connect  r$i.CLK           to clock3
  connect  pkt_switch.in$i   to pkt_in$i
  connect  pkt_switch.out$i  to pkt_out$i
}
```

*Figure 3.9.* CIL listing for a 4 ports packet switch composition

the signals' split-level interface will pick the appropriate type (`Pkt`) among all possible implementation types in the library. The same process is done for the packet switch, where in this case it will pick the implementation with four ports that processes the `Pkt` packet format. The types for the senders and receivers will also be inferred to transmit and process the `Pkt` types. It is required that these types be available in the libraries for the split-level interface to instantiate them. Because of the regularity of the packet switch structure with respect to the number of ports and their types, the usage of the CIL provides the following advantages for flexibility and abstraction:

1 *Control flow parameterization for regular structures*:
The CIL imperative control structures are used to build the system architecture. In this example, the design structure is parameterized with respect to the number of ports. The `for` loop instantiates and connects sender and receiver components and surrounding signals for every port of the switch.

2 *Name expansion for regular structures*:
Component names are expanded by the interpreter with interpreted variable values. Names for signals, senders and receivers are expanded such

like in `pkt_in$i` variable `i` is replaced with the value of the iteration counter to `pkt_in0`, `pkt_in1`, etc.

3 *Type inference*:
Types are assigned to partially specified components. The components and connections are introspected by the environment and the split-level interfaces select a full type. In this example, the components will be picked by the environment to process the `Pkt` data type for the switch, signals, senders and receivers.

The compactness of the CIL listing is due to the parameterization capabilities that enable a separation of concerns of type compatibility from concerns of composition of the architectural structure. Using control statements and name completion can be useful to complete component type names as well as instance names. Unlike code generation approaches, it also has the advantage of avoiding recompilation cycles when changing parameter values. The same topology can also be built using only C++ with all sub-types and parameters explicitly specified. This yields longer descriptions that are less flexible as parameters changes (i) may need to be done throughout the design, and (ii) necessitate re-compilation. An XML-based approach would also have yielded a lengthy description. The CIL is like an imperative ADL where checks are done on the composition as the design is constructed. A difference from most ADLs is that the CIL describes architectures without explicitly enumerating every component, connection or type in the topology. It is actually very close to a domain-specific language for system-level design. This shortens description sizes and helps the designer manage the complexity of describing architectures.

## 6. Conclusion and Future Work

A good component composition framework (CCF) provides a composition language, and capabilities for dynamic composition, simulation and verification. One of the novelties of the BALBOA environment is the separation between component definition and architecture elaboration through split programming to take advantage of weaker typing dependencies for typing abstractions at the architectural level. Typing abstraction means that it is possible to reduce the type dependencies of the strongly typed compiled C++ layer, through careful type management at the wrapper level. The split-level interfaces can implement the type inference to keep the CIL description focused on component instantiations, compositions and connections.

We are currently working to extend the composition theory and framework for temporal and behavioral types re-introduce the UML front-end in the framework.

## References


[1] R. Allen and D. Garlan. A Formal Basis for Architectural Connection. *ACM Transactions on Software Engineering and Methodology*, July 1997.

[2] D. Harel and P.S. Thiagarajan. Message sequence charts. In Luciano Lavagno, Grant Martin, and Bran Selic, editors, *UML for Real: Design of Embedded and Real-Time Systems*, chapter 4, pages 77–106. Kluwer Academic Publishers, 2003.

[3] L. de Alfaro and T. A. Henzinger. Interface Theories for Component-Based Design. In *First International Workshop on Embedded Software*, volume 2211 of *Lecture Notes in Computer Science*. Springer, October 2001.

[4] J. A. Rowson and Alberto Sangiovanni-Vincentelli. Interface-Based Design. In *Proc. IEEE/ACM Design Automation Conf.*, 1997.

[5] B. Selic and J. Rumbaugh. Using uml for modeling complex real-time systems. Rational whitepaper, available online at http://www.rational.com/media/whitepapers/umlrt.pdf, 1998.

[6] R. Chen, M. Sgroi, L. Lavagno, G. Martin, A. Sangiovanni-Vincentelli, and J Rabaey. Uml and platform-based design. In Luciano Lavagno, Grant Martin, and Bran Selic, editors, *UML for Real: Design of Embedded and Real-Time Systems*, chapter 5, pages 107–126. Kluwer Academic Publishers, 2003.

[7] N. Medvidovic and R. N. Taylor. A Classification and Comparison Framework for Software Architecture Description Languages. *IEEE Trans. on Software Engineering*, January 2000.

[8] D. Garlan, R. Allen, and J. Ockerbloom. Architectural Mismatch: Why Reuse Is So Hard. *IEEE Software*, November 1995.

[9] F. Doucet, R. Gupta, M. Otsuka, P. Schaumont, and S. Shukla. Interoperability as a Design Issue in C++ Based Modeling Environments. In *Proc. Int. Symposium on System Synthesis*, 2001.

[10] R. A. Bergamaschi, S. Bhattacharya, R. Wagner, C. Fellenz, M. Muhlada, F. White, W. R. Lee, and J.-M. Daveau. Automating the Design of SOCs Using Cores. *IEEE Design and Test of Computers*, 18(5):32–44, September-October 2001.

[11] W. Cesario, A. Baghdadi, L. Gauthier, D. Lyonnard, G. Nicolescu, Y. Paviot, S. Yoo, A. A. Jerraya, and M. Diaz-Nava. Component-based design approach for multicore socs. In *Proc. IEEE/ACM Design Automation Conf.*, 2002.

[12] R. Siegmund and D. Muller. Automatic synthesis of communication controller hardware from protocol specification. *IEEE Design and Test of Computer*, 19(4):84–95, July-August 2002.

[13] P. Gerin, S. Yoo, G. Nicolescu, and A. A. Jerraya. Scalable and Flexible Cosimulation of SoC Designs with Heterogeneous Multi-Processor Target. In *Proc. Asia-South Pacific Design Automation Conf.*, 2001.

[14] The Ptolemy 2 Project at UC Berkeley, available online at URL: http://ptolemy.eecs.berkeley.edu/.

[15] C. Szyperski, D. Gruntz, and S. Murer. *Component Software: Beyond Object Oriented Programming*. Addison-Wesley, 2nd edition, 2002.

[16] F. Doucet, S. Shukla, and R. Gupta. BALBOA: A Component-Based Design Environment for Composition and Simulation of System Level Models. *IEEE Trans. on Computer-Aided Design of Integrated Circuits and Systems*, 22(12):1597–1612, December 2003.

[17] J. K. Ousterhout. Scripting: Higher-Level Programming for the 21st Century. *IEEE Computer*, March 1998.

[18] L. Breslau, D. Estrin, K. Fall, J. Heidemann, A. Helmy, P. Huang, S. McCanne, K. Varadhan, Y. Xu, and H. Yu. Advances in Network Simulation. *IEEE Computer*, May 2000.

[19] F. Buschmann, R. Meunier, H. Rohnert, P. Sommerlad, and M. Stal. *Pattern Oriented Software Architecture: A System of Patterns*. John Wiley and Sons, 1996.

[20] T.-R. Chuang, Y. S. Kuo, and C.-M. Wang. Non-intrusive object introspection in C++: Architecture and application. In *Proc. Int. Conf on Software Engineering*, 1998.

[21] F. Doucet, S. Shukla, and R. Gupta. Introspection in System-Level Language Frameworks: Meta-level vs. Integrated. In *Proc. Design Automation and Test in Europe Conf.*, 2003.

[22] B. C. Pierce. *Types and Programming Languages*. MIT Press, 2002.

[23] M. R. Garey and D. S. Johnson. *Computers and Intractability: A Guide to the Theory of NP-Completeness*. W. H. Freeman and Company, 1979.

[24] J. D. Ullman. *Principles of Database and Knowledge-base Systems: Vol I and II*. Computer Science Press, 1989.

[25] Thorsten Groetker, Stan Liao, Grant Martin, and Stuart Swan. *System Design with SystemC*. Kluwer Academic Publishers, 2002.

Chapter 4

# TRULY HETEROGENEOUS MODELING WITH SYSTEMC

Hiren D. Patel
*FERMAT Research Lab.*
*Center for Embedded Systems for Critical Applications*
*Virginia Polytechnic Institute and State University*
*Blacksburg, Virginia, 24060, USA*
hiren@vt.edu

Sandeep K. Shukla
*FERMAT Research Lab.*
*Center for Embedded Systems for Critical Applications*
*Virginia Polytechnic Institute and State University*
*Blacksburg, Virginia, 24060, USA*
shukla@vt.edu

**Abstract** Among the many factors contributing to the industrial "productivity gap", the use of inappropriate abstraction levels for design entry, nonexistence of a standard high-level design methodology, modeling frameworks & tools, and insufficient methodologies and tools for hardware & software co-design are often blamed by industry experts. In our effort to address these specific issues, our research focuses on designing an appropriate framework that allows designers to express their specification of a complex hardware-software system in appropriate abstraction level by supporting multiple models of computations (MOC) and seamless simulation of the models comprising such heterogeneity. In this short chapter, we motivate the need for such frameworks, and discuss briefly how we have been extending SystemC's simulation kernel to support easier heterogeneous modeling and efficient simulation of such models. We illustrate this in particular through our implementation of a Synchronous Data Flow model of computation within SystemC kernel, and by showing the resulting efficiency gains in simulation.

**Keywords:** System Level Design Languages, Heterogeneous, Simulation, SystemC, Models of Computation, Embedded System Design, Synchronous Data Flow, Discrete-Event.



*R. Gupta et al. (eds.), Formal Methods and Models for System Design,* 83-101.

# 1. Motivation

Designs of complex systems challenges existing design methodologies and tools in managing the development of large and complex applications. Embedded system designers face stringent time-to-market deadlines and the task of producing designs of great complexity and heterogeneity, but lack the support from design methodologies and tools. These productivity trends are undoubtedly expected to continue furthering the need for a resolution of this productivity crisis.

One of the proposed solution by many to mitigate this productivity crisis is raising levels of abstraction for design entry modeling and simulation. The use of System Level Design Languages (SLDL) for SOC and embedded system design have gained considerable attention and importance in the research community and industry alike, due to the rising complexity of designing such systems. Heterogeneous and hierarchical system modeling and the ability to exploit these facets during all phases of the design including modeling and validation is difficult. The success of an SLDL is predicated upon its adoption as a standard design entry language by the industry and resulting closure of the productivity and verification gaps. We believe SystemC [14] is poised as one of the strong contenders for such a language. SystemC is an open-source scheme that allows designers to simulate their system level models in a similar fashion to VHDL [22] and Verilog [21].

However, we argue that going to a higher level abstraction for closure of the productivity gap, while keeping the Discrete-Event simulation semantics of SystemC are contradictory goals. In fact, most system models for SOCs are heterogeneous in nature, and encompass multiple Models Of Computation in its different components as mentioned earlier. As a result, we have to start thinking about simulation semantics of such a langauge away from the semantics of hardware simulation only, and we have to be able to provide a way to express and simulate other Models Of Computation.

This chapter describes our plans and ongoing efforts in managing such complex designs while reducing the productivity gap specifically related to the modeling and simulation frameworks that play a crucial role in validation of systems. In particular we focus on the importance of providing a System Level Design Language with sufficient simulation efficiency and modeling expressiveness to realize various Models Of Computation (MOC), and to realize behavioral hierarchy cutting across different MOCs. Note that most hardware description languages are well endowed with structural hierarchy, but the right model of behavioral hierarchy is absent from most of them. In our view, there lies the greatest challenge in providing a high level modeling and simulation framework for such systems. We broadly classify the most challenging requirements for such frameworks as follows:

1 The ability to express different Models Of Computation in the language and modeling framework to express heterogeneity, and behavioral hierarchy and

2 To provide efficient simulation framework that can exploit the different MOCs and hierarchy.

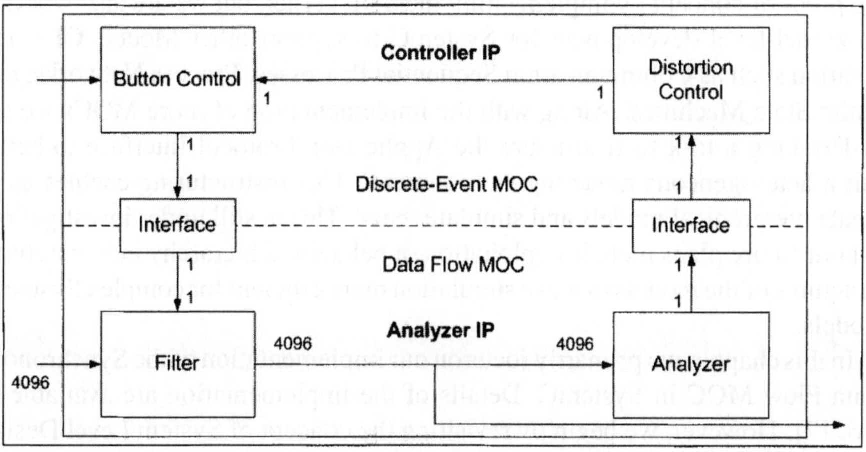


*Figure 4.1.* Example of Digital Analyzer using Model of Computations [8]

Most complex SOC embedded systems have varied functionalities which often are modeled as a hierarchy and composition of distinct Models Of Computation [9, 11]. Models Of Computation encapsulate the behavior of a particular system by defining the nature of communication between different computing components and the semantics for executing a model. The computing components themseleves encapsulate computation and the semantics define when the computation is to proceed as well as the communication between components. Figure 4.1 displays a digital analyzer using the Discrete-Event and Data Flow Model of Computation.

In addressing this challenge to mitigate the productivity crisis, we describe our implementation of an alternative Model of Computation in SystemC [14]. As a proof of concept we have implemented the Synchronous Data Flow (SDF) MOC kernel in SystemC. Our new SDF kernel is interoperable with SystemC's Discrete-Event (DE) kernel providing designers with the capability to model and simulate complex heterogeneous models constructed from a collection of DE and SDF models. It is natural that by extending the simulation kernel, we not only improve the modeling expressiveness of SystemC to naturally construct models for SDF, but we also make simulation more efficient. Since now we can express SDF models better in SystemC, the kernel can interpret these SDF models differently by exploiting the use of static scheduling as opposed to dynamic scheduling that the DE kernel employs.

We acknowledge that implementations of such kernels in SystemC can vary due to the vast expressiveness of the underlying C++ framework. However, no efforts to our knowledge have attempted at implementing kernel-level solutions for multi-MOC modeling for SystemC and we believe to be the first at investigating this issue. Nonetheless, we do not plan to stop by simply showing the proof of concept by implementing the SDF kernel, but we are also working on kernel-level development for SystemC to support other Models Of Computation such as Communication Sequential Processes, Process Networks, and Finite State Machines. Along with the implementation of more MOCs we are undertaking a task to restructure the Application-Protocol Interface to better suit a heterogeneous modeling environment. This restructuring enables us to create hierarchical models and simulate them. This is still under investigation, but our future plans include exploitation of behavioral hierarchy in hierarchical structures of the models to make simulation more efficient for complex SystemC models.

In this chapter, we primarily focus on our implementation of the Synchronous Data Flow MOC in SystemC. Details of the implementation are available in [16, 17]. However, we begin by revisiting the concept of System Level Design Languages, our choice of SLDL, the existing simulation kernel and the case study where we implement the Synchronous Data Flow Model of Computation kernel for SystemC, followed by experiments and results.

## 2. System Level Design Languages and Frameworks

In this section, we review the concept of System Level Design Languages and Frameworks and how SystemC represents an embedded and hybrid system design language and framework.

### What are SLDLs

System Level Design Languages provide users with a collection of libraries of data types, kernels and components, accessible through either graphical or programmatic means to model systems and simulate the system behavior. A simulation framework is a collection of core libraries that simulates a system and a modeling framework is a representation to create models for simulation. A successful SLDL must have its respective modeling and a simulation framework.

An SLDL's modeling expressiveness refers to the constructs, language and design guidelines that facilitate complete description of a system with appropriate MOCs, whereas simulation efficiency is measured by the time a model takes to simulate from start to end.

## Examples of Some SLDLs

In the past few years, we have seen the introduction of many System Level Design Frameworks and Languages. There also is an effort to lift the abstraction level in the hardware description languages, exemplified in the efforts to standardize SystemVerilog [20]. Numerous other system level modeling and simulation frameworks are developed for specific purposes and systems. Examples of recently introduced system level modeling languages are SpecC, SystemC, Ptolemy, etc. [19, 14, 6]. Another breed of system level modeling languages are Opnet, and NS-2 which are network oriented modeling and simulation frameworks [13, 1]. These SLDLs are usually application-specific. For example, tools such as NS-2 and Opnet are predominantly used to create network models and simulate network protocols whereas VHDL [22] for RTL designs and SystemC for RTL designs and system level designs.

**About SystemC.** SystemC [14] is poised as one of the strong contenders for a standard design entry language. SystemC is a C++ based open-source SLDL. SystemC's advantage is that it has free simulation libraries to support the modeling framework. This provides designers with the ability to construct models of their system and simulate them very similar to VHDL and Verilog [22, 21]. Since it also has the full capability of C++, higher levels of abstraction, software modeling (SystemC 3.0), analog-mixed signal modeling (SystemC 4.0), all are possible. Also, since software is designed in C++/C, if the hardware model is also in the same language, the hardware-software co-simulation is much more efficient. Another consideration is that in the industry, lot of the times, high level performance models for hardware, Instruction Set Simulator or Transaction Level System Models are better done in C++, so SystemC provides the possibility of one language for all abstraction levels. However, we believe without the possibility of heterogeneous MOC modeling and co-simulation and behavioral hierarchy, SystemC will not succeed in the industry.

**Why not use existing SystemC.** SystemC's framework is based on a simulation kernel which is essentially a scheduler. This kernel is responsible for simulating the model with a particular behavior. The current SystemC has a Discrete-Event based kernel that is very similar to the event-based VHDL simulator. However, the strive for a higher level of abstraction for closure of the productivity gap, while keeping the Discrete-Event simulation semantics are highly diametric goals. As a result, it is important and necessary to think about simulation semantics of such a language diverging from semantics of pure Discrete-Event based hardware models and towards semantics to express multiple Models Of Computation. The inspiration of such a system is drawn upon the success of the Ptolemy II framework in specification and simulation of embedded heterogeneous software systems [6]. However, since SystemC is targeted to be the language of choice for semiconductor manufacturers as well as system designers, as opposed to Ptolemy II, its goals are more ambitious.

**Ptolemy II.** There is a Java-based system modeling framework developed by U.C. Berkeley called Ptolemy [6]. Ptolemy II is a hierarchical and heterogeneous modeling framework. One might raise a question as to the need for enhancing SystemC framework when Ptolemy II exists. One can find a description of Ptolemy II and its capabilities in [6]. However, we summarize our reasons for pursuing with SystemC as our investigated SLDL. First, Ptolemy II is designed for embedded software modeling and software synthesis. It has an extensive infrastructure supporting different Models Of Computations encapsulated as domains, and hierarchical modeling and simulation. However, the needs for software design, and system design consisting of hardware/software are quite distinct. For example, it may often be important not to design some system from scratch, but to use existing IP libraries to build a system. Some of these IP libraries may represent hardware, some real-time O/S and some software. SystemC language was built as C++ libraries with this kind of modeling applications in mind. And our goal is to help that by allowing IP libraries encapsulating different MOCs to interoperate without any special tricks by the model integrators, and just by the power of the framework itself Needless to day that Ptolemy II has been an inspiration for us in our project to build similar capabilities in SystemC or similar languages.

In the recent years however, SystemC has gained considerable traction in the industry. Nonetheless, the full potential of SystemC is yet to be realized because most modeling examples and tools for SystemC are still targeting RTL level of design abstraction. The reason why SystemC is not yet successful as a language of choice for designers of complete embedded systems is because the SystemC kernel is inherently Discrete-Event simulation based, suitable for hardware models. Also, as we mentioned before, lack of behavioral hierarchy and exploiting the hierarchy for simulation efficiency are important gating factors for SystemC efficiency with respect to other HDL alternatives. Ptolemy II has its niche in embedded software modeling and synthesis. With recent advances in hardware discrete event domain, Ptolemy II can model hardware as well, it is usually not used for such purposes.

SystemC can model at any arbitrary level of abstraction, almost all kinds of hierarchies, and all kinds of MOCs. Solving the problem of actually bringing these capabilities in the modeling expressiveness and the SystemC kernel, will make a huge difference in SystemC's acceptability as a language of choice for embedded system design.

Interestingly, there is a considerable industry momentum in tool offerings for SystemC, including transaction level simulation accelerators, sequential equivalence checking, behavioral synthesis etc. Ptolemy II being an embedded software generating tool does not share this momentum. So, the Embedded System Design community will certainly be helped much more if SystemC can become the language of choice at all levels of abstractions, and can also be used for seamless IP integration and simulation, efficient modeling and simulation, and for synthesis from appropriate levels of abstraction.

Another issue is that Ptolemy II has no validation facilities other than simulation. Sequential equivalence checking tools for SystemC are being promised to the market, also in a related project we have been working on automated model extraction from SystemC for Formal Property Verification. These are also motivators for attempting to make SystemC the choice for EHS designers and validators.

We believe that we are opening up an avenue through which SystemC can be developed into a heterogeneous and hierarchical modeling and simulation environment for better construction of complex designs. Our initial efforts will be followed by releasing our enhanced SystemC kernel to the SystemC user community. We also plan to submit our changes and proposed changes to the SystemC standards committee for review.

## 3. SystemC as a Simulation and Modeling Framework

There are major issues that need consideration when choosing an SLDL for modeling and simulation for any complex design. Many times these issues are overlooked by designers who begin constructing models in SLDLs that do not directly support the model's required behavior. Such an approach requires certain restrictions and almost always causes degradation in simulation efficiency. An example is attempting to model Synchronous Data Flow (SDF) designs with the existing version of SystemC [5, 12] that explicitly only support models exhibiting behaviors of the Discrete-Event MOC. Sometimes the reason for such a choice is the lack of an SLDL that does support the required behavior and other times the modeler's comfort levels in already knowing a certain SLDL and its semantics. Matching the design to an SLDL is analogous to matching the required behavior of the design to the MOCs supported by the SLDL. It is of immense importance that the behavior of the design be appropriately matched with the MOCs supported by the SLDL. The simulation framework must also adhere to the MOC being modeled. In addition, very few designs are homogeneous in functionalities and mostly all complex designs require interaction between various MOCs. This indicates the need for heterogeneity in SLDLs. Heterogeneous and hierarchical modeling and simulation environments offer support for constructing designs that exhibit different behaviors and that can be explicitly simulated by the simulation framework. SLDLs that support heterogeneous modeling and simulation increase their modeling expressiveness that can in turn result in simulation efficiency. Our proof of concept implementation of an SDF MOC in SystemC is an involved process, requiring system design, software design patterns, software architecture, as well as algorithmic innovation of heuristics.

## Simulation Efficiency Versus Expressiveness of MOCs

Lee and Sangiovanni-Vincentelli developed semantic framework to compare Models Of Computation for common systems and in particular embedded systems [9, 11]. An MOC describes the underlying behavior of a system on top of which a model following that particular behavior is simulated. Many MOCs for different behaviors for systems can be employed when targeting a model. Simulation efficiency is a major consideration when trying to incorporate systems that are inherently not of that behavior. The systems expected behavior must be modeled appropriately by the modeling framework, otherwise an unnatural modeling scheme has to be used to impose the behavior and the simulation efficiency will almost always suffer. When we describe a simulation framework supporting the Model Of Computation, we particularly refer to kernel-support for that MOC. Our motivation involves modeling and simulating SDF designs in SystemC by introducing an SDF simulation kernel. Although meant to be a system level language, the current SystemC simulation kernel uses a non-deterministic DE [11, 9, 14] kernel, which does not explicitly simulate SDF models in the way SDF models could be simulated more efficiently.

Since SystemC is planned as a language at higher levels of abstraction, notably at the system level, such Discrete-Event based simulation is not the most efficient for system level simulation. Unfortunately most industry inventors of SystemC claim that since SystemC has the full power of C++, it is capable of system level modeling. We consider that as a myth. SystemC provides no support for heterogeneous simulation making it inefficient to model systems that exhibit behaviors as other MOCs other than DE. To simulate a heterogeneous model the kernel must support each of the MOCs interlaced in the user model. This allows designers to put together heterogeneous models without worrying about the target simulation kernel and discovering methodologies to enforce certain MOCs under a single MOC kernel. Figure 4.1 shows an example of a digital analyzer [8] that employs two MOCs, one being an un-timed Data Flow MOC and the other being a timed model based on the DE MOC. The Controller IP and Analyzer IP could be possibly obtained from different sources requiring the Interfaces to allow for communication between the two. Support for modeling these IPs requires the simulation framework to provide the underlying MOCs, otherwise programmatic alterations are required to the original IPs for conformation. We believe it is crucial for modeling frameworks to allow designers to express their models in a simple and natural fashion. This can be achieved by providing MOCs via the kernel and sufficient guidelines to allow the kernels to interpret a more natural method of designer implementation. Another important inadequacy of current SystemC is lack of behavioral hierarchy representation. The structural hierarchy in SystemC is flattened during simulation, which leads to extreme inefficiency.

## Simulation Kernels and MOCs

A simulation model's responsibility is to capture the behavior of the system being modeled and the simulation kernel's responsibility is to simulate the correct behavior for that system. In essence, the Model Of Computation dictates the behavior which is realized by the simulation kernel.

We present a pictorial example of how SDF models are implemented in [5, 12] in Figure 4.2a and using our SDF kernel in Figure 4.2b. This example shows a Finite Impulse Response (FIR) model with *sc_fifo* channels and each process is of *SC_THREAD* type. This model employs the existing SystemC Discrete-Event kernel. Figure 4.2b describes our approach where each block in the model is an SDF *SC_METHOD* process and the communication is performed by SDF-specific channels.

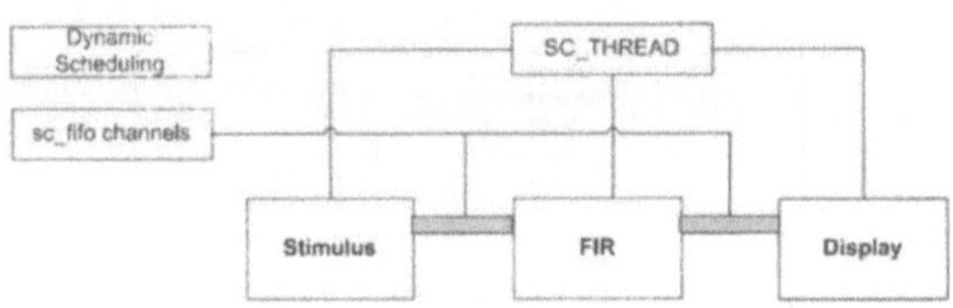


(a) FIR in DE

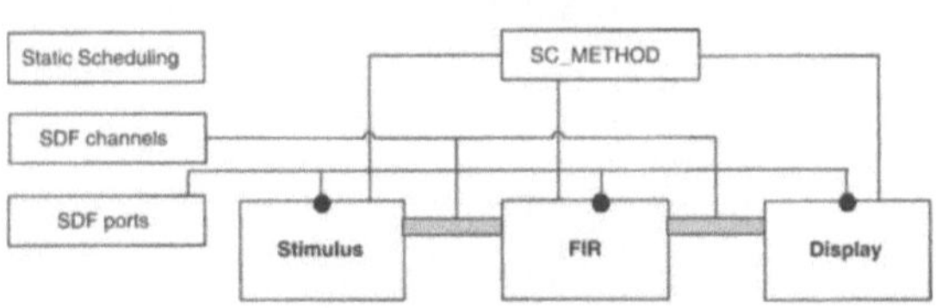


(b) FIR in SDF

*Figure 4.2.* Modeling Frameworks

**The Current SystemC DE kernel.** One of the important constraints in working on building a framework on top of an existing industry accepted framework is to not break any functionality or facility provided by the framework already, while enhancing it. As a result, all our work on bringing multiple MOCs and hierarchy has to make sure the current SystemC works in the same way, since SystemC is already on its way to become a standard.

The existing SystemC scheduler employs the Evaluate-Update paradigm to process the modules in the model as shown in Figure 4.3. The Evaluate phase is responsible for resuming execution of ready-to-run processes whereas the Update phase calls the *update()* function for the primitive channels that requests an update to propagate their values and update events. During initialization, the Evaluate phase schedules all processes that are ready-to-run, though the order selection of the execution of the processes is unspecified. The executing process can cause other processes to be ready-to-run in the same delta cycle, due to immediate event notifications or in future delta cycles using delayed notifications. When all the ready-to-run processes for that specific delta cycle have been evaluated, the scheduler enters the Update phase by updating the signal values calculated in the Evaluate phase. The end of simulation is realized when there are no timed notifications, but if there are pending timed notifications, then the scheduler advances the current simulation time to the earliest pending timed notification and continues to iterate through the Evaluate-Update phases until all timed notifications have been processed.

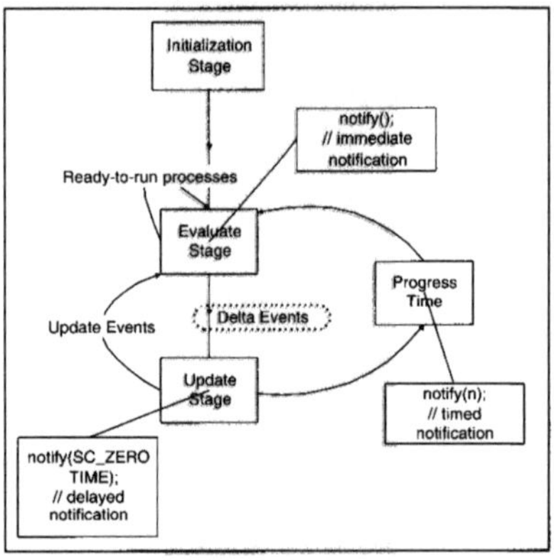


*Figure 4.3.* Discrete-Event Kernel

## 4. Examples of Models of Computation

Currently, models which are best represented with different MOCs can be modeled and simulated in SystemC. However, the cost is paid in simulation efficiency. One can see this from the examples in [5, 12] for the SDF MOC. The lack of expressiveness in SystemC for MOCs other than Discrete-Event, makes it difficult to model these MOCs. Modeling at a higher level of abstraction requires the kernel to accommodate the abstracted system rather than having the designer to force the abstracted system on to the DE kernel. So far, SystemC is not a heterogeneous modeling and simulation framework, lacking the capability to easily express and simulate MOCs other than the DE MOC. We are interested in providing design ease through better modeling expressiveness and simulation efficiency in SystemC.

A brief definition of some of the MOCs are presented below taken from [3].

**Component Interaction (CI)** The basis of the Component Interaction (CI) domain follows the construct for a data-driven or demand-driven application. An example of data-driven application is an airport departure and arrival display where information about flight departures, arrivals, delays, times, number, etc. are displayed for the public. Whenever a status changes for a particular flight, changes are made on the main server raising an event signalling the displays to perform a refresh or update.

Demand-driven computation on the other hand refers to the more common web-browsing environment where an HTTP request to a server acts as the demand resulting to an event on the server for a webpage. This event causes computation returning the requested webpage (given sufficient privilege for access) to the client requesting the webpage through a network protocol such as TCP.

**Communicating Sequential Processes (CSP)** In the CSP domain, actors are concurrently executing processes. The communication between two or more CSP actors is carried out by atomic and instantaneous actions called rendez-vous [3, 7]. If a process is ready to communicate with another process, then the ready-to-communicate process is stalled until the other process is ready to communicate, and once the processes are both ready the communication occurs simultaneously in one non-interruptible quantum, hence atomic.

**Continuous Time (CT)** As the name suggests, in the Continuous Time (CT) domain, the interaction between actors is via continuous-time signals. The actors specify either linear or nonlinear algebraic / differential equations between their inputs and outputs. The director finds a fixed point in terms of a set of continuous-time functions that satisfy the equations.

**Discrete-Events (DE)** In the Discrete Event (DE) domain, actors communicate through events on a real time line. An event occurs on a time stamp and contains a value. The actors can either fire functions in response to an event or react to these events. This is a common and popular domain used in specifying digital hardware. Unlike other DE simulators, Ptolemy provides deterministic semantics to simultaneous events by analyzing the data dependency graph to discover the order of execution.

**Finite-State Machine (FSM)** The FSM domain differs from the actor-oriented concept in Ptolemy II in that the entities in the domain are represented by states and not actors. The communication occurs through transitions between the states. Furthermore, the transitions can have guard conditions that dictate when the transition from one state to the other can occur. This domain is usually useful for specifying control state machines. This domain encapsulates the semantics of Kahn Process Networks.

**Process Networks (PN)** In this domain, processes communicate via asynchronous message passing through channels where the messages are buffered between the processes in a FIFO based structure. This implies that the receiver does not have to be ready to receive for the sender to send, unlike the CSP domain. The buffers are assumed to be of infinite size. The PN MOC defines the arcs as a sequence of tokens and the entities as function blocks that map the input to the output through a certain function.

**Data Flow (DF)** DF domain is a special case of Process Networks where the atomic actors are triggered only when input data is available. Otherwise data is queued.

# 5. Implementation of an SDF Kernel for SystemC

In this section we provide a brief description of the problems we solved in order to incorporate the Synchronous Data Flow Model of Computation with the existing framework in SystemC.

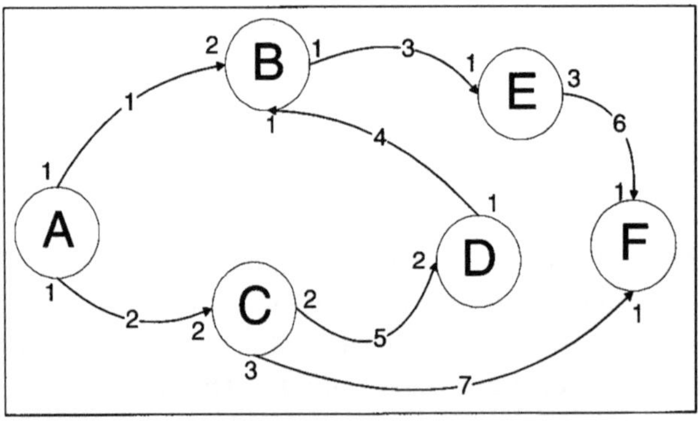


*Figure 4.4.* Synchronous Data Flow Graph Example

The Synchronous Data Flow MOC is a subset of the Data Flow paradigm. This paradigm dictates that a program is divided into blocks and arcs, representing functionality and data paths, respectively. The program is represented as a directed graph connecting the function blocks with the data arcs. Figure 4.4 shows an example of an SDF graph (SDFG) where the numbers at the head and tail of the arcs represent the production and consumption rates respectively, and the numbers in the middle represent an identification number for the arc. The production and consumption rates of an SDFG have to be known prior to its simulation. A Synchronous Data Flow MOC imposes further constraints by defining the block to be invoked only when there is sufficient input samples available to carry out the computation by the function block. SDFGs are susceptible to static scheduling [10]. There are two problems to be solved when statically scheduling an SDFG for execution. They are:

1 Computing the number of times each SDF block has to be fired that we refer to as the *repetition vector*.

2 Finding the order in which the SDF nodes have to be executed, which we term *firing order*.

Solution to the above two problems results in an *executable schedule* for an SDFG.

## Computing the Number of Times Each node needs to run

The first issue of creating a valid schedule is discussed in [10] where Lee et al. describe a method whereby static scheduling of these function blocks can be computed. The method utilizes the predefined consumption and production rates to construct a set of linear Diophantine [4] homogeneous system of equations and represent it in the form of a topology matrix $\Gamma$. It was shown in [10] that in order to have a solution, $\Gamma$ must be of rank $s - 1$ where $s$ is the number of blocks in the SDFG. Reducing the solution of this system of equations to an integer-valued solution produces the *repetition vector* for the SDF model.

We employed an algorithm finding Hilbert's basis [4] that solves linear Diophantine equations using the Completion procedure [15] to provide integer-valued Hilbert's basis. To better suit our needs for the calculation of the *repetition vector* we used a modification of this algorithm to create a hueristic specific for SDFGs. The details of the heuristic algorithm is available in [17].

## Computing the Order of the Schedule

The second issue of determining the *firing order* of execution of these function blocks is resolved by employing an algorithm from [2]. This algorithm handles the *firing order* of the SDF blocks and can appropriately schedule cyclic SDFGs as well. The algorithm defines a *delay* for every arc in an SDFG. This *delay* represents the number of tokens the modeler has to inject into the SDF model into that arc in order for the model to startup. The concept of *delay* is necessary when considering cyclical SDFGs due to the additional constraint set by the SDF paradigm that every node or block will only execute when it has sufficient input tokens on its input arcs. Thus, a cycle without the *delay* causes the SDFG to be inconsistent. The *delay* acts as an initial token on that arc to allow the simulation to begin in these circumstances. This algorithm in [2] will only yield a *firing order* if there is a valid *repetition vector*.

Along with solving these two integral problems for SDF scheduling, there was a need to construct a data structure to represent an SDF graph. Furthermore, there can be any number of SDFGs in a system hence there must be a capability to encapsulate multiple SDFGs in a model. Introducing an alternate kernel for SystemC also requires an Application Protocol Interface (API) through which kernels can "release" and "gain" control of the simulation. By "release" we imply the current executing kernel being able to invoke another kernel which "gains" control for a particular purpose. Suppose an SDFG is embedded in a Discrete-Event model, then the DE kernel invokes the SDF kernel once an SDF model is to be executed after which the DE kernel resumes its execution.

The details of data structure representation and implementation are documented in [17].

**An Example of Heterogeneous Model.** To further illustrate our heterogeneous modeling environment we construct a heterogeneous model of an image converter shown in Figure 4.5. This system begins by downloading encrypted images that are decrypted and converted to a specific type and then encrypted again to be uploaded back to the source. This is a multi-MOC model where the first DE block is responsible for downloading encrypted images and decrypting them. These images are passed onto the SDF block that performs the Sobel edge detection algorithm on the image. This is our preferred conversion type. Output from Sobel is sent to the final DE block that encrypts the converted image and uploads it back to the download source.

This model has SDF specific ports and channels through which the designer can transfer data within the SDF model. However, interaction external to the SDF block is dependent on the designer. SystemC signals and channels may be used to channel data to the top-level module of the SDF model through which they can be retunnelled through SDF specific channels. SDF specific ports and channels are required primarily because the SDF MOC describes a FIFO interconnect between modules and since the SDF model executes within one execution cycle (or in other words in multiple delta cycles) there is a need for immediate updates on signal values. This is unlike the signal implementations of SystemC whereby events are the underlying implementation for *sc_signal*. In addition, static scheduling eradicates the need for Evaluate-Update of signals within an SDF model, further supporting our decision to introduce SDF specific channels and ports.

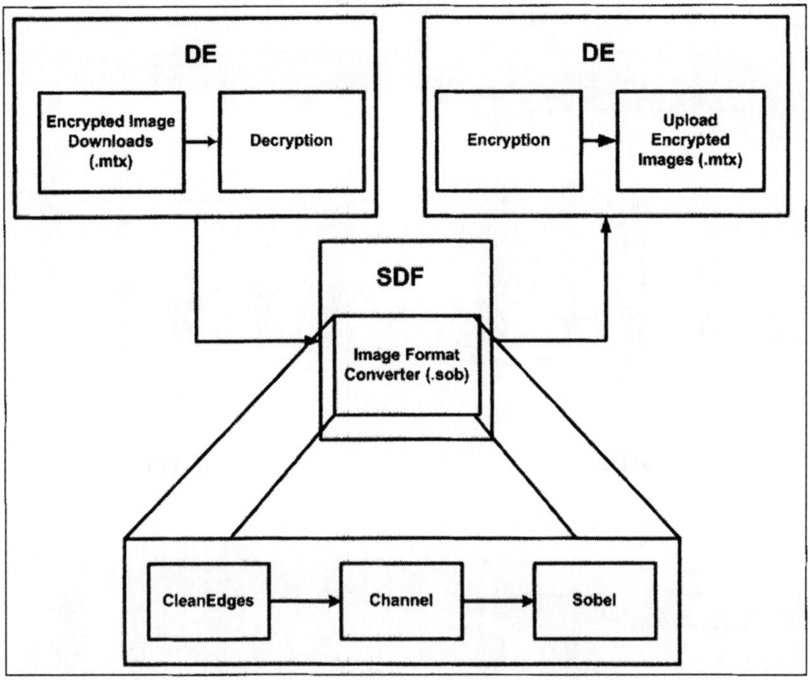


*Figure 4.5.* Image Converter Diagram

## Experiments and Results

We present some examples that uses our modeling guidelines to use the SDF kernel instead of the reference implementation kernel for SDF models.

In order to evaluate the efficiency enhancement by our SDF kernel we set out to experiment with a few models that are amenable to SDF style modeling. In this section, we show the results of simulating the same models for three distinct modeling styles, for different sample sizes of data. [18] showed around 50% or more improvement in simulation efficiency over threaded models have been reported. Our aim has been to improve upon those results reported in [18], which we call "Non-Threaded" models.

Four systems are modeled using the SDF kernel and the SDF modeling style. They are Finite Impulse Response Filter (FIR), Fast Fourier Transform (FFT), a Sobel edge detection system and a heterogeneous model that we call the Converter.

As seen in the graphs, every model showed significant improvement in the amount of simulation time that was taken over the original model and the non-threaded model. The three bars on each chart refer to the time taken in seconds for the entire model to be executed in their respective modeling environments. The leftmost bar being the original and middle bar being non-threaded are

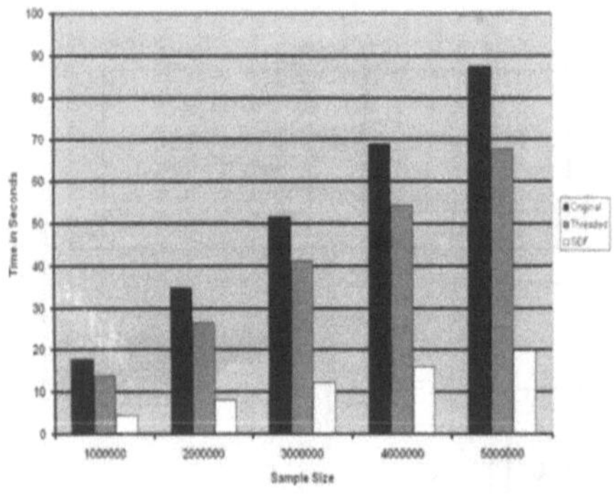


(a) FIR

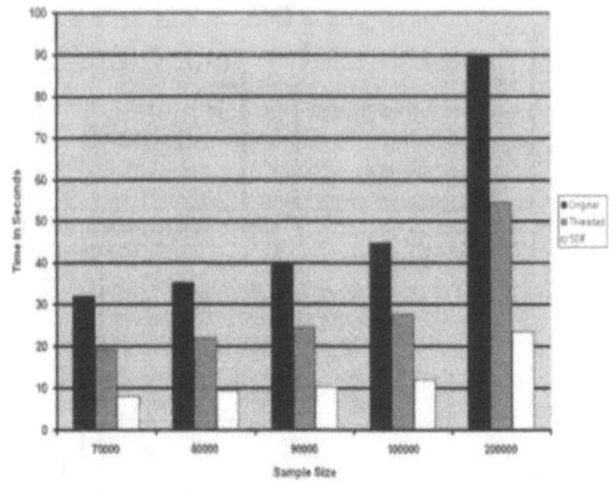


(b) FFT

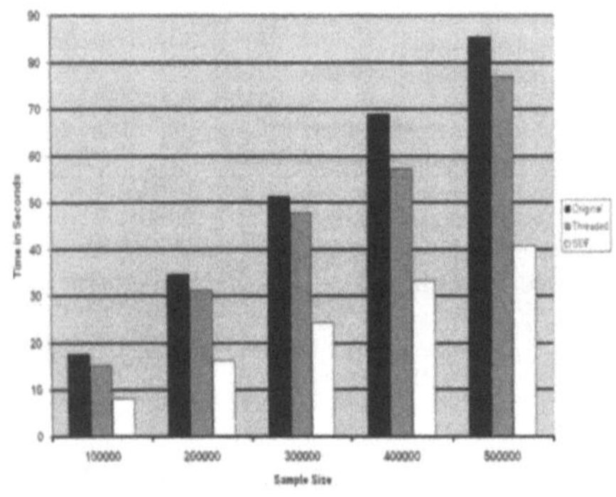


(c) Sobel

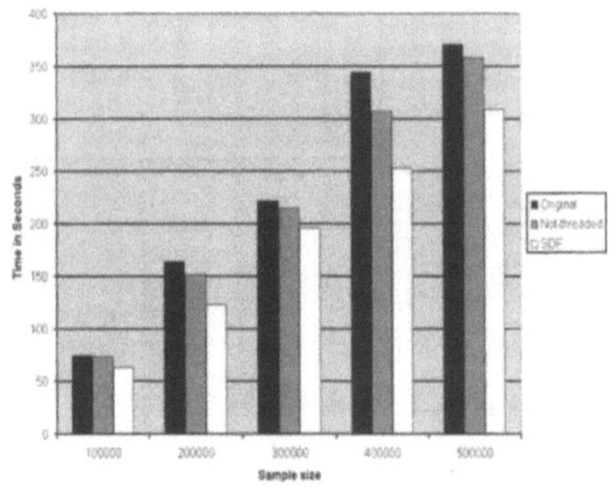


(d) Converter

(Black = Original DE models, Dark Gray = Non-Threaded models, White = SDF models)

*Figure 4.6.* Results from Experiments

modeled using the discrete event kernel and the rightmost bar is the synchronous data flow using the SDF kernel. The bar charts show that increasing the number of sample size will still preserve the efficiency presented by the set of collected data. The FIR and FFT yielded approximately 75% improvement in simulation time compared to the original model and the Sobel yielded a 53% improvement in simulation time. When the results from the SDF kernel is compared to the non-threaded models, the FIR, FFT, Sobel and Converter, showed 70%, 57%, 47% and 13% improvement in simulation time respectively. We attribute the decrease in simulation efficiency to Amdahl's law. For example, the SDF block in the converter model serves only a small portion of the entire system allowing for only that much improvement in total simulation performance. This means that when we construct models with numerous SDF blocks, the simulation efficiency of that model would be significantly larger.

## 6. Conclusion

Compiling a library of kernels describing different MOCs, enables the designer to employ them in either a timed or untimed fashion, depending on what MOC is being used. Furthermore, they can be used in unison to build heterogeneous models. As a proof of concept we have presented an extension to the SystemC standard by implementing a Synchronous Data Flow kernel that is interoperable with the reference implementation kernel providing designers with flexibility in constructing SDF-like models. Addition of other commonly used Models of Computations are being investigated for implementation of which some are Communicating Sequential Processes and Kahn Process Networks. The constricted event management of SystemC raises issues when considering implementation of some of these MOCs. We are currently addressing these issues while preparing SystemC as a possible system level design language that preserves behavioral hierarchy opposed to its current structural hierarchy.

The long term effort is in providing the design automation community with an open source multi-domain modeling and simulation framework. This chapter introduces our argument for selecting SystemC as our expected design entry System Level Design Language and lay a foundation for displaying that a heterogeneous modeling environment is possible in SystemC. However, the standardization of SystemC language is still underway, but as for the kernel specifications, SystemC only provides the Discrete-Event simulation kernel. We are working towards laying a foundation for an open source multi-domain modeling framework based on the C++ programming language.

**Acknowledgement:** This research is partially supported by NSF grants CCR-0237947 and NGS-0204028 and an SRC Integrated Systems Program.

## References

[1] Network Simulator 2, *Network Simulator 2*, Website: `http://www.isi.edu/nsnam/ns`.

[2] S. Bhattacharyya, P. Murthy, and E. Lee, *Software Synthesis from Dataflow Graphs*, Kluwer Academic Publishers, 1996.

[3] Shuvra S. Bhattacharyya, Elaine Cheong, John Davis II, Mudit Goel, Christopher Hylands, Bart Kienhuis, Edward A. Lee, Jie Liu, Xiaojun Liu, Lukito Muliadi, Steve Neuendorffer, John Reekie, Neil Smyth, Jeff Tsay,

Brian Vogel, Winthrop Williams, Yuhong Xiong, Yang Zhao, and Haiyang Zheng, *Heterogenous Concurrent Modelling and Design in Java: Volume 2 - Ptolemy II Software Architecture*, Memorandum UCB/ERL M03/28, July 2003.

[4] M. Clausen and A. Fortenbacher, *Efficient solution of linear Diophantine equations*, Journal of Symbolic Computation **8** (1989), no. 1-2, 201–216.

[5] T. Grotker, S. Liao, G. Martin, and S. Swan, *System Design with SystemC*, Kluwer Academic Publishers, 2002.

[6] Ptolemy Group, *Ptolemy II*, Website: `http://ptolemy.eecs.berkeley.edu/ptolemyII`.

[7] C. Hoare, *Communicating Sequential Processes*, Communications of the ACM, 8, vol. 21, 1978.

[8] A. Jantsch, *Modeling Embedded Systems And SOC's - Concurrency and Time in Models of Computations*, Morgan Kaufmann Publishers, 2001.

[9] L. Lavagno, A. Sangiovanni-Vincentelli, and E. Sentovich, *Models of Computation for Embedded System Design*, Website: `citeseer.nj.nec.com/lavagno98model.html`, 1998.

[10] E. A. Lee and D. G. Messerschmitt, *Static Scheduling of Synchronous Data Flow Programs for Digital Signal Processing*, In Proceedings of IEEE Transactions on Computers, NO. 1, vol. Vol. C-36, 1987.

[11] Edward A. Lee and Alberto L. Sangiovanni-Vincentelli, *Comparing Models of Computation*, In Proceedings of the International Conference on Computer-Aided Design (ICCAD), 1996, pp. 234–241.

[12] B. Niemann, F. Mayer, F. Javier, R. Rubio, and M. Speitel, *Refining a High Level SystemC Model*, Kluwer Academic Publishers, 2003, In SystemC: Methodologies and Applications, Ed. W. Muller and W. Rosenstiel and J. Ruf.

[13] OPNET, *OPNET*, Website: `http://www.opnet.com`.

[14] OSCI, *SystemC* , Website: `http://www.systemc.org`.

[15] D. Pasechnik, *Linear Diophantine Equation Solver*, Website: `http://www.thi.informatik.uni-frankfurt.de/~dima/software.html`.

[16] H. D. Patel and S. K. Shukla, *Towards a heterogeneous simulation kernel for system level models: A systemc kernel for synchronous data flow models*, In the Proceedings of the Great Lakes Symposium on VLSI 2004, 2004.

[17] Hiren D. Patel, *HEMLOCK: HEterogeneous ModeL Of Computation Kernel for SystemC*, Master's thesis, Virginia Polytechnic Institute and State University, December 2003, Website: http://fermat.ece.vt.edu.

[18] S. Sharad and S. K. Shukla, *Efficient Simulation of System Level Models via Bisimulation Preserving Transformations*, Tech. report, FERMAT Lab Virginia Tech., 2003-07.

[19] SPECC, *SpecC*, Website: `http://www.ics.uci.edu/specc`.

[20] SystemVerilog, *System Verilog*, Website: `http://www.systemverilog.org`.

[21] VERILOG, *Verilog*, Website: `http://www.verilog.com`.

[22] VHDL, *VHDL*, Website: `http://www.vhdl.org`.

Chapter 5

# MODE: A METHOD FOR SYSTEM-LEVEL ARCHITECTURE EVALUATION

Jan Romberg, Jewgenij Botaschanjan
*Systems & Software Engineering*
*TU München*
{romberg,botascha}@in.tum.de

Oscar Slotosch
*Validas AG*
*Lichtenbergstr. 8*
*85748 Garching, Germany*
slotosch@validas.de

**Abstract** Formalisms for system-level design of embedded HW/SW systems face several challenges: in order to be amenable to systematic formal analysis based on state-space exploration, a modeling notation with a simple formal semantics is desired. Early evaluation of design alternatives demands notations which concentrate on certain aspects of system functionality, while other aspects (such as communication and scheduling) are implicitly encoded in the language semantics, and later realized by HW components, operating systems, and protocol stacks. We describe a system-level design methodology targeted for automotive control applications. Models in a simple graphical component-based input language are compiled into complex system models incorporating abstractions for hardware, operating systems, and inter-processor communication. System models are based on the time-synchronous AutoFOCUS notation and are used as a basis for formal analysis such as systematic worst-case response time analysis. This paper describes the supported platforms for implementation, the MoDe design notation, and the translation to system models along with a case study describing the analysis procedure.

**Keywords:** System-level design methodologies, Performance estimation



*R. Gupta et al. (eds.), Formal Methods and Models for System Design,* 103-133.

# 1. Introduction

With as many as 80 electronic control units running distributed applications in modern premium-class vehicles, the complexity of software applications for automotive control is advancing towards the orders of magnitude known from the business information system and telecommunication domains. At the same time, non-functional properties such as timing constraints, per-unit costs, power consumption, and space play a predominant role in the domain. To tackle complexity, the predominant focus in automotive development on subsystem-level ECU design is shifting towards a purely functionality-oriented perspective, where the corresponding functions are distributed across several subsystems. However, for non-functional properties, finding a design optimized on the system level is typically much harder than optimizing a single subsystem.

The goal of the MoDe (*Mo*del Based *De*ployment) approach is to give early guidance for design decisions using SW components to model functionality, and simple abstractions of HW units. In the following, we will restrict ourselves to those early decisions that require a performance model of the overall system. Examples for such decisions are: choice of scheduling strategy to maximize utilization, mapping of SW components to HW units, choice of processor for minimal HW cost, or examination of bus load,

The MoDe approach is based on a design notation with a simple and extensively supported formal semantics, AutoFOCUS (Huber et al., 1997). AutoFOCUS is used at two levels: Firstly, the AutoFOCUS formalism is extended for modeling both the functional aspects of the application and the platform aspects of the HW architecture. Secondly, AutoFOCUS is used in order to represent an overall semantic model of the system used for detailed analysis. The use of system models with a clear formal semantics opens MoDe to a wide variety of formal analysis options based on state-space exploration.

## 1.1 Deployment

The term deployment in our context denotes the distribution of system functionality onto physical components. *Model-based* deployment uses abstractions of the deployed system to perform analysis such as validation and simulation. We have identified two possibilities of employing AutoFOCUS models for model-based deployment:

1 Use a given AutoFOCUS model directly as the behavioral specification of the system. In this case, the main difficulty for deployment is ensuring behavioral consistency between the model and its implementation as a combination of hardware, operating system, and software.

2 Use additional (and possibly non-deterministic) abstractions for communication and scheduling in an overall AutoFOCUS model of the system.

> The system model is preferably compiled from separate notations for functionality and HW architecture, and standard models for scheduling and communication are automatically inserted.

For MoDe, we choose the second option, using a separate input language and AutoFocus system models. Let us briefly discuss the reasons for this choice: If the first option is chosen, and the system's functionality is specified in a time-synchronous language such as AutoFocus, the platform must guarantee some assumptions made by the formalism. These assumptions, like the existence of a global time base, tend to be rather strong for globally synchronous models. For distributed communication, strongly deterministic architectures such as Time-Triggered Architectures (Kopetz, 1997) are definitely a good candidate for the first approach, and are currently being evaluated for use in safety-critical automotive applications.

On the other hand, the majority of applications to this date are based on event-triggered, weakly deterministic architectures such as the Controller Area Network (CAN) protocol for inter-processor communication, queued events and shared variables for intertask communication, and preemptive scheduling with intertask synchronization primitives. When such a design is being evaluated for early design decisions, much of the detailed performance characteristics of the implementation are not known. If one wants to analyze a design with formal models, e.g. for checking safety properties or for simulation, one possibility is to drop relative timing altogether, and an asynchronous system model results. Unwinding the possible executions of asynchronous models leads to a large number of possible process interleavings. As a consequence, methods based on state-space exploration quickly become impractical, despite some encouraging progress in the area (Godefroid, 1990).

In MoDe, functional models of the system are enriched with actual platform and implementation information; the system model which is subjected to analysis incorporates these informations to improve analyzability. Consequently, system models are tied to properties of the platform: the possible executions of the model closely reflect possible executions in the implementation in order to ensure the validity of model-based analysis. By offering automated support for compiling platform abstractions into the system model, MoDe offers a flexible method for evaluating the same functionality for different platforms.

In the remainder of this article, we will give an overview of our methodology and the design flow in Section 2. The supported target platforms are characterized as the reference platform in Section 3. In Section 4, we will give a definition of the graphical notation used to describe HW/SW architectures in MoDe. A case study in Section 5 presents an application example and describes the currently available analysis techniques as well as some experimental results. Finally, Section 6 gives a summary and an outlook discussing possible future extensions of our method.

# 2. Overview

## 2.1 Design Flow

Figure 5.1 shows the design flow for the MoDe approach. The designer starts out with a functional view of the application, the *logical model*, a view of the technical architecture, the *technical model*, and an additional *synthesis descriptor*. The MoDe system model generator reads the input representations, checks them for mutual consistency, and generates an AutoFOCUS *system model* used for simulation, verification, and performance analysis.

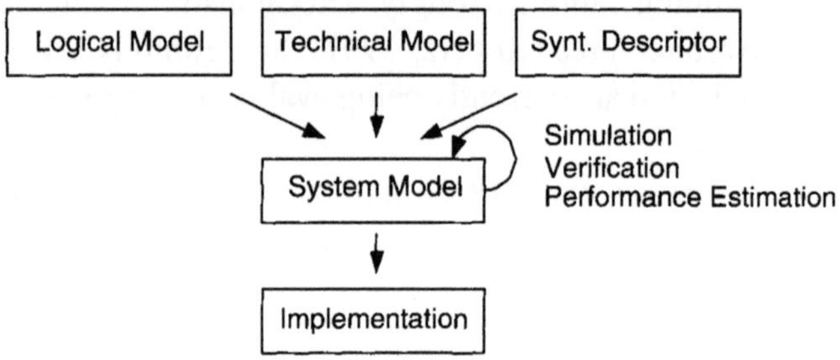


*Figure 5.1.* Design Flow

**Logical Model.** The logical model captures the software architecture in terms of *software components*, component *interfaces*, and *communication dependencies* between components. Each component is specified in terms of its functional and temporal behavior, the latter of which depending on properties of the technical model described below. The logical model abstracts from the actual deployment, i.e. communication protocols and scheduling are not part of the logical specification.

**Technical Model.** The characteristics of the platform are specified within the *technical model*. The technical model captures the structure of the underlying hardware, the characteristics of the different nodes of the platform, the characteristics of the links between nodes, and parameters of the operating system such as supported scheduling policies and I/O drivers. In a way, the technical model fills the behavioral and structural "blanks" left unspecified by the logical model. Note that the technical model is intended to be an independent and self-contained view of the platform that is reusable across several designs. The concepts of the technical model are supported by *component libraries* which define the components required to generate the system model.

**Synthesis Descriptor.** As the logical and technical models alone are not expressive enough to perform an automated synthesis of system models, some additional mapping information has to be provided by the developer in the form of a *synthesis descriptor*. It is required to ensure an unambiguous translation from logical and technical models to system models. The synthesis descriptor captures information that depends both on the logical and technical models; this typically includes the mapping from logical components to technical nodes, and additional information required for scheduling such as static schedules or priorities, depending on the scheduling policy.

The separation into three distinct descriptions ensures, to the greatest possible extent, a separation of concerns between application and platform aspects, and enables the developer to separately focus on desired functionality and technical realization. We will refer to the combination of logical model, technical model, and synthesis descriptor as the *deployment model* in the remainder of this paper.

**System Model.** The MoDe system model generator compiles the logical and technical models into an AutoFocus system model. The synthesis step uses the mapping from logical to technical components in the synthesis descriptor to synthesize the system model as a complete representation of the system behavior relevant to analysis. All analysis is based on the system model; if several different deployments are evaluated, a corresponding number of analysis cycles are required.

## 2.2 Analysis

There is a wide variety of analysis options supported by AutoFocus which are all applicable to MoDe system models: a visual simulation interface using precompiled test vectors or user interaction as input (Huber et al., 1997), formal verification using either the symbolic BDD-based model checker SMV or the bounded-depth solver SATO( Wimmel et al., 2000), and automated test case generation using a Constraint Logic Programming (CLP) environment ( Pretschner et al., 2004).

In the following sections, we will focus on system-level response time analysis. The current framework supports this kind of analysis both through the AutoFocus simulation interface, and by systematic quantitative analysis described in Section 5. In an iterative systems development process, the results of the analysis may serve as a basis for modifications of the deployment model in the next iteration.

# 3. Supported Platforms

## 3.1 Inter-Processor Communication

The reference platform for MoDe consists of a number of single-processor boards (*nodes*) communicating over a number of communication *links*. In MoDe system models, properties of the processor are modeled by service timing functions (see Section 4.4). For modeling inter-processor communication, MoDe system models incorporate abstractions for *drivers* and *synchronization*:

**Drivers.** In a layered communication architecture, the logical model typically specifies the application layer. For distributed communication, models of *drivers* are required to specify the behavior of layers between the application layer and the layer modeling the actual link in the system model. Note that this layer is not necessarily the physical layer. The choice depends on the level of detail required for the analysis.

**Transport medium and synchronization.** This includes synchronization between multiple nodes (e.g. for buses), potential message loss, etc. For instance, a synchronization component for a bus with priority-based arbitration yields the bus to the requesting driver with the highest priority for a given arbitration cycle.

Models for drivers and synchronization are imported through a library mechanism. The MoDe communication library currently supports two types of links:

**RS232 serial communication.** RS232 is a simple unidirectional serial protocol. In the MoDe library, RS232 is modeled with comparatively simple driver components; a separate synchronization component is not necessary for point-to-point links.

**CAN bus communication.** CAN is a local area network protocol with priority-based arbitration geared at real-time and automotive applications. The MoDe library includes both a driver model representing the sub-application-layer functionality, and a synchronization component modeling arbitration.

## 3.2 Inter-Task Communication

Nodes are assumed to run a real-time operating system with support for multitasking, mutual exclusion protection, and I/O drivers. The mapping of software components to nodes is assumed to be fixed during run-time, i.e. migration of software components is not considered.

The MoDe approach imposes some constraints on the OS services used for implementation:

- The OS primitives used are *threads*, *message queues*, and *shared resources*.
- All computations are done by threads. Threads share memory with other threads, and inter-thread communication is performed either via message queues, or by shared-variable communication through protected memory resources.

- Access to protected resources is restricted to one thread at a time (mutex protected). A running thread may acquire a free resource or release a locked resource at any time.
- Queues realize FIFO queues of unbounded size. Messages are never lost.

The abstractions currently used for scheduling and inter-task communication correspond to the OSEK OS 2.2 standard (OSEKOS, 2001). OSEK OS is an open operating system standard for applications with small and statically predictable resource consumption, and was developed by a consortium of European automobile manufacturers.

## 3.3 Scheduling

OSEK standard defines two kinds of tasks: *basic* and *extended* tasks. Basic tasks can be either in *suspended*, *ready*, or *running* state, while extended tasks have an additional fourth state, *waiting*, similar to the ready state, in which the stack context is saved. MoDe currently supports two options for scheduling:

**Cyclic (offline) scheduling, basic tasks.** Threads are activated according to a fixed cycle. Static scheduling can be enforced in OSEK's fixed-priority framework by either using a cyclic executive within one OSEK task, or by ensuring a deterministic cyclic sequence of task activations.

**Fixed priority scheduling, no preemption, basic tasks.** Threads are dynamically scheduled according to their priorities. Scheduling is restricted to the OSEK basic task model and non-preemptive scheduling.

## 3.4 The Mine Pump Example

For the following explanations, we will use an example mine pump system whose specification was adapted from (Joseph, 1996). Though not from the automotive domain, the example exhibits some typical properties of automotive control applications, namely *real-time requirements*, *safety requirements*, *distribution*, and *resource constraints*.

The mine pump is used to pump water out of the bottom of a mine (Fig. 5.2). It is equipped with two water level sensors (high, low) and a gas sensor. A pump controller switches the pump on when the water reaches the high water level and turns it off when the water is below the low water level. As an additional operational requirement, the mine pump may only operate if the methane ($CH_4$) level is below a critical threshold. If a methane alarm occurs, the pump has to be switched off. If the methane sensor fails, the mine pump has to be transferred into a fail-safe state. More formally, the following requirements can be derived:

1 A high methane value reading shall cause an emergency shut down of the mine pump within 30 milliseconds.

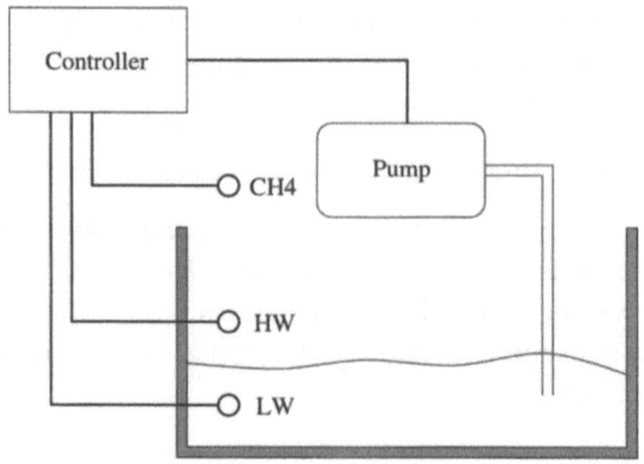


*Figure 5.2.* Mine Pump Example

2 A methane sensor failure shall be recognized and lead to a shutdown of the pump within 65 milliseconds.

3 After the methane value is back to normal, the pump shall be initialized (ready to pump) within 100 milliseconds.

4 Under the condition of normal methane values, a high water level reading shall cause the pump to be turned on within 100 milliseconds.

5 If the low water level has been reached, the pump shall be turned off within 75 milliseconds.

In the following section, we will illustrate how the MoDe approach allows flexible modeling of the mine pump controller system and how different architectural alternatives can be evaluated using MoDe.

## 4. Notations and Tools

This section presents the notations used for representing deployment and system models in MoDe, and gives a brief description of the tool used for editing and translating deployment models.

### 4.1 Logical Model

The visual notation for logical models is loosely based on the AutoFOCUS notation. In AutoFOCUS, systems are specified as hierarchical component networks, where components communicate via typed and directed channels. Similar in style to UML-RT( Selic et al., 1994), component networks are specified in System Structure Diagrams (*SSDs*). Rectangles represent components, arrows between components represent channels, and interface of a component to a channel is called port. Outgoing ports may be connected to several channels, while incoming ports are connected to one channel. Components may be defined by other SSDs, or by STDs introduced below. All visual elements may be extended by UML-like *stereotypes* of the form ≪`ID`≫, and *tags* of the form {`KEY=VALUE`}. AutoFocus models may be extended by datatype and function definitions in a functional language. The corresponding text documents are called Data Type Definitions (*DTDs*).

Figure 5.3 shows the SSD for the logical model for the mine pump example. Ignoring the stereotypes for now, the system is decomposed into the following logical components:

- A pump controller (`Controller`)
- Two components handling high and low water events, respectively (`WaterLowComponent`, `WaterHighComponent`)
- One component monitoring the methane level (`MethaneSensorComponent`)

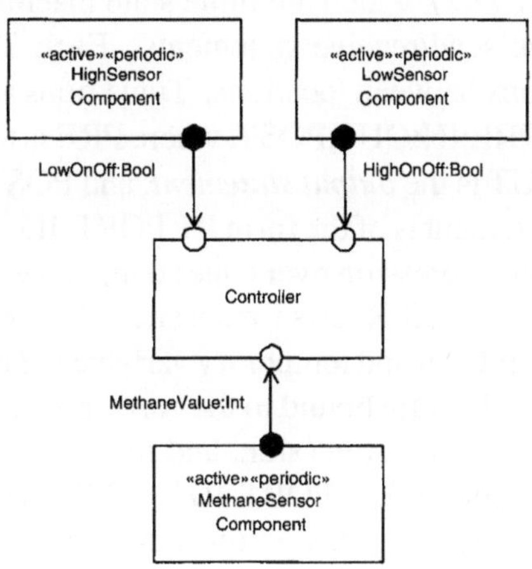


*Figure 5.3.* SSD for Logical Model

A component may be refined by a network of sub-components defined in another SSD. In the mine pump example, the `Controller` component is defined by a sub-SSD (Figure 5.4). The interface ports of `Controller` appear as external ports; there are three components refining `Controller`'s behavior:

- A safety component acting as a watchdog (`SafetyComponent`)
- A component managing the state of the pump (`PumpController`)
- A component holding the current methane level, and putting timestamps on incoming new values (`MethaneValueComponent`)

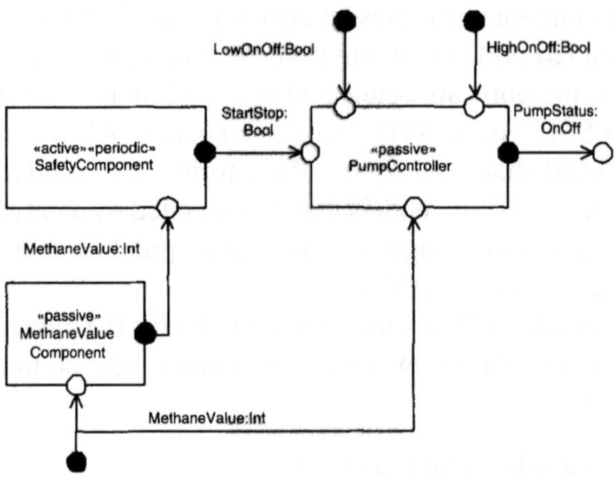


*Figure 5.4.* SSD for `Controller`

Leaf components in AutoFOCUS component hierarchies are defined by State Transition Diagrams (*STDs*), which are finite state machines extended by local variables and message send/receive statements. Each STD has a number of *locations* and *transitions* between locations. Transitions are labeled; the syntax of transition labels is PRE:IN:OUT:POST, where PRE is the *precondition*, IN is an *input statement*, OUT is the *output statement*, and POST is the *postcondition* (action). The input statement is of the form IN_PORT_ID ? EXPR, where EXPR is a functional language expression over (free) temporary variables and (bound) local variables. An empty EXPR tests for absence of a message. The guard is a boolean expression over local and temporary variables. The transition is chosen (and the temporary variables are bound to the corresponding values in the input) if the input matches the input expression, and the guard evaluates to true. An output statement of the form OUT_PORT_ID ! EXPR sends the message EXPR to an output port, and postconditions perform assignments to local variables.

Figure 5.5 shows a simplified STD for the `PumpController` without timing annotations. The pump controller's STD has three locations: the pump is turned off, the pump is turned on, and the pump has been stopped by the safety component or the methane sensor component. All three locations are *final* locations, marked by a dot at the right hand side: in MoDe, this is used to indicate interruption points for scheduling. Location `PumpOff` is the *initial* location (dot on left hand side) activated at the start of the system. The controller switches between the `PumpOn` and `PumpOff` locations whenever the high and low water levels are reached, respectively. If either `Stop` message on the `StartStop` channel or a high methane value is encountered, the controller switches to the `PumpStopped` location. It is switched back to `PumpOff` when a `Start` message is detected.

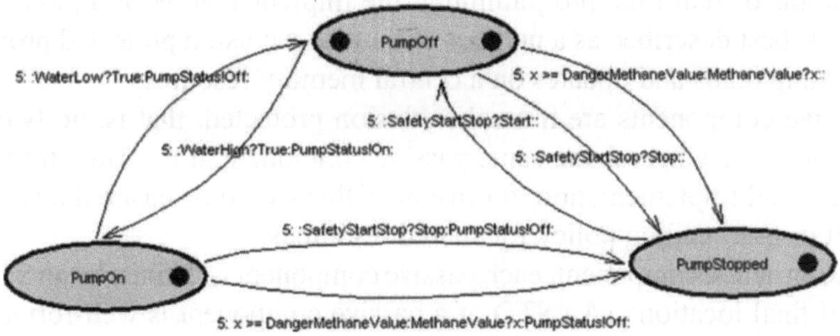


*Figure 5.5.* Simplified STD for `PumpController`

In logical models, as opposed to standard AutoFOCUS models, leaf components are further differentiated along their mode of activation as *active* or *passive* components.

In the mine pump example (Figure 5.3), the `PumpController` and the `MethaneValueComponent` are stereotyped as passive, while all other components, such as `SafetyComponent` or the sensor monitoring components, are active.

**Active Components.** Active components are triggered by external events such as interrupts or periodically elapsing timers. Each active component corresponds to a lightweight task (thread) on the implementation level. The additional stereotype «`periodic`» defines that the activation is periodic, i.e. time-triggered, while the stereotype «`sporadic`» indicates that the activation of a component is triggered by incoming messages. Note that in combination with a cyclic (offline) scheduling policy, the «`periodic`» and «`sporadic`» stereotypes are ignored, and a cyclic activation results.

The behavior of an active component is specified by an STD. The STD defines an initial location and a number of final locations indicating interruption points for the scheduler. Well-formed STDs for active components have at least one final location; this is checked by the generator. The active component keeps its local control and data state between subsequent activations.

Messages passed to other active components will be queued and processed in FIFO order by the receiving component, while messages to passive components are immediately processed (see below).

**Passive Components.** Passive components are activated whenever the component receives a message from another component. The activation ends whenever a final state is reached.

In terms of real-time programming, the implementation of a passive component is best described as a number of mutual exclusion protected procedures performing reads and updates on a central memory resource.

Passive components are mutual exclusion protected, that is, no two active components may access the same passive component at the same time. In an OSEK-based implementation, for instance, this would be ensured by using the default priority ceiling policy for shared resources.

Like an active component, each passive component is defined by an STD with a set of final locations. An STD of a passive component is well-formed if all outgoing transitions from final locations have exactly one input statement with a nonempty expression, and all other transitions do not have input statements. Note that transitions may not contain more than one send statement to a passive component, so sequentialization of calls is explicitly reflected in the calling component's automaton.

**Timed and Clocked STDs.** MoDe system models are interpreted as discrete-time models where each system step corresponds to a fixed time period. A logical component delays the overall computation by remaining in a non-final state for a number of steps. In principle, the expressiveness of STDs is sufficient for modeling delays; however, the resulting models tend to be somewhat clumsy. The MoDe tool provides two extensions of the basic STD notation, *clocked STD*s and *timed STD*s. Timed STDs can be translated to clocked STDs; clocked STDs, in turn, translate to basic STDs.

Clocked STDs allow local variables to be qualified as *clocks*. Clocks are of (nonnegative) integer type and are increased by one with each system step; like local variables, clocks are controlled by their components. Each of the locations is labeled with an additional *invariant* over the component's clocks, with the conjunction of negated outgoing guards as the default; the invariant defines which clock valuations are permissible for the respective location. Clearly, the resulting behavior may be nondeterministic. In combination with preconditions over clock valuations for outgoing transitions, invariants allow for the specification of *counting locations* modeling execution time for computations. In order to omit violations of invariants, the totality check for the transition relation has to be performed before the system model can be generated.

Timed STDs have the same syntax as basic STDs; in addition, transitions may have *timing labels* of the form ≪`time(MIN;MAX)`≫ or ≪`time(VAL)`≫, where `MIN`, `MAX`, and `VAL` are expressions over the component's variables of integer type. The `(MIN;MAX)` notation specifies a nondeterministic interval, while the `(VAL)` syntax refers to a deterministic delay.

Figure 5.6 shows the timed STD of the component `SafetyComponent`. The safety component may perform one of the following two actions:

- If the methane value is within a safe range (`x <= DangerMethaneValue`), a `Start` command is issued to the pump controller
- If the methane value exceeds the maximum (`x > DangerMethaneValue`), or if no methane value has been available for a certain time interval (`t - lastTime > TMaxSilent:MethaneValue?::`), then a `Stop` command is sent to the pump controller.

We assume that the execution time for either of the three transitions is between `tCheckingMin` and `tCheckingMax`, which are predefined constants. The generator will translate the timed STD to a clocked STD. In the translation step, a new clock $c$ is added to the component. Each timed transition $\delta$ with label $\text{PRE}_\delta : \text{IN}_\delta : \text{OUT}_\delta : \text{POST}_\delta : \ll\texttt{time(MIN;MAX)}\gg$ is replaced with

- a counting location with invariant $c < \text{MAX}$ modeling the ongoing computation,
- a transition labeled $\text{PRE}_\delta : \text{IN}_\delta :: \text{POST}_\delta \wedge (c = 0)$ from the source location to the counting location,
- a transition labeled $c \geq \texttt{MIN} :: \text{OUT}_\delta :$ from the counting location to the destination location.

Examples for clocked STDs resulting from the translation can be found in section 5.

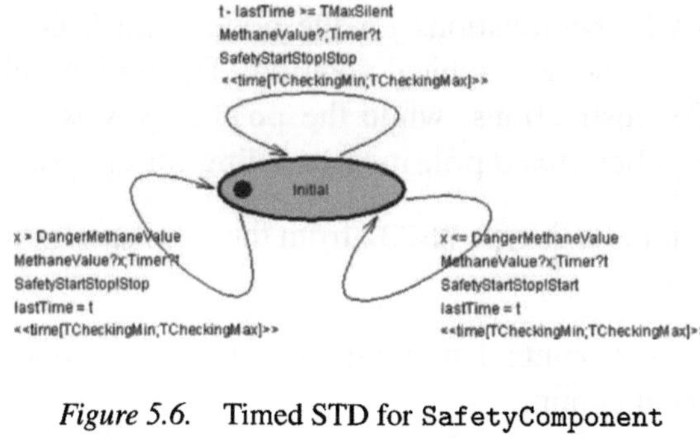


*Figure 5.6.* Timed STD for `SafetyComponent`

## 4.2 Technical Model

The properties of the technical platform are specified in the technical model. The technical model is an independent and self-contained view of the system.

Technical models consist of networks of *nodes*, *links* between nodes, and *connectors* on nodes. *Nodes* represent physical computers with their associated I/O interfaces and their operating system. Specification of node-specific properties like scheduling policy, operating system, etc. is done by a number of tags on the respective nodes. In our example, all three nodes use the OSEK operating system with a (non-preemptive) Fixed Priority Scheduling policy.

*Links* denote communication links between nodes which can be either directed or undirected. Every directed link may be connected with one source connector and arbitrarily many destination connectors (broadcast). For undirected links there is no inherent restriction on the number of connectors, but restrictions on the number of connectors may be part of the associated *link type* definition. The behavior of the link is defined by associated library components. The corresponding drivers are also identified by link types. In the mine pump example, both links are of type `RS232`, and the corresponding driver components are taken from the MoDe communication library.

*Connectors* denote those "access points" of a node where links can be attached. Connectors, like links, are typed; for directed link types, the connector carries an additional source/destination flag. Naturally, the connectors of a given node are restricted to those protocols that are supported by libraries for the component.

Figure 5.7 shows the technical model for the mine pump example in a simple graphical notation. The technical model consists of the following entities:

- Three nodes for processing signals from the water high/low sensors (`WaterSensorProcessor`), processing signals from the gas sensors (`GasSensorProcessor`), and a central control unit (`ControlUnit`), respectively.

- Additional OS specifications for the nodes. Each OS specification is shown as a tag in the graphical notation. The `os` key selects the library used for OS abstractions, while the `policy` key is set to the desired scheduling policy (fixed priority scheduling for the example).

- Two directed links of type `RS232` from the two signal processing units to the control unit.

- Four connection points defining the interfaces from nodes to links – not shown in the diagram.

## 4.3 Synthesis Descriptor

The third element of the deployment model is the synthesis descriptor. The synthesis descriptor currently consists of two parts: a `mapping` part mapping components to nodes, and a `schedule` part specifying the relative order of activations (in case of cyclic scheduling) or priorities (in case of fixed-priority scheduling) of the active components. A cyclic schedule is written in the form

$$\texttt{N(}<ac\_id_1>\texttt{,N(}<ac\_id_2>\texttt{,}\ldots\texttt{,Finished)}\ldots\texttt{)}$$

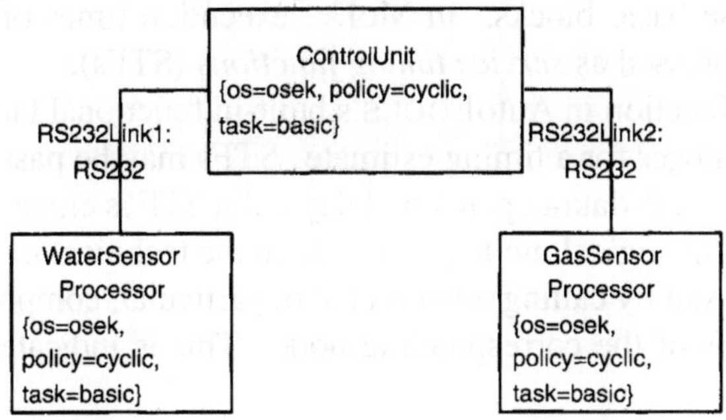


*Figure 5.7.* Technical Model

where <*ac_id*$_1$>, <*ac_id*$_2$>, ... are the identifiers of active components. Figure 5.8 shows the synthesis descriptor for the example. Though conceptually the synthesis descriptor is considered a separate document, the MoDe tool facilitates navigation within the model by supporting direct annotations to diagrams of the logical and technical models.

```
syntdesc {
  mapping {
    {PumpController, MethaneValueComponent,
      SafetyComponent} mapsto ControlUnit
    MethaneSensorComponent mapsto GasSensorProcessor
    {HighSensorComponent, LowSensorComponent}
      mapsto WaterSensorProcessor
  }
  schedule {
    {ControlUnit, N(SafetyComponent, Finished)},
    {GasSensorProcessor, N(MethaneSensorComponent, Finished)},
    {WaterSensorProcessor, N(HighSensorComponent, N(LowSensorComponent, Finished))}
  }
}
```

*Figure 5.8.* Synthesis Descriptor

## 4.4 Service Timing Functions

MoDe models the timing of complex computations in terms of the delay of simple computations (basic blocks). Timing analysis in MoDe is based on the assumption that execution time analysis based on these basic blocks is sufficiently precise for analysis. Basic blocks typically correspond to simple, short algorithms, or even single processor instructions. The execution times of more complex software algorithms may be described as a (possibly data-dependent) function of these basic blocks. In MoDe, execution times of algorithms, or *services*, are expressed as *service timing functions* (STFs).

An STF is a function in AutoFOCUS's built-in functional language yielding a nonnegative integer for a timing estimate. STFs may be passed a number of parameters to express data-dependent delays. An STF is either associated with a component in the logical model, or a node in the technical model. STFs may compute their result by calling other STFs; in particular, component STFs may refer to the STFs of the corresponding node. This is indicated by the prefix `$Node$`.

In the mine pump example, component `MethaneSensorComponent` reads a value from an external sensor port `ActualMethane` and forwards its value to the `MethaneValue` port. MethaneSensorComponent filters the sensor signal in order to eliminate noise. The STF `tFilter` is used to derive an estimate for the worst case execution time of the filter; note that the model abstracts from the actual filter operation.

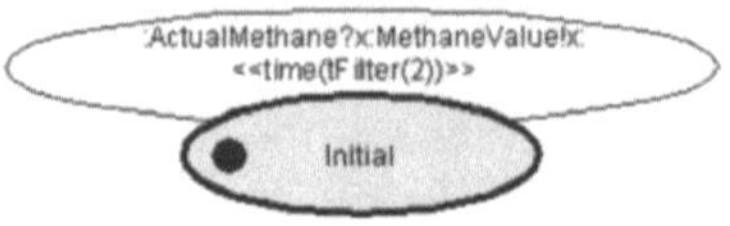


*Figure 5.9.* Timed STD for `MethaneSensorComponent`

The component's STF `tFilter` models the worst case execution time for a discrete Butterworth filter of order $n$:

```
MethaneSensorComponent_tFilter: Int -> Int;
fun MethaneSensorComponent_tFilter(n:Int) =
  $Node$_tIntAdd(16) * (2*n)
+ $Node$_tIntMult(16) * (2*n+1);
```

The corresponding node, `GasSensorProcessor`, defines two service timing functions, `tIntAdd` and `tIntMult`, giving some estimates for the worst case execution time of integer additions and multiplications on its processor:

```
GasSensorProcessor_tIntAdd: Int -> Int;
fun GasSensorProcessor_tIntAdd(16) = 2 |
  GasSensorProcessor_tIntAdd(32) = 4;
GasSensorProcessor_tIntMult: Int -> Int;
fun GasSensorProcessor_tIntMult(16) = 13 |
  GasSensorProcessor_tIntMult(32) = 22;
```

The STF `tFilter(2)` evaluates to $2 \times (2 \times 2) + 13 \times (2 \times 2 + 1) = 73$. Consequently, `MethaneSensorComponent`'s transition will be delayed by the same number of steps.

## 4.5 System Model

The system model expresses the semantics of the deployment model as a network of basic AutoFOCUS components. Each node in the technical model is an abstract processor which executes software components sequentially. The passing of control between different components must be modeled appropriately in the AutoFOCUS notation, which uses parallel lock-step composition for components.

There are two ways control is passed between components running on the same abstract processor: *sequential* and *scheduled*. Sequential passing of control closely resembles a procedure call in the implementation; it applies whenever a message is sent to a passive component. For preemptive scheduling policies, scheduled passing of control may happen at any point in time during execution of the system. For nonpreemptive scheduling, active components allow scheduled passing of control whenever one of their final locations are reached. Note that the passive components may be accessed by several active components.

The translation proceeds roughly along the following steps:

1 Create a system model. Each node in the technical model corresponds to a top-level AutoFOCUS component in the system model. Figures 5.10, 5.11 and 5.14 show the top-level structure and a part of the model for the `ControlUnit` node of the mine pump, respectively.

2 Insert communication handler components for distributed communication through RS232 or CAN.

3 Insert message queue components for communication between active components mapped to the same node.

4 Depending on the scheduling policy, insert a predefined scheduler component for each node, and insert appropriate communication channels to the active components for scheduled passing of control. Each active component's STD is extended with a special `Suspended` location that models those states where the component is inactive.

5 Insert acknowledgement channels and waiting locations in the calling component's STD for sequential passing of control with passive components.

6 Translate timed STDs to clocked STDs, and clocked STDs to basic STDs.

## 4.6 System Model Generator

The MoDe method is supported by a tool extension of the publicly available AutoFOCUS tool (AutoFocus, 2003). As an input notation for deployment

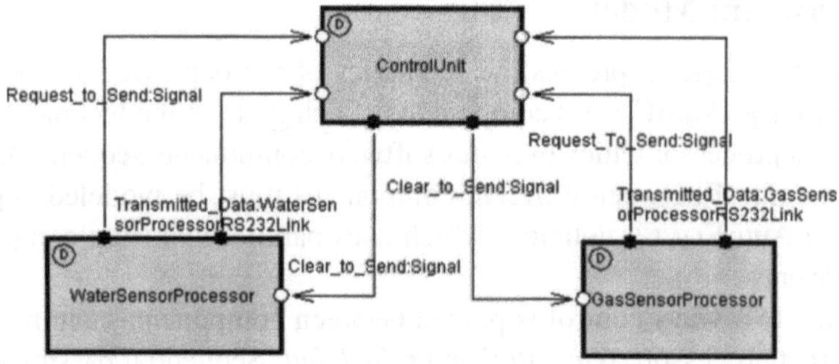


*Figure 5.10.* SSD for System Model (Top Level View)

models, the MoDe generator currently uses annotated AutoFOCUS models. The generator then applies a well-formedness check on the deployment model, and automatically generates system models according to the above procedure, if the deployment model is found to be correct. The system model is used for simulation or quantitative real-time analysis. Note that automatic translation of MoDe deployment models is currently limited to cyclic scheduling.

## 5. Case Study

Section 4.5 has described the deployment model for the mine pump example. According to Fig. 5.10, the resulting system model is decomposed into three components at the top level for controller, water sensor processor, and methane sensor processor. Sections 5.1–5.3 describe the translation of the sensor nodes, the `ControlUnit` node, and the scheduler component for the `ControlUnit` node.

**Translation and Simplification.** The system model for the cyclic scheduling policy was generated from the deployment model described in Sections 4. Because the system model generator currently supports cyclic scheduling only, system models for fixed priority scheduling were created by manually modifying the scheduler component, and by adding appropriate interrupt handler components as event sources (see Section 5.3 for an overview). In order to reduce the state-space of the generated system model for the analysis, some parts of the model were further optimized, without a change to the observable behavior. We will indicate simplifications at the appropriate points in the text.

### 5.1 WaterSensorProcessor and GasSensorProcessor

The water and methane sensors are modeled as simple periodic active components with a period of 40 steps (`HighSensorComponent`, `LowSensorComponent`) and a period of 70 steps (`MethaneSensorComponent`), respectively.

Because nothing is known about the behavior of the environment (methane concentration and water level), we use nondeterministic abstractions of the environment: any sensible combination of sensor inputs may occur at the scheduled execution times. In our abstraction, the case where the water level sensors indicated both high and low water level at the same time is excluded.

**Simplifications.** Instead of having separate AutoFOCUS components for scheduler components, active components, and protocol handlers, the overall behavior of both `WaterSensorProcessor` and `GasSensorProcessor` was encoded in one STD for each node. As a further simplification, the integer value for the methane level sent by `MethaneSensorComponent` was encoded as a combination of Boolean values (Table 5.1).

*Table 5.1.* Abstractions for `MethaneSensorComponent`'s data

| Event | Coding |
|---|---|
| $CH_4$High | `True` |
| $CH_4$Low | `False` |

## 5.2 ControlUnit

The controller consists of three active components (see Figure 5.11): `SafetyComponent`, `RS232InHandlerGSP` and `RS232InHandlerWSP`. The latter two components are RS232 serial communication protocol drivers for the communication with the methane and water sensor processors, respectively. The `SafetyComponent` is responsible for the emergency stop of the mine pump if the methane values are too high or the methane sensor fails. Moreover this component reactivates the pump once the methane value reading goes back to normal. The active components listed above use a number of passive components.

**ControllerComponent.** The mine pump is controlled by a passive component, `ControllerComponent`. It captures the actual state of the pump, specified in the logical model (see Fig. 5.5): `PumpOn`, `PumpOff`, `PumpStopped`. The execution times specified in the timed STDs yield counting locations in the clocked STD. The simulated delays correspond to the values of the `MIN` and `MAX` expressions in the timing labels. The `ControllerComponent` offers to its callers the possibility to stop or start the pump as well as to start pumping (assuming that the pump is on).

In order to model sequential passing of control, acknowledgment channels of type `Signal` are inserted by the system model generator. Consider for example

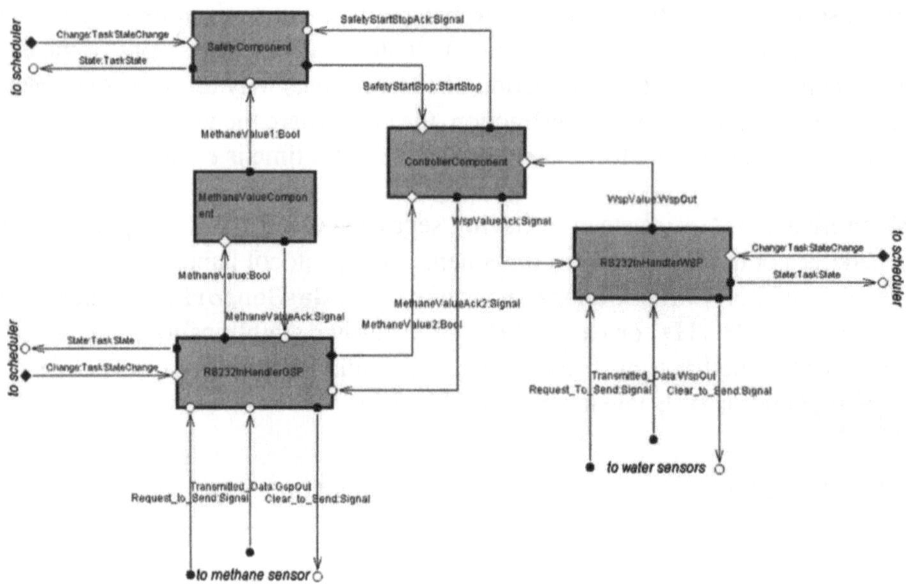


*Figure 5.11.* Active Components and Passive Components of the Controller Node

the channel WspValueAck on Figure 5.11. A signal on an acknowledgment channel emulates return point from the procedure. The caller of the passive component is blocked until the signal arrives.

**Drivers.** RS232InHandlerGSP and RS232InHandlerGSP are standard predefined components which model the RS232 drivers.

The RS232InHandlerGSP receives methane values from the sensor processor. If the methane value is too high, it stops the pump by calling the ControllerComponent with this value as an argument. Afterwards, the MethaneValueComponent is called, which stores the last received methane value together with its arrival time in the shared memory of the controller.

The RS232InHandlerWSP component calls the ControllerComponent if the water level received from sensors is too high or too low.

**Safety Component.** The SafetyComponent analyzes the environment of the pump and decides whether the usage of the pump can be considered as safe or not (see Fig. 5.12). The ControllerComponent is called with a Start message if the pump operation is safe, and with a Stop message if not. The operational environment of the pump is considered as unsafe if the methane value is high or the last received methane value is too old. The last case indicates the possible failure of the methane sensor. For this purpose the time stamp of the last methane value is compared with the current time.

The difference may not exceed a specific threshold, called `Freshness` which is set to the send period of the methane sensor (see table 5.1) plus the worst case execution time (WCET) of the `RS232InHandlerGSP` component. Thus, `Freshness` represents the maximum delay between two sensor readings if the sensor is operational.

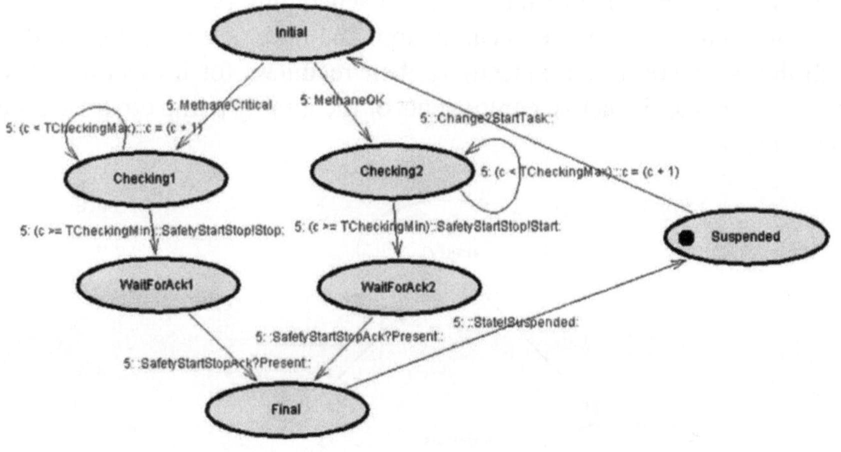


*Figure 5.12.* Clocked STD for `SafetyComponent`

## 5.3 Scheduler for ControlUnit

The `Scheduler` component for the `ControlUnit` in the system model, shown in Fig. 5.14 represents the OS scheduler responsible for the scheduled passing of control flow between active components. The scheduler is generated automatically and realizes the scheduling policy chosen for the node in the technical architecture. In the case study two different scheduling strategies were explored: *cyclic scheduling* and *fixed-priority non-preemptive scheduling*.

**Cyclic Scheduling.** In the case of cyclic scheduling, the `Scheduler` component invokes the three active components in a fixed order by sending them an `ActivateStartTask` message via `StateChange` channels. Active components suspend themselves by transitioning to a special `Suspended` location, which is generated automatically for every active component in the system model. It has a set of incoming transitions with the final locations as their sources and one outgoing transition with the start location as its destination. The control is given back to the scheduler, by sending a `Suspended` message through the corresponding `State` channel.

**Fixed-Priority Scheduling.** For each round, the fixed-priority scheduling invokes the active component with the highest priority, which is ready for execution. The scheduler, which realizes this policy (Fig. 5.14), stores the operating state of every active component (task) and updates it on corresponding events. The task states and transitions between them correspond to the *basic state model* of the *OSEK* standard (OSEKOS, 2001) (Fig. 5.13). Every active component is assigned an integer *priority*, where small numbers indicate low priorities. In the realization of the described scheduling policy, the priorities were encoded into the scheduler's automaton, and have determined the order, in which the tasks are tested concerning their readiness for invocation. The priorities assigned to the active components of the mining pump controller are listed in Table 5.2.

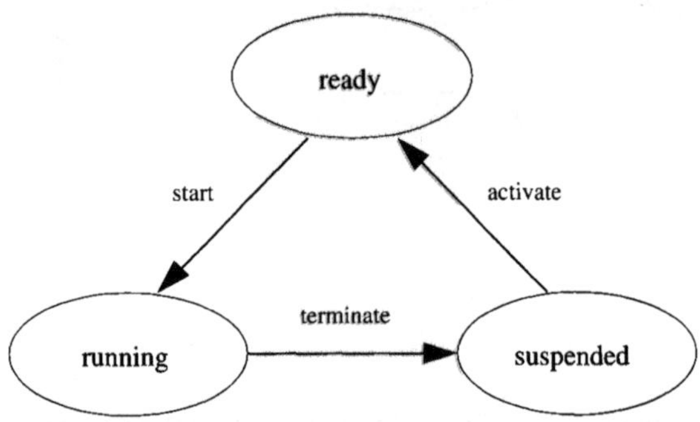


*Figure 5.13.* OSEK basic task state model

The active components are *activated* by specific external *invocation events*, like timers or certain values arriving on the controller's input ports. The events cause hardware interrupts, which update within their *interrupt service routines* (*ISRs*) the corresponding task states. `SafetyComponent` and `RS232InHandlerGSP` are periodically activated. Their periods are listed in Table 5.2. The `RS232InHandlerGSP` component was implemented as a sporadic component, activated by the value `Present` on the `RequestToSend_DCE_WSP` port, which indicates the initiation of the send protocol, originated by the water sensor processor. The arrival rate of this event defines the period of the sporadic component.

The deadlines of the controller's tasks were set to their periods in the presented case study. By this the priority assignment to tasks were accomplished in the *rate monotonic* order (Fig. 5.2), i.e. the task's priority is proportional to its rate of activation.

Note that the interrupt service routines (ISRs) were added as components running in parallel to the rest of the system model (see Fig. 5.14). This corresponds to the assumption of zero execution time for interrupt service routines.

*Table 5.2.* Active Components's periods and priorities for `ControlUnit`

| Active Component | Type | Priority | Period |
|---|---|---|---|
| RS232InHandlerGSP | periodic | 8 | 40 |
| SafetyComponent | periodic | 5 | 55 |
| RS232InHandlerWSP | sporadic | 3 | 70 |

We expect this assumption to be valid for applications with short ISRs vs. long application tasks, and small context switch times for executing ISRs.

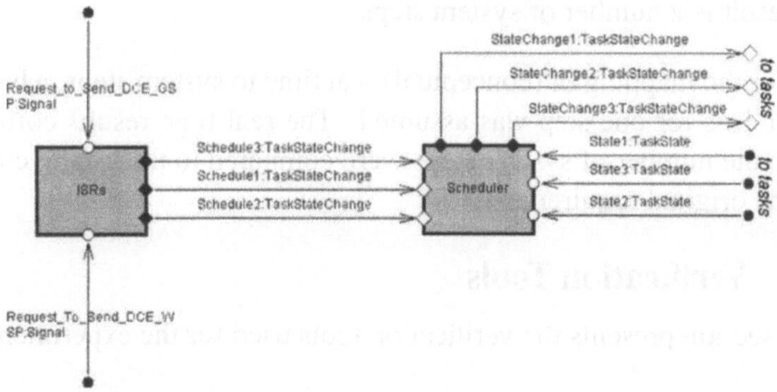


*Figure 5.14.* Scheduler and ISRs of the Controller Node

## 5.4 Analysis

In architecture-level design, the following questions are of major interest for analysis:

- Are the system-level timing requirements fulfilled for the given architectural choices?
- What is an optimal scheduling and partitioning of the system with respect to timing requirements and robustness?
- What is the *best/worst case response time* (*BCET/WCET*) of the application/ its components?
- How flexible is the system with respect to addition of further functionality?

In the next paragraphs, the experiments accomplished within the scope of this case study are described and their results are listed. The *schedulability* and conformance with the *time constraints* formulated in Section 3.4 were the two emphases of the analysis. These two aspects were investigated for the both scheduling policies of the controller node.

The analysis of the mine pump system model was performed according to the following scenario:

- The properties of the model were formulated as sets of evaluation traces in the terms of the system model (e.g. using logical formulas over component STD locations and/or local variables/ports valuations).

- These properties were quantified (e.g. the length of longest/shortest evaluation traces were measured) using the *model checking* technique. The result is a number of system steps.

- For the mapping of (conceptual) real time to system steps, a basic period of 1ms for one step was assumed. The real-time results corresponding to the number of system steps were compared to the timing condition of the original requirement.

## 5.5 Verification Tools

This section presents the verification tools used for the experiment.

**NuSMV.** NuSMV ( Cimatti et al., 2002) is an extended re-implementation of the BDD-based model checker SMV (J.R. Burch et al., 1990). NuSMV requires a description of a finite-state system and a set of properties as input. It checks the fulfillment of these properties for the given system. The properties are formulated as (RT)CTL or LTL formulas. If the properties are not fulfilled by the system the model checker delivers a counter-example trace.

Moreover quantitative information about the system's behavior can be computed by NuSMV. Using the upper and lower bound algorithms from (Campos et al., 1994), NuSMV measures the maximum/minimum delay (expressed as the number of system steps) between sets of states characterized by logical formulas. For this purpose the COMPUTE statement is offered as the part of the input language syntax:

```
compute_decl ::=
        'COMPUTE' ( 'MIN' '[' rtctl_expr ','
                              rtctl_expr ']'
                  | 'MAX' '[' rtctl_expr ','
                              rtctl_expr ']' ) [';']
```

- `MIN [start,final]` computes the set of states reachable from *start*. If a state from *final* is reached, the minimum number of steps taken to reach this step is returned. Otherwise, if the iteration reaches a fixed point without visiting any state from the *final* set, *infinity* is returned.

- `MAX [start,final]` computes the longest path from the states in *start* to the states in *final* and returns its length. If there exists a path from a state in *start* that never reaches a state in *final*, *infinity* is returned.

**Quest.** Quest is a framework, which integrates AutoFocus models with model checkers (SMV), SAT-solvers (SATO), theorem provers (VSE) or test case and code generation tools. Using this framework, the AutoFocus system model of the mine pump controller was automatically translated into the input language of NuSMV (Wimmel et al., 2000).

## 5.6 Cyclic Scheduling

**Schedulability.** An important schedulability requirement for the mine pump is that none of its tasks shall miss its deadline. For this purpose the worst/best case execution times of every task were computed using the `COMPUTE` statement described above. The maximum duration of a task's execution can be measured as the number of steps between a `StartActivateTask` message sent by the scheduler, and the corresponding `Suspended` message sent after termination. Measurement of the scheduler execution times is vice versa, so the interval between `Suspended` and `StartActivateTask` is measured. For example, the evaluation of the following statement delivers the worst case execution time of the `SafetyComponent` task as result:

```
COMPUTE MAX [ Scheduler.StateChange1 == StartActivateTask,
              SafetyComponent.@ == Suspended ]
```

The special expression *<comp_id>*.@ refers to the current location of component *<comp_id>*'s STD.

Moreover the minimal and maximal *invocation rates* (e.g. time between two succeeding invocations) for every task were measured. The results are listed in Table 5.3. Note that the execution time of passive components is included in WCET estimation for each task.

We observe that the minimal invocation period of a task is the sum of best case execution times of all tasks plus three invocations of the scheduler in between. The maximal/minimal invocation rate of the scheduler is observed to be the maximal/minimal execution time of tasks (e.g. `SafetyComponent` and `RS232InHandlerGSP` respectively). These observations are explained by the nature of the deployed scheduling policy.

**Performance.** The verification results for the timing requirements in section 3.4, are summarized in Table 5.4. They were measured using the `COMPUTE` statement, as demonstrated by the following example:

*Table 5.3.* Schedulability for Cyclic Policy

| Subsystem | Invocation Rate Min | Invocation Rate Max | Execution Time Min | Execution Time Max |
|---|---|---|---|---|
| SafetyComponent | 18 | 27 | 9 | 14 |
| RS232InHandlerGSP | 24 | 34 | 3 | 12 |
| RS232InHandlerWSP | 21 | 35 | 6 | 11 |
| Scheduler | 3 | 14 | 3 | 3 |

```
COMPUTE MAX [ GasSensorProcessor.TransmittedDataDTE == True,
              ControllerComponent.@ == PumpStopped ]
```

The statement computes the longest path between the high methane value transmitted by the gas sensor (see Table 5.1) and the transition of the `ControllerComponent` to the location `PumpStopped`, which represents an emergency stop of the pump. Thus, the evaluation result of the presented statement has to be compared with the timing condition of the first requirement from Table 5.4.

*Table 5.4.* Real-Time Behavior for Cyclic Scheduling

| | Requirement | Transaction Time Min | Transaction Time Max |
|---|---|---|---|
| 1. | A high methane value reading shall cause an emergency shut down of the mine pump within 30 milliseconds | 35 | 40 |
| 2. | A methane sensor failure shall be recognized and lead to a shut-down of the pump within 65 milliseconds. | 6 | 43 |
| 3. | After the methane value is back to normal, the pump shall be initialized (e.g. shall be ready to pump) within 100 milliseconds. | 6 | 85 |
| 4. | Under the condition of normal methane values, a high water level reading shall cause the pump to be turned on within 100 milliseconds. | 32 | 46 |
| 5. | If the low water level has been reached, the pump shall be turned off within 75 milliseconds. | 32 | 46 |

The cyclic schedule meets the original requirements, except the first one. Thus, the cyclic scheduling policy is not applicable for the given implementation of the mine pump specification.

**Schedulability.** Besides the best/worst execution times and periods of invocation, there are a number of further factors relevant for the schedulability of the system under the fixed-priority policy:

- The *worst/best case response time* (*WCRT*/*BCRT*) is defined as the maximum/minimum time between the arrival of an invocation and the completion of the computation for that invocation. For a task WCRT may not exceed its deadline.

- The invocation rate is the time between two succeeding invocations of a task. The *maximum invocation rate* indicates whether a task can be executed within every scheduling period.

These factors of schedulability were measured for every task and are summarized in table 5.5. Note that the execution times of the tasks are the same as in

*Table 5.5.* Schedulability for Fixed-Priority Policy

| Subsystem | Response Times | | Invocation Rates | |
|---|---|---|---|---|
| | Min | Max | Min | Max |
| RS232InHandlerGSP | 3 | 26 | 27 | 35 |
| SafetyComponent | 9 | 22 | 41 | 49 |
| RS232InHandlerWSP | 6 | 32 | 44 | 66 |
| Scheduler | 1 | 1 | 4 | 51 |

the case of cyclic scheduling (see Table 5.3).

We observe that the worst case response times are below the corresponding task deadlines listed in table 5.2. The further observation is that invocation rates do not exceed the periods. Thus, the tasks of the mine pump controller are schedulable with the presented fixed-priority policy.

**Performance.** The performance of the system is explored by the table 5.6. All requirements are met by the system for the applied scheduling policy.

# 6. Results Summary

## 6.1 Application Examples

We have used our framework to model both the mine pump example and a simplified HW/SW architecture of a car periphery supervision system by an automotive supplier.

*Table 5.6.* Real-Time Behavior for Fixed-Priority Scheduling

| | Requirement | Transaction Time Min | Transaction Time Max |
|---|---|---|---|
| 1. | A high methane value reading shall cause an emergency shut down of the mine pump within 30 milliseconds | 26 | 29 |
| 2. | A methane sensor failure shall be recognized and lead to a shut-down of the pump within 65 milliseconds. | 6 | 42 |
| 3. | After the methane values get normal again, the pump shall be initialized (e.g. shall be ready to pump) within 100 milliseconds. | 16 | 96 |
| 4. | Under the condition of normal methane values, a high water level reading shall cause the pump to be turned on within 100 milliseconds. | 43 | 86 |
| 5. | If the low water level has been reached, the pump shall be turned off within 75 milliseconds. | 49 | 68 |

## 6.2 Related Work

The concept of active and passive components is inspired by Burns and Wellings's model as described in (Joseph, 1996). In contrast to the programming-language and scheduling-theoretic foundations of Burns and Wellings, MoDe is intended to be susceptible to formal analysis using assertional methods to prove real-time properties.

In the POLIS approach( Balarin et al., 1997), HW/SW systems are modeled as an asynchronous composition of components defined either by an FSM-like formalism, or a composition of Esterel processes. Verification of timing properties in POLIS is mainly simulation-based; as opposed to MoDe, the combined HW/SW representation is not made explicit as a system model.

There are several current approaches that support the use of synchronous languages for distributed implementation( Benveniste et al., 2002, Caspi et al., 2001). As opposed to our approach, the timing and relative order of events in the implementation is assumed to be unknown, and the focus is on ensuring the robustness of synchronous components under such circumstances.

## 6.3 Conclusion and Future Work

We have presented a concept for architectural-level design and analysis of real-time systems using a formal notation. Our approach is aligned with the predominant platforms in the domain of automotive control applications.

A more synthesis-oriented use of MoDe models is straightforward and can be built on existing technology. Being a synchronous formalism, AutoFOCUS leads itself naturally to both software( Blotz et al., 2002) or hardware synthesis, similar to other synchronous formalisms such Esterel( Berry, 1991), or Statecharts (Drusinsky and Harel, 1989).

The currently used AutoFOCUS simulator shows animated views of the rather complex system model. We expect the evaluation of simulation runs to be simplified by instrumenting the design with the logical and technical views on the system.

Quite obviously, the method described here requires some further experimental evaluation. This should include combination with existing approaches for BCET and WCET analysis( Kirner and Puschner, 2002), and a detailed comparison of analysis results with the actual timing of an implementation.

Lastly, we feel that quantitative real-time analysis is, at least on subsystem level, a promising technology, though subject to the same scalability issues as other formal methods based on full state-space exploration.

**Acknowledgements.** Thanks to Gabor Hahn for contributions to the MoDe tool.

## References

AutoFocus (2003). *AutoFocus Homepage*. TU München. URL: `http://autofocus.in.tum.de`.

Balarin, F., Chiodo, M., Giusto, P., Hsieh, H., Jurecska, A., Lavagno, L., Passerone, C., Sangiovanni-Vincentelli, A., Sentovich, E., Suzuki, K., and Tabbara, B. (1997). *Hardware-Software Co-Design of Embedded Systems: The POLIS Approach*. Kluwer, Boston.

Benveniste, A., Caspi, P., Guernic, P. Le, Marchand, H., Talpin, J.-P., and Tripakis, S. (2002). A protocol for loosely time-triggered architectures. In *Proceedings of EMSOFT 2001*. Springer-Verlag.

Berry, G. (1991). A hardware implementation of Pure Esterel. Technical Report 06/91, INRIA, Sophia-Antipolis, France.

Blotz, A., Huber, F., Lötzbeyer, H., Pretschner, A., Slotosch, O., and Zängerl, H.-P. (2002). Model-based software engineering and Ada: Synergy for the development of safety-critical systems. In *Proc. Ada Deutschland Tagung*, Jena.

Campos, S., Clarke, E., Marrero, W., Minea, M., and Hiraishi, H. (1994). Computing quantitative characteristics of finite-state real-time systems. In *Proceedings of IEEE Real-Time Systems Symposium*.

Caspi, P., Mazuet, C., and Paligot, N. R. (2001). About the design of distributed control systems: The quasi-synchronous approach. In *Proceedings of SAFECOMP 2001*, pages 215–226.

Cimatti, A., Clarke, E. M., Giunchiglia, E., Giunchiglia, F., Pistore, M., Roveri, M., Sebastiani, R., and Tacchella, A. (2002). NuSMV 2: An open source tool for symbolic model checking. In *Proceedings of the International Conference on Computer-Aided Verification (CAV 2002)*, Copenhagen, Denmark.

Drusinsky, Doron and Harel, David (1989). Using statecharts for hardware description and synthesis. *IEEE Transaction on Computer-Aided Design of Integrated Circuits and Systems*, 8(7):798–807.

Godefroid, P. (1990). Using partial orders to improve automatic verification methods. In *Proceedings of the 2nd Workshop on Computer-Aided Verification*, LNCS 531, pages 176–185. Springer.

Huber, F., Schätz, B., and Einert, G. (1997). Consistent Graphical Specification of Distributed Systems. In Fitzgerald, J., Jones, C.B., and Lucas, P., editors, *Industrial Applications and Strengthened Foundations of Formal Methods (FME'97)*, LNCS 1313, pages 122–141. Springer Verlag.

Joseph, Mathai, editor (1996). *Real-time Systems: Specification, Verification and Analysis*. Prentice Hall.

J.R. Burch, E.M. Clarke, K.L. McMillan, D.L. Dill, and L.J. Hwang (1990). Symbolic Model Checking: $10^{20}$ States and Beyond. In *Proceedings of the Fifth Annual IEEE Symposium on Logic in Computer Science*, pages 1–33, Washington, D.C. IEEE Computer Society Press.

Kirner, Raimund and Puschner, Peter (2002). International workshop on WCET analysis - summary. Research Report 12/2002, Technische Universität Wien, Institut für Technische Informatik, Vienna, Austria.

Kopetz, H. (1997). *Real-Time Systems: Design Principles for Distributed Embedded Applications*. Kluwer, Boston.

OSEKOS (2001). *OSEK/VDX Operating System Version 2.2*. OSEK VDX consortium.

Pretschner, A., Lötzbeyer, H., and Philipps, J. (2004). Model based testing in incremental system development. *Journal of Systems and Software*, 70(3):315–329.

Selic, B., Gullekson, G., and Ward, T. (1994). *Real-Time Object-Oriented Modeling*. John Wiley & Sons, New York.

Wimmel, G., Lötzbeyer, H., Pretschner, A., and Slotosch, O. (2000). Specification based test sequence generation with propositional logic. *Journal of Software Testing, Verification & Reliability (STVR): Special Issue on Specification Based Testing*.

II

# MODELS AND METHODS FOR SYSTEM VALIDATION

Chapter 6

# A VERIFICATION METHODOLOGY FOR CONCURRENT SOFTWARE WITH SYNCHRONOUS COMMUNICATION*

Christoph Sprenger
*Ecole Polytechnique Fédérale de Zürich*
Christoph.Sprenger@inf.ethz.ch

Krzysztof Worytkiewicz
bluewin.ch

**Abstract** The verification methodology studied in this paper stems from investigations on respectively deduction-based model checking and semantics of concurrency. Specifically, we consider imperative programs with CSP-like communication and use a *categorical semantics* as foundation to extract from a program a *control graph* labelled by *transition predicates*. This logical content acts as system description for a *deduction-based model checker* of LTL properties. We illustrate our methodology with a concrete realisation in the form of the **Mc5** verification tool written in Ocaml and using the theorem prover PVS as back-end.

## 1. Introduction

Formal verification of concurrent software has been studied for more than two decades and is still considered far from being simple. The difficulties stem essentially from the conjunction of two characteristics most concurrent systems exhibit: non-deterministic behaviour on one hand and huge state spaces on the other.

*Research done while both authors were members of the Computer Networking Laboratory of the Swiss Federal Institute of Technology in Lausanne, Switzerland



*R. Gupta et al. (eds.), Formal Methods and Models for System Design,* 137-157.

The first characteristic makes pure *proof-theoretic* approaches tedious and the second, commonly known as *state space explosion*, eliminates the use of state-of-the-art automatic *model-checking* for realistic applications. One remedy for this situation is to get the best of both proof-theoretic and *model-theoretic* worlds by using model-checking algorithms as *decision procedures* within a theorem prover (Owre et al., 1996). We advocate an alternative approach where model checkers are devised to operate by application of *deduction rules*. The latter are designed to incrementally construct a *proof object*, possibly associated to some *validity conditions* of logical nature. The approach is still *model-based* since the deduction rules in question are formulated with respect to a class of models. Model checkers based on deductive methods are able to cope with systems of arbitrary size. However, they require expert knowledge to be taken advantage of.

In our case, the models are graphs representing control, with edges labelled by *transition predicates*. They are extracted from a program text written in our concurrent programming language **Mc** including imperative features and synchronous communication infrastructure à la CSP (Hoare, 1985). We apply categorical semantics of the language to extract the transition predicates in question, a novel approach to the best of our knowledge. The individual *processes* and communication channels occurring in a program determine a *diagram* in a category (Mac Lane, 1971) where the objects are essentially graphs with edges labelled by spans, i.e. pairs of functions with common domain. The composed behaviour is calculated as *limit* of this diagram. Finally, the transition predicates are calculated from the spans by *image factorisation*, the categorically inclined reader will identify the whole construction as a *reflector*. This kind of semantics scales up to more complex languages (Worytkiewicz, 2000) and has the advantage of being *compositional*.

Given a model, our set of deduction rules is designed to construct a *proof structure* i.e. a graph with nodes labelled with *sequents* associating *control points* with logical content in form of predicates and LTL formulas (Manna and Pnueli, 1992). Once constructed, a proof structure *per se* does in general *not* represent a *proof*. In order to do so, it needs to be validated by proving *side conditions* and *discharge conditions*. The former guarantee the *coherence* of the proof structure with respect to the model while the latter guarantee that no system run is a counterexample to the LTL property to be proved. Discharge conditions are constructed from a Büchi automaton associated to the proof structure along with user-supplied *rankings*, a *measure* of a *failure potential*. Side conditions and discharge conditions are first-order predicates. Due to the nature of the models however, it turns out to be convenient to work in a higher-order logic framework. This temporal deduction system is novel to the best of our knowledge and scales up to CTL* and models including fairness constraints (Sprenger, 2000).

It is fair to say that this work stems from a productive blend of two generally disjoint disciplines as the concepts sketched above led us to devise the verification tool **Mc5**. We present its first design iteration and the implementation of an early prototype. The latter is written in Ocaml (Ocaml, 2000) and uses PVS (Owre et al., 1995) as back-end for proving side and discharge conditions.

**Related Work** Expressing a system of communicating processes as diagram and the composed behaviour as its limit goes back to Goguen's work in the late seventies (Goguen et al., 1977). Errington recently refined the idea by introducing twisted systems in his doctoral dissertation (Errington, 1999b) and in (Errington, 1999a). The second author of the present paper applied a slightly simplified version of the setup to the semantics of a concurrent programming language based on *active objects* and made the link with conventional operational semantics precise (Worytkiewicz, 2000).

Closest to our work on verification are the diagram-based techniques of deductive model checking (DMC) (Sipma et al., 1999) and generalised verification diagrams (GVDs) (Browne et al., 1995; Manna et al., 1998). They use automata (called *diagrams*) very similar to our discharge automata. In contrast to our method, DMC repeatedly *refines* an initial diagram by applying transformations preserving the *invariant* that each diagram accepts all potential counterexamples of the system. It is noteworthy that discharge automata satisfy this invariant (cf. Lemma 6.11) and can hence be refined using DMC transformations. The GVD method consists of the construction of an abstraction of the system, followed by automatic model checking of the property on the abstraction. Whereas GVDs can be constructed independently of the property to be verified, a proof structure can be seen as an abstraction that is constructed hand-in-hand with the tableau of the property $\phi$ and is thus *tailored* to the latter.

**Organisation** This article is structured as follows: Section 2 elaborates on the language **Mc** and its semantics, Section 3 is about the temporal deduction system, in Section 4 we describes the **Mc5** verification tool, which is applied to an example proof of a liveness property in Section 5, followed by conclusive remarks in Section 6.

## 2. A Concurrent Programming Language

The concurrent programming language **Mc** acts as input language for the temporal deduction tool to be introduced in Section 4. It is a variant of concurrent Pascal with message passing, enriched by a rigorously designed theory of types essentially reflecting the connectivity of the communication channels.

## 2.1 Syntax

**Mc**'s sequential fragment is imperative while its concurrent infrastructure consists of input and output over *synchronous channels*. The rules for well-typed **Mc** programs in-context are summarised in tables 6.1, 6.2, 6.3 and 6.4. Types are lists of *interface points* or *ip*'s, the latter being pairs associating a type to a name. The operator $\otimes$ on types performs list concatenation followed by removal of duplicates while the ok-judgement (on types as well as on contexts) says that no name occurs twice. Finally, notice that the ground types as well as the ground function symbols are taken in an *algebraic theory* (Crole, 1993).

$$\textbf{Asg}\ \frac{\Gamma, \mathtt{x} : \tau \vdash \mathtt{t} : \tau}{\Gamma, \mathtt{x} : \tau \vdash \mathtt{x} := \mathtt{t} : \langle\rangle} \qquad \textbf{Nop}\ \frac{}{\vdash \mathtt{nop} : \langle\rangle}$$

$$\textbf{Out}\ \frac{\Gamma \vdash \mathtt{t} : \tau}{\Gamma \vdash \mathtt{c!t} : \langle \mathtt{c} : \tau \rangle} \qquad \textbf{In}\ \frac{}{\Gamma, \mathtt{x} : \tau \vdash \mathtt{c?x} : \langle \mathtt{c} : \tau \rangle}$$

*Table 6.1.* Well-typed atomic statements.

$$\textbf{If}\ \frac{\Gamma \vdash \mathtt{p} : \mathtt{bool} \quad \Gamma \vdash \mathit{stat}_1 : \Delta \quad \Gamma \vdash \mathit{stat}_2 : \Theta \quad \Delta \otimes \Theta : \mathtt{ok}}{\Gamma \vdash \mathtt{if\ p\ then}\ \mathit{stat}_1\ \mathtt{else}\ \mathit{stat}_2\ \mathtt{end} : \Delta \otimes \Theta}$$

$$\textbf{Whl}\ \frac{\Gamma \vdash \mathtt{p} : \mathtt{bool} \quad \Gamma \vdash \mathit{stat} : \Delta}{\Gamma \vdash \mathtt{while\ p\ do}\ \mathit{stat}\ \mathtt{end} : \Delta}$$

$$\textbf{Seq}\ \frac{\Gamma \vdash \mathit{stat}_1 : \Delta \quad \Gamma \vdash \mathit{stat}_2 : \Theta \quad \Delta \otimes \Theta : \mathtt{ok}}{\Gamma \vdash \mathit{stat}_1; \mathit{stat}_2 : \Delta \otimes \Theta}$$

*Table 6.2.* Well-typed compound statements.

$$\textbf{MCut}\ \frac{\begin{array}{ll} \Gamma \vdash \mathit{stat}_1 : \langle \Delta, \mathtt{c}_1 : \tau_1; \cdots; \mathtt{c}_n : \tau_n \rangle & \Delta \otimes \Theta : \mathtt{ok} \\ \Xi \vdash \mathit{stat}_2 : \langle \Theta, \mathtt{d}_1 : \tau_1; \cdots; \mathtt{d}_n : \tau_n \rangle & \Gamma, \Xi : \mathtt{ok} \end{array}}{\Gamma, \Xi \vdash \mathit{stat}_1 \, \| c_1 \asymp d_1, \ldots, c_n \asymp d_n \| \, \mathit{stat}_2 : \Delta \otimes \Theta}$$

*Table 6.3.* The composition rule.

The type system reflects the interfaces of the programs to their environment. Specifically, two ip's with matching types may form a *channel*. The **MCut**-rule expresses composition: its application creates an arbitrary number

| | |
|---|---|
| **Weak** $\dfrac{\Gamma \vdash stat : \Delta}{\Gamma, x : \tau \vdash stat : \Delta}$ | **Perm** $\dfrac{\Gamma \vdash stat : \Delta}{\pi_1\Gamma \vdash stat : \pi_2\Delta}$ |

*Table 6.4.* The structural rules.

of *private* channels between two processes, resulting in a composite one. In this sense, **MCut** roughly corresponds to *parallel composition* followed by *hiding* as encountered in conventional process calculi like CCS (Milner, 1990) or CSP (Hoare, 1985).

## 2.2 Semantics

In this subsection, we present the semantic concepts underlying **Mc**.

DEFINITION 6.1

1 *A* graph *$G$ is pair $(E, V)$ along with functions dom, cod : $E \to V$. Then $l \xrightarrow{e} m$ is a shortcut for $dom\,(e) = l \;\wedge\; cod\,(e) = m$ and* $E\,(l, m) \stackrel{def}{=} \left\{ e \in E \,|\, l \xrightarrow{e} m \right\}$

2 *A* relational *graph is a graph* $G = (E, V)$ s.t. $E \subseteq V \times V$.

3 *A* path *$\pi$ in a graph $G = (E, V)$ is a (finite or infinite) sequence* $v_1 e_1 \ldots v_i e_i v_{i+1} \ldots$ s.t. *$dom\, e_i = v_i$ and $cod\, e_i = v_{i+1}$ for all indices $i$. The* trace *of $\pi$ is the sequence $v_1 v_2 \ldots v_i \ldots$ of vertices occurring in $\pi$.*

*Twisted systems* are graphs with edges labelled by *spans* i.e. by pairs of functions with common domain. They rely upon the concept of control locations and transitions on one hand and on (possibly explicit) state and computation on the other. In line with the underlying intuition we dub the graph *control,* its vertices *control points* and its edges *transitions.* Twisted systems are a platform of choice for handling semantics of imperative programs with CSP-style input/output on channels. For instance, the **Mc**-program (without communication)

```
x:nat ⊢ x:= 20; while x > 0 do x:= x-1 end : ⟨⟩
```

can be translated to the twisted system in Figure 6.1.

In particular, the program's syntactic locations are one-to-one with the control points and each transition carries a computation reflecting an assignment or the evaluation of a branching condition. Given a twisted system arising from a meaning of an **Mc**-program, there is a set $D$ such that the labels of all transitions are of type $D \times D \to \{\mathsf{tt}, \mathsf{ff}\}$. $D$ is a product where each factor corresponds as usual to an individual program variable.

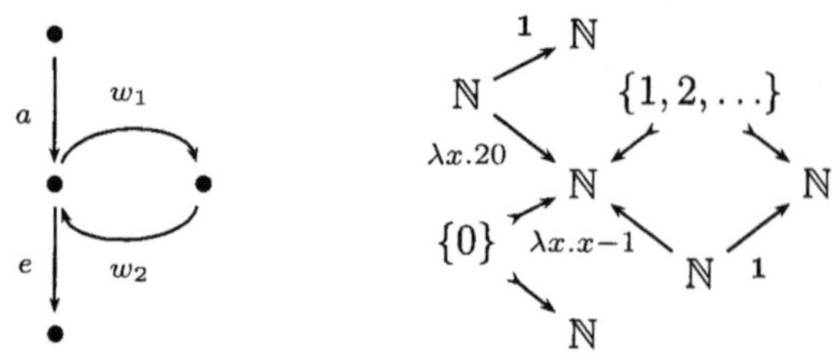


*Figure 6.1.* Twisted system composed of control and labelling with spans

Labelling of graphs with spans can be formulated in terms of graphs, functors and natural transformations determining a category $\mathbf{Ts}\,(\mathbf{Sets})$ with *finite limits*, the latter being the ingredient required for the semantics of the **MCut**-rule (Worytkiewicz, 2003; Worytkiewicz, 2002). The rest of this subsection is dedicated to sketch an equivalent category $\mathcal{T}\,(\mathbf{Sets})$ which turns out to be more convenient for the present setup. The practise-oriented reader may safely skip this part of the discussion keeping in mind that the relevant constructions can be carried out in a higher-order logic over a simple type theory as implemented in PVS.

PROPOSITION 6.2 *Let* $\mathbf{Osp} = (\mathbf{Osp}_0, \mathbf{Osp}_1, dom, cod, comp)$ *be the category having spans as objects and the obvious triples of functions as morphisms. The category* $\mathcal{T}\,(\mathbf{Sets})$ *given by the data*

- *Objects:* $(G, s)$ *with* $G = (E, V)$ *a reflexive graph and* $s : E \to \mathbf{Osp}_0$
- *Morphisms:* $(G, s) \xrightarrow{(h,\kappa)} (H, t)$ *with* $h = (h_0, h_1) : G \to H$ *a homomorphism of reflexive graphs and* $\kappa : E \to \mathbf{Osp}_1$ *s.t. everything commutes in*

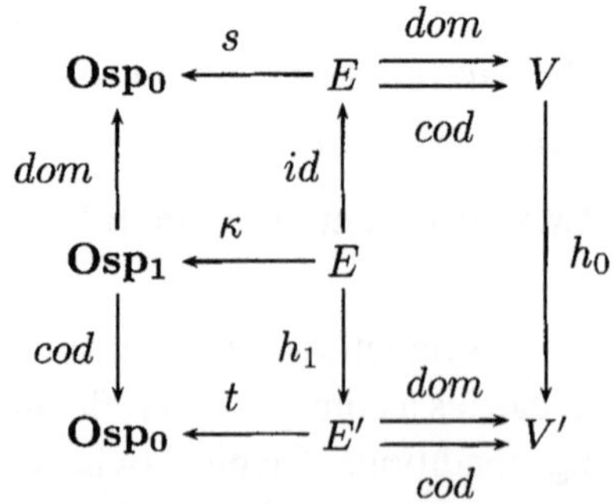


- *Composition:* $(l, \kappa') \circ (h, \kappa) = (l \circ h,\; comp \circ \langle \kappa,\, \kappa' \circ h_1 \rangle)$

- *Units:* $id_{(G,s)} = (id_G, K_1)$ *where* $K_1(e) = id_{s(e)}$

*has finite limits.*

It is not hard to see how finite limits are calculated in $\mathcal{T}(\mathbf{Sets})$ and that the constructions can be carried out in the internal language of **Sets**.

## 2.3 Systems

In this subsection we introduce the format of the systems to be used by the temporal deduction engine. Let $\mathbf{Pred}(A) = A \to \{\mathsf{tt}, \mathsf{ff}\}$ be the set of predicates over a set $A$.

DEFINITION 6.3 *A* system $\mathcal{S} = (T, C, D, dom, cod, \rho)$ *is a graph with edges labelled by predicates where* $dom, cod : T \to C$ *are the graph data and* $\rho : T \to \mathbf{Pred}(D \times D)$ *is the labelling.*

Systems determine a *reflective subcategory* of $\mathcal{T}(\mathbf{Sets})$, where the legs of the labelling spans are *jointly monic,* i.e. they determine a predicate. The basic underlying fact is that logical information can be extracted from a span by *image factorisation* using the familiar correspondence $\mathbf{Sub}(S) \cong \mathbf{Pred}(S)$ between subsets of $S$ and their *classifying predicates*:

$$A \xleftarrow{f} X \xrightarrow{g} B \quad \mapsto \quad A \times B \xrightarrow{\exists x \in X.\,(a,b)=(f(x),g(x))} \{\mathsf{tt}, \mathsf{ff}\}$$

In other words, any twisted system admits an associated system, canonically. The construction can be carried out in the internal language of **Sets**, essentially a higher-order logic over a simple type theory.

# 3. Deductive Temporal Verification

We introduce a tableau-based proof method for the verification of LTL properties of systems (stemming from **Mc**-programs). The method consists of two parts: a set of local proof rules used to construct *proof structures* and a global *discharge condition* that serves to establish that a proof structure is a proper *proof* of a property. For the remainder of this section, let $\mathcal{S} = (T, C, D, dom, cod, \rho)$ stand for an arbitrary but fixed system. We first define the notions of states and runs of a system.

DEFINITION 6.4 *A* run *of* $\mathcal{S}$ *is an infinite sequence*

$$\sigma = (l_0, d_0) \cdots (l_i, d_i) \cdots \in (C \times D)^\omega$$

*of* states *such that for all* $i \in \mathbb{N}$ *there is some* $t \in T$ *with* $l_i \xrightarrow{t} l_{i+1}$ *and* $\rho(t)(d_i, d_{i+1}) = \mathsf{tt}$. *A run* $\sigma$ *is a* $p$-run *for* $p \in \mathbf{Pred}(C \times D)$ *provided* $p(l_0, d_0) = \mathsf{tt}$. *Finally,* $\mathcal{R}(\mathcal{S}, p)$ *stands for the set of* $p$*-runs of* $\mathcal{S}$.

Next, we give a definition of Hoare triples (Hoare, 1969) adapted to our context. These will appear as a side condition of one the proof rules of our tableau proof system as well as in discharge conditions.

DEFINITION 6.5 *Let $\mathcal{S}$ be a system and $p, q \in \mathbf{Pred}\,(C \times D)$. Then*

$$\begin{aligned}\{p\}\,t\,\{q\} \;\stackrel{\mathit{def}}{=}\; & \forall\,(c, c' \in C)\,,(d, d' \in D)\,.\\ & \qquad p\,(c, d) \wedge c = \mathit{dom}\,(t) \wedge c' = \mathit{cod}\,(t) \wedge \rho\,(t)\,(d, d')\\ & \qquad \Rightarrow q\,(c', d')\end{aligned}$$

*and* $\{p\}\,U\,\{q\} \stackrel{\mathit{def}}{=} \bigwedge_{t \in U}\{p\}\,t\,\{q\}$ *are* Hoare triples *with respect to $\mathcal{S}$ for a transition $t \in T$ and for a set of transitions $U \subseteq T$, respectively.*

## 3.1 Linear-Time Temporal Logic

The formulas of linear-time temporal logic (LTL) over $X$ are given by

$$\varphi ::= p \mid \varphi_1 \vee \varphi_2 \mid \varphi_1 \wedge \varphi_2 \mid \varphi_1 \,\mathsf{U}\, \varphi_2 \mid \varphi_1 \,\mathsf{V}\, \varphi_2 \mid \mathsf{X}\,\varphi$$

where $p \in \mathbf{Pred}(X)$ is a predicate, $\mathsf{U}$ is the *Until* operator, $\mathsf{V}$ is its dual, the *Release* operator, and $\mathsf{X}$ is the *Next* operator. As usual, some useful derived connectives can be defined from the basic ones such as $\mathsf{F}\,\varphi \stackrel{\text{def}}{=} \mathsf{tt}\,\mathsf{U}\,\varphi$ (*Eventually*) and $\mathsf{G}\,\varphi \stackrel{\text{def}}{=} \mathsf{ff}\,\mathsf{V}\,\varphi$ (*Always*).

These formulas are interpreted over infinite sequences $\sigma \in X^\omega$. For $\sigma \in X^\omega$ and an LTL formula $\varphi$ over $X$, the *satisfaction relation* $\models$ is inductively defined by

$$\begin{array}{lll}\sigma \models p & \text{iff} & p(\sigma(0)) = \mathsf{tt}\\ \sigma \models \varphi_1 \wedge \varphi_2 & \text{iff} & \sigma \models \varphi_1 \text{ and } \sigma \models \varphi_2\\ \sigma \models \varphi_1 \vee \varphi_2 & \text{iff} & \sigma \models \varphi_1 \text{ or } \sigma \models \varphi_2\\ \sigma \models \mathsf{X}\,\psi & \text{iff} & \sigma^1 \models \psi\\ \sigma \models \varphi_1 \,\mathsf{U}\, \varphi_2 & \text{iff} & \exists k \in \mathbb{N}.\,(\sigma^k \models \varphi_2 \text{ and } \forall i < k.\,\sigma^i \models \varphi_1)\\ \sigma \models \varphi_1 \,\mathsf{V}\, \varphi_2 & \text{iff} & \forall k \in \mathbb{N}.\,(\forall i < k.\,\sigma^i \not\models \varphi_1) \text{ implies } \sigma^k \models \varphi_2\end{array}$$

where $\sigma(i)$ denotes the $i$th position and $\sigma^i$ the $i$th suffix of $\sigma$. For a predicate $p \in \mathbf{Pred}(C \times D)$ and an LTL formula $\phi$ over $C \times D$, the notation $p \models_{\mathcal{S}} \phi$ means that all $p$-runs of $\mathcal{S}$ satisfy $\phi$. For more details on LTL we refer the interested reader to (Clarke et al., 1999; Manna and Pnueli, 1992).

## 3.2 Proof Structures

Next, we introduce our tableau proof system used to construct graphs called proof structures. In the next section we will study criteria under which a proof structure can be considered as a proof. We start by introducing sequents and their semantics.

DEFINITION 6.6 *A sequent over $C \times D$ is a triple $l, p \vdash \Phi$ where $l \in C$ is a control point, $p \in \mathbf{Pred}(C \times D)$ is a predicate and $\Phi$ is a* finite, non-empty *set of LTL formulas over $C \times D$. By abuse of language we use $l \in C$ as predicates with $l(c,d) = \mathtt{tt}$ if and only if $l = c$. A sequent $l, p \vdash \Phi$ is called* valid *if $l \wedge p \models \bigvee_{\phi \in \Phi} \phi$. $\mathbf{Seq}\,(C, D)$ is the set of sequents over $C \times D$.*

Proof structures are essentially graphs that are constructed using a set of tableau proof rules. Some proof rules have first-order side conditions, the validity of which needs to be shown as part of the rule application. Note that we write proof rules in tableau style with the conclusion above and the premises below the line.

DEFINITION 6.7 *Let $\Pi = (V, E, v_r, \eta)$ where $(E, V)$ is a relational graph, $v_r \in V$ is a distinguished vertex called* root *and $\eta : V \to \mathbf{Seq}\,(C, D)$. $\Pi$ is a* proof structure *for the sequent $l_r, p_r \vdash \phi_r$ with respect to system $S$ provided*

$$\eta(v_r) = l_r, p_r \vdash \phi_r$$

*and for all $v \in V$*

*1* $v$ *is* reachable *from $v_r$*

*2 if $v$ has $n \geq 0$ successors $v_1, \ldots, v_n$ then $\eta(v)$ is the conclusion and $\eta(v_1), \ldots, \eta(v_n)$ are the premises of a correct instantiation of some rule $R$ of Table 6.5 and the side condition of rule $R$ holds (if any)*

*3 if $(v, v') \in E$ then rule* $\mathsf{R}(sp)$ *is not applied at both $v$ and $v'$*

*We often write $l_v, p_v \vdash \Phi_v$ for $\eta(v)$.*

The aim of a proof structure is to establish the validity of its root sequent. Clause (3) in Definition 6.7 prevents circular propositional reasoning. However, the latter is necessary but not sufficient for a proof structure to be a *proof* of the validity of the root sequent. Indeed, we have to make sure that formulas of the form $\phi_1 \,\mathsf{U}\, \phi_2$ eventually satisfy their promises $\phi_2$ which is a non-local condition that cannot be guaranteed by the construction of a proof structure alone. Therefore, we need a global criterion to identify proofs.

## 3.3 Proofs

We formulate a criterion for a proof structure $\Pi$ to be *proof*, in terms of the language accepted by a derived automaton with a generalised Büchi acceptance condition. We begin by introducing some auxiliary notions used in the definition of this automaton.

For the rest of this section we consider an arbitrary but fixed proof structure $\Pi = (V, E, v_r, \eta)$ with respect to $S$ with root sequent $\eta(v_r) = l, p \vdash \phi$. By a *trace in* $\Pi$ we mean a trace in the underlying graph.

| Rule | Premise / Conclusion | Side condition |
|---|---|---|
| $\mathsf{R}(ax)$ | $\dfrac{l, p \vdash \Phi, p}{\cdot}$ | |
| $\mathsf{R}(bsf)$ | $\dfrac{l, p \vdash \Phi, q}{l, p \vdash \Phi}$ | $(l \wedge p) \rightarrow \neg q$ |
| $\mathsf{R}(\vee)$ | $\dfrac{l, p \vdash \Phi, \phi_1 \vee \phi_2}{l, p \vdash \Phi, \phi_1, \phi_2}$ | |
| $\mathsf{R}(\wedge)$ | $\dfrac{l, p \vdash \Phi, \phi_1 \wedge \phi_2}{l, p \vdash \Phi, \phi_1 \quad l, p \vdash \Phi, \phi_2}$ | |
| $\mathsf{R}(\mathsf{U})$ | $\dfrac{l, p \vdash \Phi, \phi_1 \,\mathsf{U}\, \phi_2}{l, p \vdash \Phi, \phi_2, \phi_1 \quad l, p \vdash \Phi, \phi_2, \mathsf{X}(\phi_1 \,\mathsf{U}\, \phi_2)}$ | |
| $\mathsf{R}(\mathsf{V})$ | $\dfrac{l, p \vdash \Phi, \phi_1 \,\mathsf{V}\, \phi_2}{l, p \vdash \Phi, \phi_2 \quad l, p \vdash \Phi, \phi_1, \mathsf{X}(\phi_1 \,\mathsf{V}\, \phi_2)}$ | |
| $\mathsf{R}(\mathsf{X})$ | $\dfrac{l, p \vdash \mathsf{X}\,\Phi}{m_1, q_1 \vdash \Phi \quad \cdots \quad m_n, q_n \vdash \Phi}$ | $\{p\}\, T(l, m_i)\, \{q_i\}$ for all $i$ |
| $\mathsf{R}(sp)$ | $\dfrac{l, p \vdash \Phi}{l, q_1 \vdash \Phi \quad \cdots \quad l, q_n \vdash \Phi}$ | $(l \wedge p) \rightarrow \bigvee_{i=1}^{n} q_i$ |

*Table 6.5.* The local proof rules; in Rule $\mathsf{R}(\mathsf{X})$ we write $\mathsf{X}\,\Phi$ for $\{\mathsf{X}\,\phi \mid \phi \in \Phi\}$ and $m_1, \ldots, m_n$ is the set of successor vertices of $l$ in the control graph.

DEFINITION 6.8 *A trace $\pi$ in $\Pi$ is called* next-free *if the Next rule* $\mathsf{R}(\mathsf{X})$ *is applied at most at the last vertex of $\pi$. Let $l_\pi$ be the (unique) control point appearing in labels along a next-free trace $\pi$.*

DEFINITION 6.9 *A vertex $v$ of $\Pi$ is called $\psi$-*safe *for a $\mathsf{V}$-formula $\psi = \phi_1 \,\mathsf{V}\, \phi_2$ if $\Phi_v \cap \{\psi, \mathsf{X}\,\psi\} \neq \varnothing$ and $\phi_2 \notin \Phi_v$. A trace $\pi$ in $\Pi$ is $\psi$-*safe *if it is the case elementwise.*

The discharge automaton we define next is a condensed version of the proof structure: its states are the maximal next-free traces of the proof structure. Each $\mathsf{V}$-subformula $\psi$ of the original property formula contributes an accepting set to the generalised Büchi acceptance condition, containing the states that are not $\psi$-safe.

DEFINITION 6.10 *Let $\{\psi_1, \ldots, \psi_n\}$ be the set of $\mathsf{V}$-subformulas of $\phi$. The* discharge automaton

$$\mathcal{A}^\Pi \stackrel{\text{def}}{=} (Q, Q_0, \Delta, \mu, \mathcal{F})$$

*is given by the data*

$$
\begin{array}{rcl}
Q & \stackrel{def}{=} & \{\pi \mid \pi \text{ is a maximal next-free trace in } \Pi\} \\
Q_0 & \stackrel{def}{=} & \{\pi \in Q \mid \pi(0) = v_r\} \\
\Delta & \stackrel{def}{=} & \{(\pi, \pi') \in Q \times Q \mid (\pi(|\pi| - 1), \pi'(0)) \in E\} \\
\mu(\pi) & \stackrel{def}{=} & l_\pi \wedge \bigwedge_{0 \le k < |\pi|} p_{\pi(k)} \\
\mathcal{F} & \stackrel{def}{=} & \{F_1, \ldots, F_n\} \\
 & & \text{where } F_i \stackrel{def}{=} \{\pi \in Q \mid \pi \text{ is not } \psi_i - \text{safe}\} \text{ for } 1 \le i \le n
\end{array}
$$

*A maximal trace $r$ in $(\Delta, Q)$ with $r(0) \in Q_0$ is called a* run of $\mathcal{A}^\Pi$. *Such an $r$ is a run over a sequence $\sigma \in (C \times D)^\omega$ if $\mu(r(i))(\sigma(i)) = \mathrm{tt}$ for all $i < |r|$. The sequence $\sigma$ is* accepted *by $\mathcal{A}^\Pi$ if there is a run $r$ of $\mathcal{A}^\Pi$ s.t. for each $F \in \mathcal{F}$ there are infinitely many $j \in \mathbb{N}$ such that $r(j) \in F$. The set of accepted sequences is called the* language of $\mathcal{A}^\Pi$ *and is denoted by $\mathcal{L}(\mathcal{A}^\Pi)$.*

Note that due to the side conditions of the Split and Next rules there is a run of $\mathcal{A}^\Pi$ over any $(l \wedge p)$-run of the system. An accepting run of $\mathcal{A}^\Pi$ corresponds to a trace $\pi$ in $\Pi$ containing infinitely many $\psi$-unsafe vertices for each V-subformula $\psi$ of $\phi$. One can show that this means that some U-formula $\phi_1 \mathsf{U} \phi_2$ does not satisfy its promise $\phi_2$ along $\pi$. As a consequence, we get the following lemma.

LEMMA 6.11 *A $(l \wedge p)$-run $\sigma$ of $\mathcal{S}$ accepted by $\mathcal{A}^\Pi$ if and only if $\sigma$ is a counterexample to the validity of the root sequent $l, p \vdash \phi$ of $\Pi$.*

DEFINITION 6.12 *A proof structure $\Pi$ is a* proof *if*

$$\mathcal{L}(\mathcal{A}^\Pi) \cap \mathcal{R}(\mathcal{S}, l \wedge p) = \varnothing.$$

Definition 6.12 goes in pair with Lemma 6.11 since the latter carves effectively out those proof structures which can be considered as proofs of the property $\phi$.

## 3.4 Discharge Conditions

Proposition 6.13 below provides us with a method to establish that a proof structure is a proof. As we can apply the well-known construction to transform a generalised Büchi automaton ($|\mathcal{F}| \ge 0$) into a standard Büchi automaton ($|\mathcal{F}| = 1$) accepting the same language (Thomas, 1990), we assume here that $\mathcal{F}$ is a singleton set.

PROPOSITION 6.13 *Let $\Pi$ and $\mathcal{A}^\Pi$ be as in definition 6.10 except that $\mathcal{F} = \{F\}$. $\Pi$ is a proof if there is a well-founded domain $(W, \prec)$ and a ranking function $\delta : Q \to (D \to W)$ such that for all $(q, r) \in \Delta$ and $w \in W$ the condition*

$$\{\mu(q) \wedge \delta(q) = w\}\, T(l_q, l_r)\, \{\mu(r) \to \delta(r) \prec w\}$$

*holds if* $q \in F$ *and*

$$\{\mu(q) \wedge \delta(q) = w\}\, T(l_q, l_r)\, \{\mu(r) \rightarrow \delta(r) \preceq w\}$$

*holds if* $q \notin F$.

Intuitively speaking, the ranking function measures progress towards the non-acceptance of system runs by $\mathcal{A}^{\Pi}$: the ranking is required to decrease at accepting vertices and to not increase at other vertices. Recall that there is a run of $\mathcal{A}^{\Pi}$ over any $(l \wedge p)$-run of $\mathcal{S}$. Hence, provided that the discharge conditions hold, the well-foundedness of $(W, \prec)$ implies that no $(l \wedge p)$-run of the system can be accepted by $\mathcal{A}^{\Pi}$.

It is important to remark that the discharge automaton $\mathcal{A}^{\Pi}$ of a proof structure $\Pi$ for a *safety formula*, that is, a formula not containing any U-subformulas, always accepts the empty language. The reason is that all vertices of a given strongly connected component of $\Pi$ are $\psi$-safe for some V-formula $\psi$. In other words, a proof structure for a safety formula is always a proof.

## 3.5 Soundness and Relative Completeness

Our proof method is sound and complete *relative* to the *validity* of the side conditions of the local rules and the discharge conditions.

THEOREM 6.14 $l \wedge p \models_{\mathcal{S}} \phi$ *if and only if there is a proof structure* $\Pi$ *for* $l, p \vdash \phi$ *with respect to* $\mathcal{S}$ *and a ranking function* $\delta$ *such that the discharge conditions of Proposition 6.13 are valid for* $\mathcal{A}^{\Pi}$.

The game-theoretic proof of soundness and relative completeness presented in (Sprenger, 2000) can easily be adapted to the present proof system.

# 4. The Mc5 Verification Tool

In this section we report on our preliminary experience with a prototype tool implementing the framework described in the previous two sections. Its overall architecture is displayed in Figure 6.2. Two basic parts can be distinguished:

1 the *front-end* consisting of the **Mc** language compiler, the proof manager, the tactical language interpreter and the graphical user interface, and

2 the *back-end* theorem prover PVS (Owre et al., 1995) used to discharge first-order verification conditions.

The implementation language for the front-end is Ocaml (Ocaml, 2000). We now give a short description of each of the components of the tool front-end.

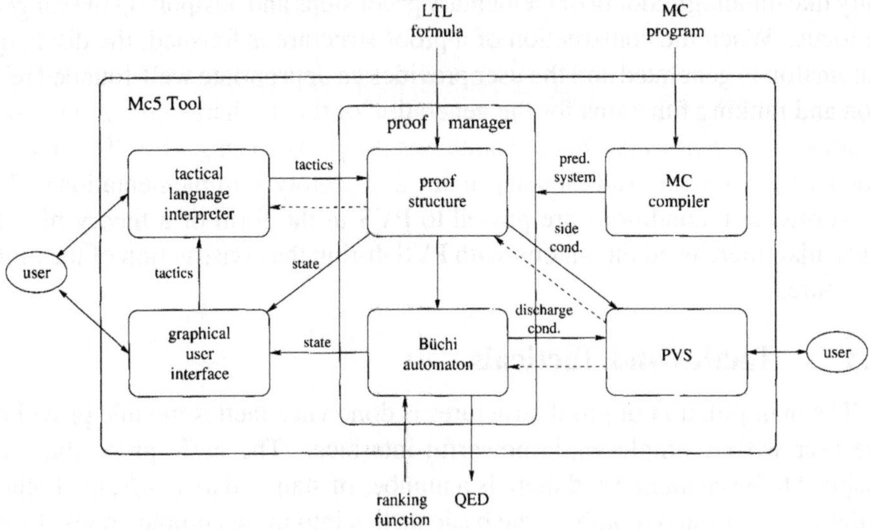


*Figure 6.2.* **Mc5** Tool Architecture

## 4.1 Mc Compiler

The functionality of the back-end with respect to a given program is to extract a system from the syntactical information collected by the parser. In section 2, we argued that all the underlying constructions can be represented in PVS. However, taking the idea *literally* from an implementation perspective would result in significant inefficiencies, since direct calculation is not possible in PVS but has to be performed by proving series of lemmas. We therefore opted for the following trade-off: graph-related calculations, e.g. the limit construction, are carried out *explicitly* while the labelling is handled in a *symbolic* way, *assembling* relevant $\lambda$-terms. So what is handed over to the proof manager is an explicit representation of the control graph and a function on edges returning the abstract syntax tree of a predicate. Not surprisingly, we had to implement "optimisations" in form of $\beta$- and $\delta$-reductions in order to obtain human-readable PVS-terms. Finally, it is worth to notice that the compilation of individual processes can be done separately and also to be decoupled from the compilation of communicating systems built with such processes. This is possible by virtue of the compositionality of semantics based on twisted systems.

## 4.2 The Proof Manager

This part handles proof structures and discharge automata by maintaining the *proof state* including the set of open goals while providing additional functionality like undoing/redoing of elementary proof steps and postponing the subgoal in focus. When the construction of a proof structure is finished, the discharge automaton is generated and the user provides an appropriate well-founded relation and ranking functions for the generation of the discharge conditions. Side conditions as well as discharge conditions are proved using PVS. The interaction with PVS is very rudimentary in this first prototype implementation in that all verification conditions are passed to PVS in the form of a theory file. In particular, there is no interaction with PVS during the construction of the proof structure.

## 4.3 Tactics and Tacticals

The manipulation of proof structures is done via a tactics module providing the user with a simple while powerful interface. The basic proof rules are mapped to basic *tactics* and there is a number of standard *tacticals*, i.e. higher-order tactics, used to combine the basic tactics into more complex ones. In the current state of development, the tactical commands are directly interpreted by the Ocaml shell. The basic tactics directly manipulate the proof structure by adding new subgoal nodes and looping back to existing ones. Thus, applying a complex tactic might add a large number of nodes to the proof structure. It might be possible to drop some intermediate proof goals, but great care needs to be taken to preserve the correct generation of the discharge automaton from the proof structure.

Counting the number $k$ of basic rules applied to the proof structure by a tactic allows us to undo the effects of the latter in case of failure by calling the undo method of the proof structure $k$ times. In case of success, $k$ is pushed onto a stack thus implementing a simple tactics-level undo facility.

## 4.4 Graphical User Interface

The daVinci graph visualisation tool (daVinci, 2001) has proved to be extremely helpful as a rudimentary graphical user interface for our tool. It is currently used only for displaying system specifications, proof structures and discharge automata. In the future, we plan to make use of daVinci's customisable graphical user interface to obtain a more interactive and user-friendly interface, where certain prover actions can be invoked via buttons and menus.

# 5. Example

As a simple example consider the program below. It is built from two processes, $P$ and $Q$, running in parallel and communicating over a synchronous channel. Control points are indicated by comments in the program.

```
P = x:nat ⊢
   while true do
     (*a*) cp!x+x
     (*b*) cp?x
   end : <cp:nat>
```

```
Q = y:nat ⊢
   while true do
     (*x*) cq?y;
     (*y*) y:=y+1;
     (*z*) cq!y
   end : <cq:nat>
```

For the parallel composition $S = \mathtt{x}, \mathtt{y} : \mathtt{nat} \vdash (P \, \|\mathtt{cp} \asymp \mathtt{cq}\| \, Q) : \langle\rangle$ we verify that the variable $y$ grows without bound whenever $x$ equals $y$ in the initial state. The liveness property of unbounded growth is expressed in LTL by the formula

$$\phi = \mathsf{F}(y \geq n)$$

where $n$ is an arbitrary but fixed parameter. We now describe the system generated by the **Mc** compiler from the program above and the proof of validity of the sequent $\mathsf{ax}, y = x \vdash \mathsf{F}(y > n)$ using the **Mc5** tool.

## 5.1 Model Generation

The control graph of the system specification generated from the program $S$ above is shown in Figure 6.3. There are three control points called $\mathsf{ax}$, $\mathsf{by}$ and

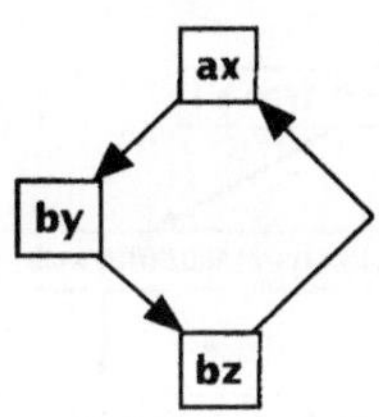


*Figure 6.3.* Control graph of system specification

$\mathsf{bz}$, respectively, and the transition predicates labelling the edges are

$$\begin{array}{lll} \mathsf{ax} \to \mathsf{by} & \mapsto & (x' = x) \wedge (y' = x + x) \\ \mathsf{by} \to \mathsf{bz} & \mapsto & (x' = x) \wedge (y' = y + 1) \\ \mathsf{bz} \to \mathsf{ax} & \mapsto & (x' = y) \wedge (y' = y) \end{array}$$

In line with traditional notation, unprimed variables represent the source state and primed variables represent the target state of a transition. The first predicate stems from $P$ sending $x + x$ to $Q$. The second one corresponds to $Q$ incrementing $y$ locally and the final one describes the transmission of $y$ back from $Q$ to $P$.

## 5.2 Proof Structure

A proof structure for this system and the sequent $\mathsf{ax}, y = x \vdash \mathsf{F}(y > n)$ is displayed in Figure 6.4. The command used for its construction was

```
byscript [
  repeat_t log_tac; inv_next_tac "y=x+x";
  repeat_t log_tac; inv_next_tac "y=(x+x)+1";
  repeat_t log_tac; inv_next_tac "y=x"
]
```

The function `byscript` takes a proof script (that is, a list of tactics) as an argument and applies them sequentially to the proof structure under construction.

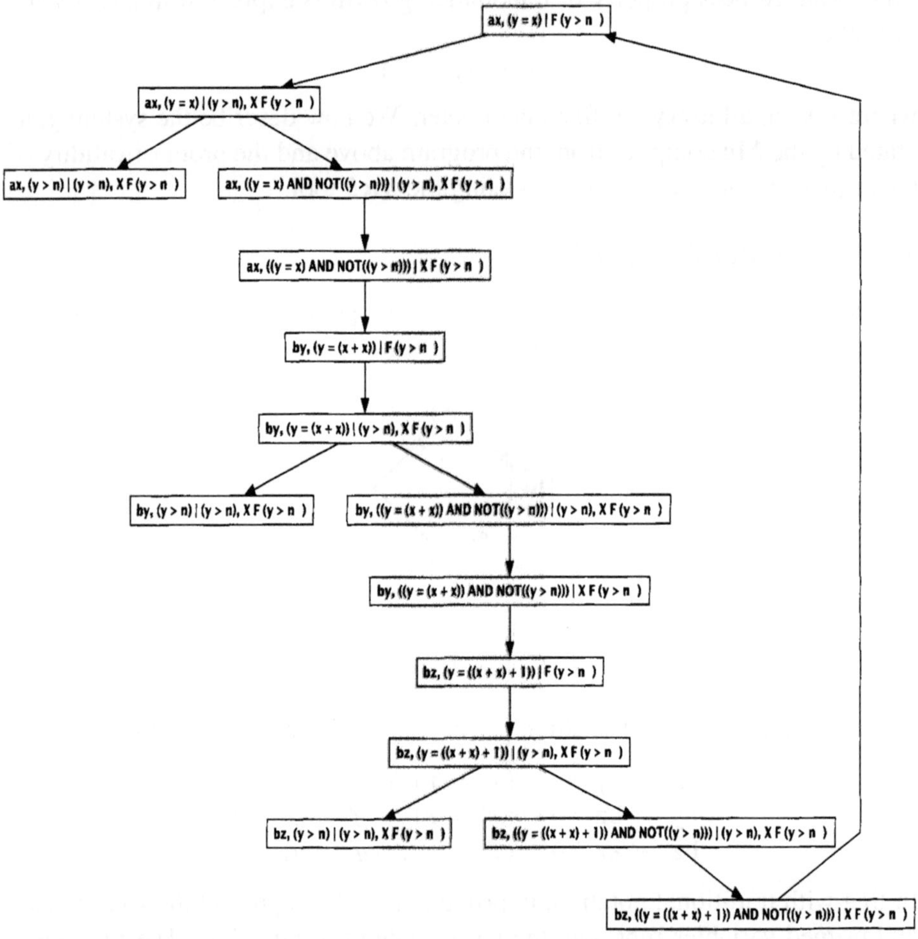


*Figure 6.4.* Proof Structure

The first tactic, applied to the root sequent, is `(repeat_t log_tac)`. It uses the tactical `repeat_t` to repeat the tactic `log_tac` as many times as possible. The latter is defined as

```
(orelse_t bool_tac (orelse_t temp_tac pred_tac))
```

and tries to apply an appropriate rule for a boolean or temporal operator (except Next) or for the elimination of a predicate, in this order of preference.

At the root sequent, tactic `temp_tac` succeeds in applying rule R(F), the obvious specialisation of rule R(U) to the Eventually (F) formula. This produces the sequent in "line" 2 of Figure 6.4. There, `bool_tac` as well as `temp_tac` fail, but `pred_tac` applies and uses in turn the Split, Axiom and Bsf rules to produce three new vertices. The new subgoal is the sequent at line 4. Here, `log_tac` fails, thus `(repeat_t log_tac)` returns and `byscript` proceeds by applying `inv_next_tac ''y=x+x''`. The latter tactic takes a predicate $p$ as an argument and applies the Next rule with $p$ as the left-hand side predicate in all successor sequents. In our case, this generates the sequent on line 5.

This sequence of tactics is repeated totally three times, each time using a different predicate with the `inv_next_tac`. The three given predicates reflect the relation of $x$ and $y$ before and after each transition. Their disjunction is an invariant of the system. All side conditions (totally 9) are proved automatically in PVS using the powerful `grind` command.

## 5.3 Discharge Automaton

Having constructed the proof structure, the next step is to generate the derived discharge automaton (see Figure 6.5). It reflects the structure of the system: each vertex corresponds to a control point and each transition to a transition of the system[1]. As there is no V-subformula in the property formula, all states of the automaton are accepting. Let us call its vertices ax, by and bz according to the associated control points.

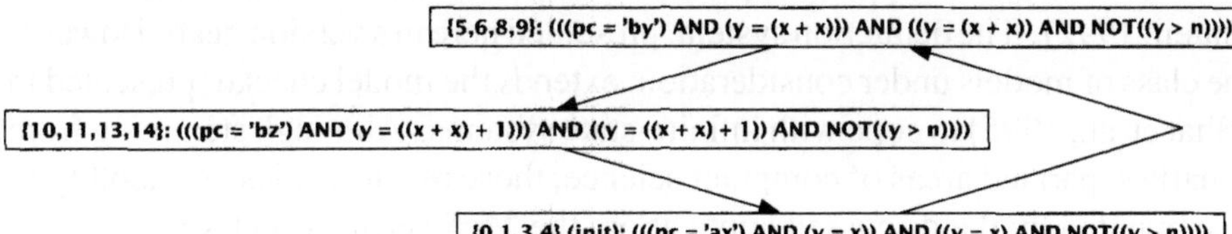


*Figure 6.5.* Discharge automaton

Clearly, all sequences of states accepted by this automaton violate our property, since $y \leq n$ holds in all states of such a sequence. Contemplating the

[1]Terminal vertices with no outgoing edges can safely be removed without changing the language accepted by the automaton.

automaton, we easily come up with the rankings

$$\begin{aligned}
\delta_{\mathsf{ax}}(x,y) &\stackrel{\text{def}}{=} \mathtt{lex2}(n-y,1)\\
\delta_{\mathsf{by}}(x,y) &\stackrel{\text{def}}{=} \mathtt{lex2}(n-y,0)\\
\delta_{\mathsf{bz}}(x,y) &\stackrel{\text{def}}{=} \mathtt{lex2}(n-y,2)
\end{aligned}$$

where `lex2` is a PVS function mapping a lexicographically ordered pair of natural numbers to (the PVS representation of) an ordinal. The first argument measures the distance between $y$ and $n$, while the second one measures progress to the next decrease of the first component. As an example, for the automaton transition from by to bz, the tool generates the PVS discharge condition[2]

```
(FORALL (pc:string),(x:nat),(y:nat),(pc':string),(x':nat),(y':nat):
  (((pc = "by") AND (y = (x + x)) AND NOT(y > n))
  AND ((x = x') AND (y + 1 = y'))
  AND ((pc' = "bz") AND (y' = x' + x' + 1) AND NOT(y' > n))
  IMPLIES (lex2(n - y, 0) > lex2(n - y', 2))))))))
```

which is easily seen to be valid. All of the totally three discharge conditions are proved in PVS without user intervention using the command `grind`.

## 6. Conclusions

We propose a verification methodology building on original results in the relevant fields of semantics of concurrency respectively deduction-based model checking. Model generation relies on a novel categorical semantics of processes with synchronous communication. The semantics is compositional, which has to be contrasted with other approaches known in the literature (Manna and Pnueli, 1992). The deduction system, presented here in a version geared towards the class of models under consideration, extends the model checker presented in (Bhat et al., 1995) to cope with infinite-state systems. Although stemming from usually separated areas of computer science, those results coalesce smoothly as documented by the discussed application, the **Mc5** verification tool.

It is obvious that a lot of work remains to be done. The most apparent limitation of the current prototype is its lack of automation. Automatic invariant generation techniques (Bjørner et al., 1997) could be applied with great benefit to produce the predicates to be supplied to the Next rule. These need to be strong enough for the discharge conditions to be provable. At the same time the discharge conditions could be weakened by replacing the acceptance condition by an equivalent (language-preserving) one with less accepting states[3].

[2]slightly edited to improve readability

[3]cf. the example in Section 5, where a single accepting state would be sufficient and obviate the need for the second component in the ranking

Another issue concerns the condition for looping back to an existing vertex in the proof structure, where the current implementation requires that a sequent generated by a rule *equals* a sequent already present. Equality is a much too strong condition on predicates and should be replaced by *implication.* Each new subgoal should be checked automatically against loop-back candidates using for example PVS' batch mode.

The connection with PVS should become more interactive. In particular, side conditions could be submitted to PVS immediately in an attempt to prove them automatically and feedback about their status could be handed back from PVS to the proof manager which keeps track of proved and open conditions.

## References

Bhat, G., Cleaveland, R., and Grumberg, O. (1995). Efficient on-the-fly model checking for CTL*. In *Logic in Computer Science, LICS 95*, pages 388–397.

Bjørner, Nikolaj S., Browne, Anca, and Manna, Zohar (1997). Automatic generation of invariants and intermediate assertions. *Theoretical Computer Science*, 173(1):49–87.

Browne, I. A., Manna, Z., and Sipma, H. B. (1995). Generalized temporal verification diagrams. In *Foundations of Software Technology and Theoretical Computer Science, FSTTCS '95, Bangalore, India*, volume 1026 of *Lecture Notes in Computer Science*, pages 484–498. Springer-Verlag.

Clarke, Edmund M., Grumberg, Orna, and Peled, Doron (1999). *Model Checking*. The MIT Press.

Crole, Roy L. (1993). *Categories for Types*. Cambridge University Press.

daVinci (2001). daVinci home page. `http://www.tzi.de/daVinci`. University of Bremen, Germany.

Errington, Lindsay (1999a). On the semantics of message passing processes. In *Proceedings of CTCS99*.

Errington, Lindsay (1999b). *Twisted Systems*. PhD thesis, Department of Computing, Imperial College, London.

Goguen, J.A., Thatcher, J.W., Wagner, E.G., and Wright, J.B. (1977). Initial algebra semantics and continuous algebras. *Journal of the ACM*, 24:68–95.

Hoare, C. A. R. (1985). *Communicating Sequential Processes*. Prentice Hall International Series on Computer Science. Prentice-Hall.

Hoare, C.A.R. (1969). An axiomatic basis for computer programming. *Communications of the ACM*, 12(10).

Mac Lane, Saunders (1971). *Categories for the Working Mathematician*. Springer.

Manna, Zohar, Browne, Anca, Sipma, Henny B., and Uribe, Tomás E. (1998). Visual abstractions for temporal verification. In *AMAST '98*, volume 1548 of *Lecture Notes in Computer Science*, pages 28–41. Springer-Verlag.

Manna, Zohar and Pnueli, Amir (1992). *The Temporal Logic of Reactive and Concurrent Systems*. Springer Verlag.

Milner, Robin (1990). Operational and algebraic semantics of concurrent processes. In van Leeuwen, Jan, editor, *Handbook of Theoretical Computer Science*, volume B: Formal Models and Semantics, chapter 1202–1242. Elsevier Science Publishers.

Ocaml (2000). *The Objective Caml system, release 3.00*. INRIA.

Owre, S., Rajan, S., Rushby, J.M., Shankar, N., and Srivas, M.K. (1996). PVS: Combining specification, proof checking, and model checking. In Alur, Rajeev and Henzinger, Thomas A., editors, *Computer-Aided Verification, CAV '96*, volume 1102 of *Lecture Notes in Computer Science*, pages 411–414, New Brunswick, NJ. Springer-Verlag.

Owre, Sam, Rushby, John, Shankar, Natarajan, and von Henke, Friedrich (1995). Formal verification for fault-tolerant architectures: Prolegomena to the design of PVS. *IEEE Transactions on Software Engineering*, 21(2):107–125.

Sipma, Henny, Uribe, Tomás E., and Manna, Zohar (1999). Deductive model checking. *Formal Methods in System Design*, 15:49–74.

Sprenger, Christoph (2000). *Deductive Local Model Checking – On the Verification of CTL* Properties of Reactive Systems*. Ph.D. Thesis no. 2215, Department of Computer Science, Swiss Federal Institute of Technology, Lausanne, Switzerland.

Thomas, W. (1990). Automata on infinite objects. In van Leeuwen, J., editor, *Handbook of Theoretical Computer Science*, volume B, pages 133–191. Elsevier Science Publishers , Amsterdam.

Worytkiewicz, Krzysztof (2000). *Components and Synchronous Communication in Categories of Processes*. Ph.D. Thesis no. 2131, Department of Computer Science, Swiss Federal Institute of Technology, Lausanne, Switzerland.

Worytkiewicz, Krzysztof (2002). Concrete process categories. In Kurz, Alexander, editor, *Electronic Notes in Theoretical Computer Science*, volume 68. Elsevier Science Publishers.

Worytkiewicz, Krzysztof (2003). Paths and simulations. In Richard Blute, Philipp Scott and Selinger, Peter, editors, *Electronic Notes in Theoretical Computer Science*. Elsevier Science Publishers. To appear.

# Chapter 7

# HIGH LEVEL VERIFICATION OF CONTROL INTENSIVE SYSTEMS USING PREDICATE ABSTRACTION*

Edmund Clarke,[1] Orna Grumberg,[2] Muralidhar Talupur,[1] and Dong Wang[3]

[1] *School of Computer Science*
*Carnegie Mellon University*
*Pittsburgh, PA 15213*
*USA*

[2] *Computer Science Department*
*TECHNION - Israel Institute of Technology*
*Technion City, Haifa 32000*
*Israel*

[3] *Electrical and Computer Engineering Department*
*Carnegie Mellon University*
*Pittsburgh, PA 15213*
*USA*

**Abstract** Predicate abstraction has been widely used for model checking hardware/software systems. However, for control intensive systems, existing predicate abstraction techniques can potentially result in a blowup of the size of the abstract model. We deal with this problem by retaining important control variables in the abstract model. By this method we avoid having to introduce an unreasonable number of predicates to simulate the behavior of the control variables.

*This research is sponsored by the Semiconductor Research Corporation (SRC) under contract no. 99-TJ-684, the Gigascale Silicon Research Center (GSRC), the National Science Foundation (NSF) under Grant No. CCR-9803774. Any opinions, findings and conclusions or recommendations expressed in this material are those of the authors and do not necessarily reflect the views of SRC, GSRC, NSF, or the United States Government.



*R. Gupta et al. (eds.), Formal Methods and Models for System Design,* 159-179.

We also show how to improve predicate abstraction by extracting useful information from a high level representation of hardware/software systems. This technique works by first extracting relevant branch conditions. These branch conditions are used to invalidate spurious abstract counterexamples through a new counterexample-based lazy refinement algorithm. Experimental results are included to demonstrate the effectiveness of our methods.

**Keywords:** predicate abstraction, high level description, control intensive systems, abstraction refinement

## 1. Introduction

**Background.** Abstraction based model checking has been widely accepted as a valuable method for the verification of large hardware-software systems. *Predicate abstraction* [1–3, 10, 11, 13, 16, 18, 19], in particular, is one of the most successful abstraction techniques. In predicate abstraction, the concrete system is approximated by only keeping track of certain predicates over the concrete state variables. Each predicate corresponds to an abstract boolean variable. Any concrete transition corresponds to a change of values for the set of predicates and is subsequently translated into an abstract transition. Using predicate abstraction, it is possible to not only reduce the complexity of the system under verification, but also, for software systems, to extract finite models that are amenable to model checking algorithms.

Predicate abstraction is a special case of existential abstraction [5, 9, 15], which is a conservative approach for model checking universal temporal logic [9] properties (we only consider safety properties in this paper). That is, the correctness of any universal formula on an abstract system automatically implies the correctness of the formula on the concrete system. However, a counterexample on an abstract system may not correspond to any real path, in which case it is called a *spurious* counterexample [6]. To get rid of a spurious counterexample, the abstraction needs to be made more precise via refinement. *Counterexample guided abstraction refinement* [7, 6, 13, 22] (CEGAR) automates this procedure. It works as follows: For a given system, an abstract model that is guaranteed to include all behaviors of the original system is built. Model checking is then applied to the abstract model. If the property holds, it is true of the concrete model and verification terminates. In case the property is violated on the abstract model a counterexample is generated. This abstract counterexample is checked against the concrete model. If the abstract counterexample corresponds to a concrete execution path, the property is proved to be false and verification terminates. Otherwise, the abstract counterexample is *spurious* and is used to guide the refinement of the abstract model. The above procedure repeats until the property is confirmed or refuted.

**Motivation.** It is usually the case that verification effort is focused more on the control logic than the data computation because most bugs exist in designing the control logic. Traditional predicate abstraction techniques can perform badly when verifying hardware/software systems which contain extensive control structure (*control intensive systems*). The control logic usually consists of concurrent state machines. Each of these state machines may depend on several *control variables*, that encode the change of state. Since the behavior of a control intensive system is determined to a large extent by the control variables, the number of predicates over the control variables that are needed can be much larger than the number of control variables. In such a case, it is better to use the control variables as predicates, (called *variable predicates*), instead of the original predicates (called *original or formula predicates*). We propose a clustering based heuristic to identify important control variables and retain these control variables in the abstract model. By doing this we also circumvent to a certain extent the problem of building the abstract model. This method works extremely well in practice.

It is usually the case that different predicates are not independent. We describe efficient methods to compute constraints between predicates, which are added as invariants to the abstract model to make it more accurate.

Another issue that we address in this paper is the following: current predicate abstraction methods do not make use of information available in the high level descriptions of the system under verification. Most hardware/software codesign tools use high level design languages, such as ESTEREL, graphical FSMs, RTL Verilog/VHDL, C/C++ etc. But most model checking engines and existing verification tools use the bit level representation of the design under verification. There is much useful information that is relevant to verification in the high level representation, which is lost once it is translated to bit level representation. To retain this information, we extract the branch conditions in RTL Verilog (the language considered in this paper) and use them as predicates. This technique can be easily adapted to other design languages.

For a given design, there are usually many branch conditions that we can extract. Not all of them are relevant to the verification of a given property. We propose a lazy counterexample based refinement algorithm to efficiently identify the branch conditions that are relevant.

**Outline of the paper.** In the next section we introduce predicate abstraction and other relevant theory. In Section 3, we give a clustering based heuristic to identify control variables and present a modified localization reduction algorithm to bound the size of the abstract model. Algorithms to compute accurate abstract models are also discussed in the same section. Section 4 gives the predicate extraction and refinement algorithm. Some related work is discussed in Section 5. In Section 6, we describe our experiments. Section 7 concludes the paper.

## 2. Preliminary

In this section, we review the theory of *existential abstraction*. We then present *predicate abstraction* and the *localization reduction* as special cases of this concept.

### 2.1 Existential Abstraction

We model circuits and programs as transition systems. Given a set of atomic propositions, $A$, let $M = (S, S_0, R, L)$ be a *transition system* (refer to [9] for details).

DEFINITION 7.1 *Given two transition systems $M = (S, S_0, R, L)$ and $\hat{M} = (\hat{S}, \hat{S}_0, \hat{R}, \hat{L})$, with atomic propositions $A$ and $\hat{A}$ respectively, a relation $\rho \subseteq S \times \hat{S}$, which is total on $S$, is a* simulation relation *between $M$ and $\hat{M}$ if and only if for all $(s, \hat{s}) \in \rho$ the following conditions hold:*

- $L(s) \bigcap \hat{A} = \hat{L}(\hat{s}) \bigcap A$
- *For each state $s_1$ such that $(s, s_1) \in R$, there exists a state $\hat{s_1} \in \hat{S}$ with the property that $(\hat{s}, \hat{s_1}) \in \hat{R}$ and $(s_1, \hat{s_1}) \in \rho$.*

We say that $\hat{M}$ *simulates* $M$ through the simulation relation $\rho$, denoted by $M \preceq_\rho \hat{M}$, if for every initial state $s_0$ in $M$ there is an initial state $\hat{s_0}$ in $\hat{M}$ such that $(s_0, \hat{s_0}) \in \rho$. We say that $\rho$ is a *bisimulation relation* between $M$ and $\hat{M}$ if $M \preceq_\rho \hat{M}$ and $\hat{M} \preceq_{\rho^{-1}} M$. If there is a bisimulation relation between $M$ and $\hat{M}$ then we say that $M$ and $\hat{M}$ are *bisimilar*, and we denote this by $M \equiv_{bis} \hat{M}$.

THEOREM 7.2 *(Preservation of ACTL* [9])*
*Let $M = (S, S_0, R, L)$ and $\hat{M} = (\hat{S}, \hat{S}_0, \hat{R}, \hat{L})$ be two transition systems, with $A$ and $\hat{A}$ as the respective sets of atomic propositions and let $\rho \subseteq S \times \hat{S}$ be a relation such that $M \preceq_\rho \hat{M}$. Then, for any ACTL* formula $\Phi$ with atomic propositions in $A \cap \hat{A}$*

$$\hat{M} \models \Phi \textit{ implies } M \models \Phi.$$

In the above theorem, if $\rho$ is a bisimulation relation, then for any CTL* formula $\Phi$ with atomic propositions in $A \cap \hat{A}$, $\hat{M} \models \Phi \Leftrightarrow M \models \Phi$.

Let $M = (S, S_0, R, L)$ be a concrete transition system over a set of atomic propositions $A$. Let $\hat{S}$ be a set of abstract states and $\rho \subseteq S \times \hat{S}$ be a total function on $S$. Further, let $\rho$ and $L$ be such that for any $\hat{s} \in \hat{S}$, all states $s \in S$ that satisfy $\rho(s, \hat{s})$ have the same labeling over a subset $\hat{A}$ of $A$. Then an abstract transition system $\hat{M} = (\hat{S}, \hat{S}_0, \hat{R}, \hat{L})$ over $\hat{A}$ which simulates $M$ can be constructed as follows:

$$\hat{S}_0 = \exists s.\ S_0(s) \wedge \rho(s, \hat{s}) \tag{7.1}$$

$$\hat{R}(\hat{s}, \hat{s}') = \exists s\ s'.\ \rho(s, \hat{s}) \wedge \rho(s', \hat{s}') \wedge R(s, s') \tag{7.2}$$

$$\text{for each } \hat{s} \in \hat{S},\ \hat{L}(\hat{s}) = \bigcap_{\rho(s,\hat{s})} (L(s) \cap \hat{A}) \tag{7.3}$$

This kind of abstraction is called *existential abstraction* [5, 15].

## 2.2 Predicate Abstraction

Predicate abstraction can be viewed as a special case of existential abstraction. In predicate abstraction a set of predicates $\{P_1, \ldots, P_k\}$, including those in the property to be verified, are identified from the concrete program. These predicates are defined on the variables of the concrete system. They also serve as the atomic propositions that label the states in the concrete and abstract transition systems, that is, the set of atomic propositions is $A = \{P_1, P_2, .., P_k\}$. A state in the concrete system will be labeled with all the predicates it satisfies. The abstract state space has a boolean variable $B_j$ corresponding to each predicate $P_j$. So each abstract state is a valuation of these $k$ boolean variables. An abstract state will be labeled with predicate $P_j$ if the corresponding bit $B_j$ is 1 in that state. The predicates are also used to define a total function $\rho$ between the concrete and abstract state spaces. A concrete state $s$ will be related to an abstract state $\hat{s}$ through $\rho$ if and only if the truth value of each predicate on $s$ equals the value of the corresponding boolean variable in the abstract state $\hat{s}$. Formally,

$$\rho(s, \hat{s}) = \bigwedge_{1 \leq j \leq k} P_j(s) \Leftrightarrow B_j(\hat{s}) \tag{7.4}$$

Note that $\rho$ is a total function because each $P_j$ can have one and only one value on a given concrete state and so the abstract state corresponding to the concrete state is unique. Using this $\rho$ and the construction given in the previous subsection, we can build an abstract model which simulates the concrete model. We now define the *concretization function* $\gamma$, which maps a set of abstract states to the corresponding set of concrete states. Formally, let $\hat{f}$ be a propositional formula over abstract state variables,

$$\gamma(\hat{f}) = \hat{f}[B_j \leftarrow P_j]. \tag{7.5}$$

In predicate abstraction [19], the abstract initial states $\hat{S}_0$ and the abstract transition relation $\hat{R}$ are defined as

$$\hat{S}_0 = \bigwedge\{\hat{Y}_1 \mid S_0 \to \gamma(\hat{Y}_1)\} \tag{7.6}$$

$$\hat{R} = \bigwedge\{\hat{Y} \to \hat{Y}' \mid (R \wedge \gamma(\hat{Y})) \to \gamma(\hat{Y}')\} \tag{7.7}$$

where $\hat{Y}$ ($\hat{Y}_1$) is an arbitrary conjunction (disjunction) of the literals of the current state variables $\{B_1, B_2, \ldots, B_k\}$ and $\hat{Y}'$ is an arbitrary disjunction of literals of the next state variables $\{B'_1, B'_2, \ldots, B'_k\}$. It can be shown that (7.6) is equivalent to (7.1) and (7.7) is equivalent to (7.2) (refer to [23] for detailed proofs).

Equations (7.6) and (7.7) can be used to compute abstract models for both hardware and software verification. To determine the validity of the proof obligations involved, a general theorem prover, such as Simplify [17], is used. For hardware verification, a SAT solver, such as zChaff, can be more efficient. In practice, heuristics are used to reduce the number of calls to the theorem prover [1, 19]. In this paper, to reduce the abstraction time, we restrict $\hat{Y}_1$ and $\hat{Y}'$ to be at most one literal, and restrict $\hat{Y}$ to include at most two literals. The model so obtained will be an over-approximation of the abstract model. We rely on refinement to compute a precise enough abstract model when necessary.

## 2.3 Localization Reduction

Localization reduction [14] is also a special case of existential abstraction. In localization reduction, a set of important state variables, called *visible* variables, are retained in the abstract model; while the rest, called *invisible* variables, are dropped (Their values are assigned nondeterministically). The abstract transition is obtained by conjuncting the transition relations for the visible variables. Formally, let $V$ be the set of concrete state variables, and $S$ be the concrete state space. The value of a variable $v \in V$ in state $s \in S$ is denoted by $s(v)$. Given a set of variables $U = \{u_1, u_2, \ldots, u_k\}, U \subseteq V$, let $s^U$ denote the portion of $s$ that corresponds to the variables in $U$, i.e., $s^U = (s(u_1)s(u_2)\ldots s(u_k))$. Let $U$ be the set of visible variables. The set of abstract states for localization reduction is $\hat{S} = D_{u_1} \times D_{u_2} \ldots \times D_{u_k}$. The simulation relation is $\rho(s, \hat{s}) = (s^U \equiv \hat{s})$.

We also assume that neither the concrete transition relation nor the set of initial states is described as a single formula. Instead, for each individual variable $v \in V$, the transition relation of $v$ is represented as a propositional formula $R_v$ and the set of initial states of $v$ is represented as a propositional formula $I_v$. Thus the abstract initial states $\hat{S}_0$ and the abstract transition relation $\hat{R}$ are defined as

$$\hat{S}_0 = \wedge_{v \in U} I_v \tag{7.8}$$

$$\hat{R} = \wedge_{v \in U} R_v \tag{7.9}$$

It is usually the case that $\hat{R}$ depends not only on current and next state variables on $U$, but also some invisible variables (precisely those invisible variables that occur in some $R_v$ or $I_v$). In the abstract model, these invisible variables are treated as primary inputs. In general, the abstract model for localization reduction can be computed very easily, but the size of the abstract transition relation may be large since it is directly copied from the concrete model.

# 3. Clustering Based Predicate Abstraction

In this section, we show how to use clustering based heuristics to identify control variables. We present an algorithm to build an abstract model by combining localization reduction with predicate abstraction. This procedure ensures that the size of the abstract model is bound by the size of the concrete model. We also show how to use correlations between predicates and control variables to make the abstract model more accurate.

## 3.1 Identifying Control Variables

Predicate abstraction is suitable for handling variables with large domains. Such variables are usually called *data variables*. By replacing important formulas over concrete data variables with abstract predicates, it is possible to reduce the complexity of verification significantly. Besides data variables, there are other variables with small domains (e.g., boolean variables) that control the behavior of the system to be verified. These variables are called *control variables*. Abstracting control variables does not give much advantage. Because control variables typically have small domains, the amount of reduction obtained by replacing a predicate over several control variables with an abstract boolean variable is not very significant.

We propose a clustering-based heuristic to identify the important control variables for the verification of the given property. Let $\{P_1, \ldots, P_k\}$ be the set of predicates. Each predicate $P_i$ is a boolean formula over a set of concrete state variables, called the *supporting variables* of $P_i$. We partition predicates into small clusters. Initially, each predicate is a cluster. We merge two clusters if the intersection of their supports crosses a certain threshold (the support of a cluster is the union of the supporting variables for each predicate in the cluster). We continue this process until no more clusters can be merged. Thus, the clusters we create partition the predicates into disjoint sets (but the supporting variables of different clusters may still overlap). Let $c$ be a cluster, the set of indexes of predicates in $c$ be $I(c)$, the supporting variables of $c$ be $v(c)$. If all the variables in $v(c)$ are finite state, each variable can be represented by several equivalent boolean variables which encode the domain of this variable. The set of boolean variables for variables in $v(c)$ is called the set of *supporting boolean variables*. For a cluster $c$, if the number of predicates is comparable to the number of supporting boolean variables, then this cluster is called a *control cluster* and the supporting variables of $c$ are regarded as control variables.

## 3.2 Combining with Localization Reduction

It is well known that, given $n$ boolean variables, the number of distinct propositional formulas over them is $2^{2^n}$. Since control variables determine the control flow of the system under verification, in order to approximate the behavior of the concrete system, many predicates over the control variables may be necessary. Each of these propositional formulas may become a predicate during predicate abstraction. Therefore, for the verification of control intensive systems, a blowup of the abstract model is likely when using existing predicate abstraction methods. Furthermore, building the abstract model using equations (7.6) and (7.7) is time consuming. Both these problems can be avoided by using our technique of *combining the localization reduction with predicate abstraction*. Using our method, it is possible to bound the size of the abstract model by that of the concrete model. We retain most the control variables in the abstract model (the criteria for retaining a control variable is discussed later in this section). The concrete transition relations for these control variables also serve as abstract transition relations after some minor modifications. So we can circumvent the problem of building abstract transition relations for all these control variables.

The modification to the concrete transition relation is as follows: for a supporting variable $v \in v(c)$, let $R_v$ be the concrete transition relation for $v$. Let $R'_v = R_v[P_j \leftarrow B_j]$, for all $j$ such that $P_j$ is a formula predicate. That is, we replace all occurrences of every formula predicate $P_j$ in $R_v$ by the corresponding abstract boolean variable $B_j$. If $R'_v$ is finite state, that is, if there are no unbounded variables or unbounded control (e.g., recursion) in it, then we use $v$ as an abstract state variable. In such a case we use $R'_v$ as the abstract transition relation for variable $v$. In the terminology of localization reduction, variable $v$ is visible and unabstracted. There is one major difference between localization reduction and our method: in localization reduction, the transition relation for a visible variable is copied from the concrete model to the abstract model, whereas in our method, we replace a subformula of the concrete transition relation if that subformula corresponds to a formula predicate. Doing this has two advantages: firstly, even if $R_v$ had unbounded variables, $R'_v$ could be finite state because of the substitutions. Secondly, the transition relations for the control variables are modified so that the abstract variables corresponding to formula predicates constrain the possible next states of the control variables. This leads to a more accurate model.

Note that the abstract model built using the localization reduction has more primary inputs (invisible variables) than the abstract model built using predicate abstraction. This can increase the size of the abstract model. Therefore, we retain unabstracted only those variables whose next state logic has a small number of inputs.

## 3.3 Correlations between Control Variables and Predicates

Our abstract model includes real predicates and control variables. In this subsection, a method to correlate predicates and control variables will be discussed. Recall from Section 3.1 that the clusters we build partition the predicates into disjoint sets (although the supporting variables of the clusters may overlap). Our method replaces the predicates in the control clusters by the supporting variables. There might be other predicates which have these control variables in their support. As an example, suppose we decide to drop a predicate cluster $\{P_1 \equiv x \vee y, P_2 \equiv x \wedge y\}$ and replace the two predicates with the variables $\{x, y\}$. Suppose also there are two additional predicates, $P_3 \equiv x \vee y \vee z$ and $P_4 \equiv x \wedge y \wedge w$ whose corresponding abstract state variables are $B_3$ and $B_4$, respectively. Thus, the abstract state variables include $x, y, B_3, B_4$. Further assume that the next state value for variable $x$ is defined as $\neg z$ in the concrete model. Note that the values of variables $x, y$ and values of $B_3, B_4$ are not independent. The following are three possible scenarios

- If we know that $B_3$ is false in an abstract state, then $x =$ false and $y =$ false in that state.
- If we know $x =$ false in an abstract state, then $B_4$ must be false in that state.
- If we know $B_3$ is false in an abstract state, then in the corresponding concrete states, $z$ is false. Therefore, in the abstract successor states, $x$ will be true.

It is desirable to incorporate the correlations/constraints between control variables and real predicates into the abstract model. This will make the abstraction more accurate. Our method does not directly compute these constraints. Instead, we selectively introduce the concrete definitions of some predicates into the abstract model as invariants. The model checking procedure will enforce any implied constraints through these invariants. Note that, only formula predicates whose supporting variables are all finite state are considered in this method. For the above example, we add $z, w$ as two additional abstract input variables and add the definitions of the two predicates as abstract invariants: $B_3 = (x \vee y \vee z)$, and $B_4 = (x \wedge y \wedge w)$. This will force the abstract model to observe any constraints between variables $x, y$ and $B_3, B_4$. Note that by doing this we have added two new variables $z, w$ to the abstract model. This could make the abstract model larger. To overcome this problem, we add the definition of a predicate to the abstract model only if most of the variables in the support of this predicate are either control variables themselves (e.g. $x$, $y$ for $B_3$) or in the support of control variables (e.g. $z$ for $x$). In this way, the added invariants will restrict the possible values of the control variables and predicates. This will ensure we only add a small number of additional variables, e.g., $z$ and $w$.

## 3.4 Correlations Between Formula Predicates

It is also possible that the predicates in a non-control cluster may not be independent, in the sense that not all possible combinations of assignments to their abstract state variables are possible. For the example in the previous paragraph, when $B_3$ = false, $B_4$ must also be false. For a given cluster $c$, let $v(c)$ be the concrete supporting variables in $c$, let $I(c)$ be the indexes of the predicates in $c$. We define $g(c)$, called the *consistent abstract states over cluster* $c$, as follows

$$g(c) = \{\hat{s} \mid \exists s \in S. \bigwedge_{j \in I(c)} (P_j(s) = B_j(\hat{s}))\} \tag{7.10}$$

It is easy to see that any $\hat{s} \notin g(c)$ does not have any corresponding concrete state and therefore it should be excluded from the abstract model checking. We represent the computed consistent abstract states for each non-control cluster as invariant in the abstract model. It is possible to compute a single set of consistent abstract states by conjuncting all predicates instead of conjuncting predicates of each cluster separately. Although this will result in a more accurate constraint, it may be computationally expensive when the number of predicates is large.

We now show how to compute $g(c)$. We have two algorithms depending on whether or not there are any unbounded variables in $v(c)$. The first algorithm is based on BDDs. It only works if all variables in $v(c)$ have finite domains. We can build BDDs for each $P_j$ and $B_j$, then $g(c)$ can be calculated by conjuncting $P_j(s) = B_j(\hat{s}), j \in I(c)$ and quantifying $v(c)$. This is not expensive because the number of predicates in a cluster is usually small. The second algorithm is based on the abstraction function [19]. Let $\hat{Y}(c)$ be a disjunction of literals over variables $B_j$, where $j \in I(c)$. It can be shown that $g(c)$ is the same as

$$\bigwedge\{\hat{Y}(c) \mid \text{true} \Rightarrow \gamma(\hat{Y}(c))\} \tag{7.11}$$

Essentially, this equation says that a formula over the abstract variables, $\hat{Y}(c)$, includes the set of consistent abstract states if the corresponding concrete formula, $\gamma(\hat{Y}(c))$, is true. The second algorithm works for variables with both finite and infinite domains. For the finite case, a SAT solver can be used; while for the other case, a general theorem prover has to be used. Since the second algorithms may require solving true $\Rightarrow \gamma(\hat{Y}(c))$ for all possible disjunctions over variables in cluster *c*, it is usually slower than the first algorithm when variables have finite domains.

## 4. Exploiting High Level Representation

In this section, we discuss how to improve predicate abstraction by using information from the high level representation of the design under verification. We first describe our method for extracting branch conditions from RTL Verilog and then we present our lazy-refinement algorithm to refine the abstract model.

### 4.1 Extracting Branch Conditions

High level design languages usually contain branch statements, such as **if**, **case** statements. The **if** statement has two branches, while the **case** statement may have multiple branches. Usually, a **case** statement can be converted to multiple **if-then-else** statements that are equivalent to it. We call the boolean predicates that determine which branch to be executed, *branch conditions*. We intend to extract the branch conditions and use them as predicates in predicate abstraction.

For the purpose of model checking, the high level representation of the system under verification is translated into a formula over the current and next state variables (referred to as the transition relation). Each extracted branch condition is translated into a subformula of the transition relation. For a branch condition, the corresponding subformula of the transition relation is called the *flattened branch condition*. The transition relation is further converted into different representations that are suitable for different model checking engines. For example, it is converted to BDDs for BDD-based model checkers, or CNF for SAT-based model checkers. For a flattened branch condition, it is straightforward to identify the corresponding representation inside the model checking engines.

We will describe a simple method to extract a set of flattened branch conditions for RTL Verilog designs. We believe it is easy to generalize this method to other design languages. One possible method is to develop a translator from RTL Verilog to gate level circuits, which can then be easily converted into a transition relation. The main disadvantage of this method is the amount of work involved in handling the semantics of Verilog, which is not formally defined [12]. In practice Verilog is interpreted by a set of standard commercial tools, such as *Synopsys Design Compiler* [20]. Our method relies on the fact that commercial synthesis tools already exist for Verilog. We first convert the RTL design into another equivalent design, where the relevant branch conditions are renamed to signals with unique names. An example is shown in Figure 7.1. We use the *continuous assignment statement* in Verilog to rename the branch

**Original design**

```
always @(posedge clk) begin
  if (mode != NO_CONF) begin
    ...
  end else if (a == b) begin
    ...
  end
end
```

**Modified design**

```
assign pred1 = mode != NO_CONF;
assign pred2 = a == b;
always @(posedge clk) begin
  if (pred1) begin
    ...
  end else if (pred2) begin
    ...
  end
end
```

*Figure 7.1.* Replace branch conditions using unique signals

conditions using unique signals, such that the modified design is equivalent to the original one. After this, a gate level circuit is generated from the modified design using Synopsys Design Compiler. We further translate the gate level circuit into a transition relation and the flattened branch conditions can be identified using the unique signal names. Our method can be easily applied to other design languages as long as there are language constructs to rename boolean predicates using new variables. Our method can take advantage of existing translators, therefore the implementation time is much shorter than building a translator from scratch.

It is usually the case that there are many branch conditions that we can extract from a high level representation of designs. Not all of them can be used as predicates to build the initial abstraction, otherwise the abstract model will become too large. We use the refinement algorithm in Section 4.2 to identify a subset of the branch conditions which are necessary to invalidate the given spurious abstract counterexample.

## 4.2 Counterexample-based Lazy Refinement

In counterexample guided abstraction refinement, a given spurious abstract counterexample is invalidated during refinement through the introduction of a set of predicates, called *invalidating predicates*, into the abstract model. Once an abstract counterexample is determined to be spurious, our algorithm identifies a subset of the flattened branch conditions as invalidating predicates.

We first introduce some notation. Let $f$ be a boolean formula, we use $\pm f$ to denote $f$ or $\overline{f}$. Let $v \in V$ be a concrete state variable, we use $v' \in V'$ to denote the corresponding next state variable. If $f$ is a boolean function over $V$, then $f'$ is the same function over $V'$.

The flattened branch conditions, which have not yet been added as predicates, are called the *candidate predicates*. A naive algorithm to compute the required set of invalidating predicates is the following: first, the set of candidate predicates is ordered according to some importance criteria. Using this order, candidate predicates can be added to the abstract model one at a time and the given counterexample can be checked on the refined abstract model. If the counterexample is invalidated, the already added candidate predicates will be the required set of invalidating predicates. This naive algorithm has two disadvantages. One is that the order of the predicates affects the size of the result. A bad order may prevent the discovery of a smaller number of invalidating predicates. Most importantly, the computation time is too high, because once a predicate is added, the abstract model has to be updated as described in Section 2.2. Instead, we have developed a new *lazy refinement* algorithm, which avoids computing the full refined abstract model at each stage. Intuitively, in this algorithm, the given abstract counterexample is extended by assigning 0, 1 or $x$ values to the abstract variables corresponding to the candidate predicates. A candidate abstract variable is given a 0 or a 1 value at time $i$ if it can be determined from the counterexample at time $i-1$ and $i$; otherwise an unknown value $x$ is given. The counterexample is invalidated if it can not be extended to the next time step. If that is the case, we perform a backward analysis from the time of failure until time 0 to identify those candidate predicates that are responsible for this failure. The predicates identified in this manner will invalidate the spurious counterexample.

Suppose there are already $m$ predicates in the abstract model. Let $ce = \langle ce_0, ce_1, \ldots, ce_n \rangle$ be a spurious abstract counterexample. Note that, each $ce_j$ is a conjunction of literals over the set of abstract state variables $B_1, \ldots, B_m$. Let $cp = \{cp_{m+1}, cp_{m+2}, \ldots, cp_{m+k}\}$ be the set of candidate predicates, which are temporarily represented by abstract state variables $\{B_{m+1}, B_{m+2}, \ldots, B_{m+k}\}$ (These candidate predicates have not been added to the abstract model yet). The example in Figure 7.2 illustrates how our algorithm works. Suppose there are 2 predicates, 3 candidate predicates and a spurious abstract counterexample of length 3. The counterexample contains values for predicates $P_1$ and $P_2$ at each time from 0 to 2. Our algorithm first determines the values for the candidate predicates at time 0. If $(S_0 \wedge \gamma(ce_0)) \rightarrow \overline{cp_4}$ is a tautology, then any valid extension of $ce_0$ must have the abstract variable corresponding to $cp_4$ set to 0.

| | time 0 | time 1 | time 2 |
|---|---|---|---|
| B1 | 1 | 0 | 1 |
| B2 | 0 | 1 | 1 |
| B3 | 1 | 1 | |
| B4 | 0 | x | |
| B5 | 0 | 1 | |

*Figure 7.2.* A refinement example

The values of other candidate predicates at time 0 can be determined similarly. The resulting extended counterexample at time 0 is denoted by $ece_0$. We then extend the counterexample at time 1 to obtain $ece_1$. For example, if we can prove that

$$(R \wedge \gamma(ce_0) \wedge cp_3 \wedge \overline{cp_4} \wedge \overline{cp_5} \wedge \gamma(ce_1)) \rightarrow cp'_3 \tag{7.12}$$

is a tautology (where $cp'_3$ is the same as $cp_3$ except that it is over the next state variables), the value of this candidate predicate must be 1. Note that we can not determine the value of $cp_4$ at time 1, therefore its value is unknown in the extended counterexample. After $ece_1$ is determined, if

$$(R \wedge \gamma(ce_1) \wedge cp_3 \wedge cp_5) \rightarrow \gamma(\overline{ce_2}) \tag{7.13}$$

is a tautology, then the counterexample can not be extended to time 2, thus it has been invalidated. Finally, we identify the set of invalidating predicates. It is possible that not all candidate predicates in the left hand side of equations (7.12) and (7.13) are necessary in showing that they are tautologies. Only those in the proof of the tautologies are necessary. Proofs can be obtained from *proof generating theorem provers* (e.g., Simplify) and *proof generating SAT solvers* [4]. Suppose, we can determine that $cp_3$, $cp_5$ in equation (7.12) and $cp_5$ in equation (7.13) are not in the respective proofs for those two implications. Then we can deduce that, of all candidate predicates, $cp_3$ alone is responsible for disabling the transition from time step 1 to time step 2 (since $cp_5$ is not needed in the proof of equation (7.13)). Moreover, of all candidate predicates, only $cp_4$ at time 0 determines the value of $cp_3$ at time step 1 (since $cp_3$, $cp_5$ do not appear in the proof of equation (7.12)). Thus the set of invalidating predicates is $\{cp_3, cp_4\}$. Note, we have worked backwards along the counterexample. We first found some invalidating predicates at time step 1 and then used that to find more invalidating predicates at time step 0. This is the basic idea of our algorithm to find the set of invalidating predicates.

We now present the lazy refinement algorithm in detail. Our algorithm is separated into three parts, the first one, which computes $ece_0$, is shown in Figure 7.3. The second one, which computes $ece_{i+1}$ making use of $ece_i$, is shown in Figure 7.4. The last one, shown in Figure 7.5, computes the invalidating predicates as a subset of the candidate predicates once the counterexample is invalidated.

**Compute_Initial**

let $ece_0 = ce_0$
**for** each candidate predicate $cp_{m+j}$
  **if** $(S_0 \wedge \gamma(ce_0)) \rightarrow cp_{m+j}$ is a tautology
    let $ece_0 = ece_0 \wedge B_{m+j}$
  **elseif** $(S_0 \wedge \gamma(ce_0)) \rightarrow \overline{cp_{m+j}}$ is a tautology
    let $ece_0 = ece_0 \wedge \overline{B_{m+j}}$
  **endif**
**endfor**

*Figure 7.3.* Algorithm to compute $ece_0$

The algorithm to compute $ece_0$ is similar to the algorithm for computing the set of abstract initial states in Section 2.2, except that we use $S_0 \wedge \gamma(ce_0)$ instead of $S_0$ alone. This makes sense because our goal is to extend the current counterexample. The idea is to determine if the set of concrete initial states $S_0$ and the concrete states corresponding to $ce_0$ can imply either the truth or falsity of each candidate predicate; otherwise the value of the candidate predicate is unknown.

Given the extended counterexample at time $i$, the algorithm in Figure 7.4 extends the counterexample to time $i+1$. It first checks whether there are any concrete transitions between $\gamma(ece_i)$ and $\gamma(ce_{i+1})$. The code for this is given in lines (1) to (4). If it is not the case, the counterexample has been invalidated by the candidate predicates, the set of invalidating predicates is calculated and returned in line (3). If it is possible to make a concrete transition from $\gamma(ece_i)$ to $\gamma(ce_{i+1})$, the algorithm will check whether a candidate predicate is guaranteed to be true/false for such concrete transitions. This is computed in line (7) and line (9) and $ece_{i+1}$ is updated. If the counterexample can be extended from time 0 until time $n$, the set of flattened branch conditions are not enough to invalidate the counterexample. We will resort to the traditional refinement methods to compute a new predicate [6] using SAT. Details can be found in [8].

If the counterexample is invalidated at line (1) in Figure 7.4, the algorithm in Figure 7.5 is called with the time $t$ and $f = (R \wedge \gamma(ece_t)) \rightarrow \gamma(\overline{ce_{t+1}})$. We use the set $np$ to hold all candidate predicates that are given a 0 or 1 value in the time steps preceding $t$ and result in the failure of the counterexample. In line (1), $np$ is initialized to all candidate predicates that are directly responsible for the failure. This is done by analyzing the proof for the failure of the counterexample. In the loop between line (2) and line (6), we go backward in time to find the set of candidate predicates that are indirectly responsible for the failure. Finally

//$i$: time to extend counterexample  
**Compute_Next**($i$)  
**if** $(R \wedge \gamma(ece_i)) \rightarrow \gamma(\overline{ce_{i+1}})$ is a tautology  
  let $f = (R \wedge \gamma(ece_i)) \rightarrow \gamma(\overline{ce_{i+1}})$  
  **return** DETERMINE_PREDICATES($i$, $f$)  
**endif**  
let $ece_{i+1} = ce_{i+1}$  
**for** each candidate predicate $cp_{m+j}$  
  **if** $(R \wedge \gamma(ece_i) \wedge \gamma(ce_{i+1})) \rightarrow cp'_{m+j}$ is a tautology  
    let $ece_{i+1} = ece_{i+1} \wedge B_{m+j}$  
  **elseif** $(R \wedge \gamma(ece_i) \wedge \gamma(ce_{i+1})) \rightarrow \overline{cp'_{m+j}}$ is a tautology  
    let $ece_{i+1} = ece_{i+1} \wedge \overline{B_{m+j}}$  
  **endif**  
**endfor**

*Figure 7.4.* Algorithm to compute $ece_{i+1}$

//$t$: the time when extending counterexample fails  
//$f = (R \wedge \gamma(ece_t)) \rightarrow \gamma(\overline{ce_{t+1}})$  
**Determine_Predicates**($t$, $f$)  
let $np = \{\langle \pm B_{m+j}, t\rangle \mid \pm\, cp_{m+j}$ is in the proof of $f\}$  
**for** $i = t - 1$ to $0$  
  let $taut(i) = \{(R \wedge \gamma(ece_i) \wedge \gamma(ce_{i+1})) \rightarrow \pm cp'_{m+q} \mid$  
    $\langle \pm B_{m+q}, i+1\rangle \in np\}$  
  let $prf = \{$ proofs for the implications in $taut(i)\}$  
  let $np = np \cup \{\langle \pm B_{m+w}, i\rangle \mid$  
    $\pm cp_{m+w}$ is in any proof in $prf\}$  
**endfor**  
**return** $\{cp_{m+j} \mid \exists 0 \leq i \leq t.\ \langle \pm B_{m+j}, i\rangle \in np\}$

*Figure 7.5.* Algorithm to compute invalidating predicates

in line (7), the set of invalidating predicates is returned. Note that, in line (3), $taut(i)$ is a subset of the tautologies we computed from the algorithm in Figure 7.4. For each implication $(R \wedge \gamma(ece_i) \wedge \gamma(ce_{i+1})) \rightarrow \pm cp'_{m+q}$ in $taut(i)$, we refine the abstract transition relation $\hat{R}$ by conjuncting it with $ece_i \rightarrow (\overline{ce_{i+1}} \vee \pm B'_{m+q})$. Therefore, our algorithm not only computes the subset of the flattened branch conditions which can invalidate the given spurious abstract counterexample but also computes the refined abstract model. Our algorithm does not build the whole refined abstract model and then test whether it invalidates the counterexample. Instead, it gradually refines the abstract model until the counterexample is invalidated. Therefore, our lazy algorithm can be more efficient than the naive algorithm.

## 5. Related Work

Some researchers have considered combining unabstracted control variables with predicate abstraction [16], but their methods are not automatic. As far as we know, no one else has considered the correlation between unabstracted control variables and predicates. Using the correlations between all predicates to constrain the abstract model has been investigated in [1]. The correlations are computed using a general theorem prover. We first partition the set of predicates into clusters based on the sharing of support sets, then correlations are computed for each cluster separately. Although our result is more approximate, the complexity of our algorithm is much less sensitive to the total number of predicates. We also give a BDD-based algorithm for the verification of finite state systems.

Exploiting high level language features for abstraction has been investigated in [6]. They extract conditions of **case** statements in the SMV language in order to build the initial abstraction. The extraction method in [6] requires modifying the *source code* of an existing translator from SMV language to transition relations, therefore it can not be applied to commercial tools. The extracted conditions are used only for the initial abstraction; while we use a new refinement algorithm to check whether branch conditions can invalidate the spurious abstract counterexample. The branch conditions become predicates only when they invalidate a spurious counterexample.

Our counterexample-based lazy refinement algorithm tries to identify the branch conditions that can invalidate the spurious abstract counterexample, before using the traditional refinement methods [8] to compute a new predicate. Therefore, our algorithm is an extension of the existing refinement algorithms. Our experiments show that this new refinement algorithm can identify the set of predicates to verify the given property much more quickly than the traditional methods alone.

Lazy abstraction for the verification of C programs has been investigated in [13]. The goals of their algorithm and ours are different. In [13], the construction of the abstract model and abstract model checking are performed only from the state where the spurious abstract counterexample fails on the concrete system. While Our refinement algorithm identifies a subset of the branch conditions that can invalidate a spurious counterexample without constructing the full refined abstract model.

## 6. Experimental Results

We have modified the zChaff SAT solver [24] to generate proofs of unsatisfiability. The techniques in this paper are implemented on top of the predicate abstraction framework reported in [8]. We compare results with and without these techniques using two sets of benchmarks: one is the integer unit (IU) of the picoJava microprocessor from Sun [21]; the other is a programmable FIR filter (PFIR) which is a component of a system-on-chip design. The size of the benchmarks is shown in Table 7.1. The first column is the name of the property. The first three properties are from the IU design; the remaining six are from the PFIR design. For all the properties shown in the first column of Table 7.2, we have performed cone-of-influence reduction before the verification. The resulting number of registers and gates are shown in the second and third columns. Most properties are true, except PFIRscr1 and PFIRprop5. The lengths of the counterexamples are shown in the fourth column.

| circuit | # regs | # gates | ctrex |
|---|---|---|---|
| IUscr1 | 4855 | 149143 | true |
| IUscr3 | 4855 | 149143 | true |
| IUscr6 | 4855 | 149143 | true |
| PFIRscr1 | 243 | 2295 | 16 |
| PFIRprop5 | 250 | 2342 | 17 |
| PFIRprop8 | 244 | 2304 | true |
| PFIRprop9 | 244 | 2304 | true |
| PFIRprop10 | 244 | 2304 | true |
| PFIRprop12 | 247 | 2317 | true |

*Table 7.1.* The benchmarks used in the experiments

All these properties are difficult for the state-of-art BDD-based model checker, Cadence SMV. Except for the two false properties, Cadence SMV can not verify any in 24 hours. The verification time for PFIRscr1 is 834 seconds, and for PFIRprop5 is 8418 seconds. In Table 7.2, we compare results obtained with and without the new techniques presented in this paper. In Table 7.2, the second to fourth columns are the results obtained without our techniques; while the last three columns are the results obtained with the techniques enabled. We compare the time (in seconds), the number of refinement iterations and the number of predicates in the final abstraction. In all cases, our new method outperforms the basic one in the amount of time used; sometimes over an order of magnitude improvement is achieved. In most cases, we use fewer refinement iterations and smaller predicate sets to verify the given properties. A detailed analysis of the PFIR results shows that the extraction algorithm extracted about 9 branch conditions from the RTL Verilog, which were later used as predicates. Without these extracted predicates, the set of predicates computed using traditional refinement algorithm was not sufficient to finish verification within 24 hours (for 3 properties).

| circuit | Basic | | | New | | |
|---|---|---|---|---|---|---|
| | time | iters | pred | time | iters | pred |
| IUscr1 | 2000 | 11 | 7 | 1265 | 7 | 18 |
| IUscr3 | 2003 | 10 | 6 | 1974 | 16 | 7 |
| IUscr6 | 9976 | 27 | 12 | 3498 | 20 | 11 |
| PFIRscr1 | 746 | 109 | 44 | 386 | 67 | 34 |
| PFIRprop5 | 1616 | 110 | 43 | 756 | 101 | 44 |
| PFIRprop8 | >24h | >276 | >80 | 159 | 40 | 25 |
| PFIRprop9 | >24h | >189 | >47 | 202 | 43 | 27 |
| PFIRprop10 | 6808 | 170 | 52 | 178 | 50 | 25 |
| PFIRprop12 | >24h | >223 | >52 | 591 | 80 | 38 |

*Table 7.2.* Compare the basic predicate abstraction [8] with our new techniques

## 7. Conclusion

We have presented two techniques to improve predicate abstraction for the verification of hardware/software systems. We give an algorithm based on localization reduction to avoid the potential blowup of the abstract models when verifying control intensive systems. This technique builds a "hybrid" abstract model, which includes predicates as well as unabstracted control variables. It is usually the case that the predicates/control variables are not independent. We give algorithms to compute correlations between them, which help to make the abstract model more accurate. We also present algorithms to exploit information in high level design languages. We give a simple method to extract branch conditions from high level design representations. Using a new counterexample-based lazy refinement algorithm, the necessary branch conditions can be added as new predicates to invalidate spurious abstract counterexamples. Experimental results demonstrate the usefulness of our methods.

## References

[1] Thomas Ball, Rupak Majumdar, Todd Millstein, and Sriram K. Rajamani. Automatic Predicate Abstraction of C Programs. In *PLDI 2001*.

[2] Thomas Ball, Andreas Podelski, and Sriram K. Rajamani. Boolean and cartesian abstractions for model checking c programs. In *TACAS 2001*, volume 2031 of *LNCS*, pages 268–283, April 2001.

[3] Saddek Bensalem, Yassine Lakhnech, and Sam Owre. Computing abstractions of infinite state systems compositionally and automatically. In *Computer-Aided Verification, CAV'98*, pages 319–331, 1998.

[4] Pankaj Chauhan, Edmund M. Clarke, Samir Sapra, James Kukula, Helmut Veith, and Dong Wang. Automated abstraction refinement for model checking large state spaces using sat based conflict analysis. In *FMCAD'02*, 2002.

[5] E. Clarke, O. Grumberg, and D. Long. Model checking and abstraction. In *POPL*, pages 343–354, 1992.

[6] Edmund Clarke, Orna Grumberg, Somesh Jha, Yuan Lu, and Helmut Veith. Counterexample-guided Abstraction Refinement. In *CAV'00*, 2000.

[7] Edmund Clarke, Orna Grumberg, Somesh Jha, Yuan Lu, and Helmut Veith. Counterexample-guided abstraction refinement for symbolic model checking. *Journal of the ACM (JACM)*, 50(5)752–794, September 2003

[8] Edmund Clarke, Muralidhar Talupur, and Dong Wang. SAT based Predicate Abstraction for Hardware Verification. In *Sixth International Conference on Theory and Applications of Satisfiability Testing*, 2003.

[9] Edmund M. Clarke, Orna Grumberg, and Doron Peled. *Model Checking*. MIT Press, 1999.

[10] Michael Colon and Tomas E. Uribe. Generating finite-state abstractions of reactive systems using decision procedures. In *CAV'98*, pages 293–304, 1998.

[11] Satyaki Das, David L. Dill, and Seungjoon Park. Experience with predicate abstraction. In *CAV'99*, pages 160–171, 1999.

[12] Michael J. C. Gordon. The semantic challenge of Verilog HDL. In *LICS'95*, pages 136–145, 1995.

[13] Thomas A. Henzinger, Ranjit Jhala, Rupak Majumdar, and Gregoire Sutre. Lazy abstraction. In *POPL*, pages 58–70, 2002.

[14] R. P. Kurshan. *Computer-Aided Verification*. Princeton Univ. Press, Princeton, New Jersey, 1994.

[15] C. Loiseaux, S. Graf, J. Sifakis, A. Bouajjani, and S. Bensalem. Property preserving abstractions for the verification of concurrent systems. *Formal Methods in System Design An International Journal*, 6(1)11–44, January 1995.

[16] Kedar S. Namjoshi and Robert P. Kurshan. Syntactic program transformations for automatic abstraction. In *CAV'00*, 2000.

[17] Greg Nelson. *Techniques for Program Verification*. PhD thesis, Stanford University, 1980.

[18] S. Graf and H. Saidi. Construction of abstract state graphs with PVS. In *CAV'97*, pages 72–83, 1997.

[19] H. Saidi and N. Shankar. Abstract and model check while you prove. In *CAV'99*, pages 443–454, 1999.

[20] Synopsys Inc.. Synopsys Design Compiler. http://www.synopsys.com.

[21] Sun Microsystems. picoJava technology. http://www.sun.com/microelectronics/communitysource/picojava.

[22] Dong Wang, Pei-Hsin Ho, Jiang Long, James Kukula, Yunshan Zhu, Tony Ma, and Robert Damiano. Formal Property Verification by Abstraction Refinement with Formal, Simulation and Hybrid Engines. In *DAC'01*, 2001.

[23] Dong Wang. *SAT based Abstraction Refinement for Hardware Verification*. PhD thesis, Carnegie Mellon University, May 2003.

[24] Lintao Zhang, Conor F. Madigan, Matthew W. Moskewicz, and Sharad Malik. Efficient conflict driven learning in a Boolean satisfiability solver. In *ICCAD'01*, 2001.

# Chapter 8

# HOW TO COMPUTE THE REFINEMENT RELATION FOR PARAMETERIZED SYSTEMS

Françoise Bellegarde, Célina Charlet and Olga Kouchnarenko
*LIFC FRE 2661 CNRS*
*INRIA group CASSIS*
*Université de Franche-Comté*
*16, route de Gray,*
*25030 Besançon, Cedex France*
{bellegar,charlet,kouchna} @lifc.univ-fcomte.fr

**Abstract** In this paper, we present a refinement verification for a class of parameterized systems. These systems are composed of an arbitrary number of similar processes. Following the approach by P. Abdulla et al., we represent the states by regular languages and the transitions by transducers over regular languages. If we can compute a symbolic model by acceleration of the actions, then we can also verify a refinement relation $\mathcal{R}$ between the symbolic models. We show that, under some conditions, if $\mathcal{R}$ is verified between two symbolic models, then refinement is verified between concrete parameterized systems. Then, we can take advantage the property (safety and $PLTL$ properties) preservation by refinement for their verification.

**Keywords:** Refinement design, parameterized systems, regular languages

## Introduction

The growing part of computer systems in critical applications points out the importance of developing techniques for verifying these systems. There are two main techniques of verification: *proof* techniques, dealing with infinite state systems but requiring human interaction, and *model-checking* techniques, completely automatic but often restricted to finite state systems [Apt and Kozen, 1986].



*R. Gupta et al. (eds.), Formal Methods and Models for System Design,* 181-209.

Complex systems verification must take into account aspects that are out of the scope of techniques and tools based on finite state systems analysis (model-checking). Among these aspects, we can distinguish: manipulation of unbounded data structures (e.g counters) or variables, parameterization (networks with an arbitrary number of components), mobility, etc.

To deal with the problem of infinite systems verification there are several approaches based on varied techniques such as abstraction, symbolic reachability analysis, etc., and using varied mathematical tools: languages and automata theory [Kesten et al., 1997, Clarke et al., 1995, Sistla and Gyuris, 1999, Abdulla and Jonsson, 1999, Bouajjani et al., 2000], logics, rewriting systems, constraints solving [Comon and Jurski, 1998, Annichini et al., 2000], etc.

In this paper, we are interested in the verification of parameterized systems. A parameterized system is a particular infinite system in the sense that each of its instances is finite but the number of states of the system depends on one or several parameters. So, taking into account the set of all possible instances of the system requires considering an arbitrary number of cases.

For classes of systems composed of an arbitrary number of processes communicating in a synchronous [Emerson and Namjoshi, 1995, Emerson and Namjoshi, 1996] or asynchronous [Emerson and Kahlon, 2000] way and eventually sharing a control process, the problem of verifying a restricted set of properties is decidable. Presently, there exists many techniques, often incomplete, for verifying parameterized systems [Annichini et al., 2000, Sistla and Gyuris, 1999, Kesten et al., 1997, Abdulla et al., 1999, Bouajjani et al., 2000, Jonsson and Nilsson, 2000, Pnueli and Shahar, 2000, Arons et al., 2001, Maidl, 2001]. An approach to apply model-checking to a parameterized system is to use a symbolic representation of the states and transitions of the system to generate a finite state abstraction of the operational model of the system.

We suppose a specification obtained by a refinement process. Several methods, like *B* [Abrial, 1996], *TLA*$^+$ [Lamport, 1999], *CSP2B* [Butler, 2000], etc. propose a refinement based development. In such a development, the system conception is done progressively by increasing the accuracy of the systems description at each specification step. The interest of a refinement based development is that it guarantees preservation of the abstract systems properties. In this paper, our refinement concept is close to the one of the *B* event systems. In the *B* refinement method, the refinement is verified by proof. Safety properties of the systems are expressed by an invariant and are verified by proof too.

In [Bellegarde et al., 2000], we express the refinement semantics as a relation between transition systems. This relation can be classified as a *divergence stability respecting completed simulation* in the van Glabbeek' spectrum. This relation can also be verified for finite state systems using an algorithmic method. This allows an algorithmic verification of the refinement by model

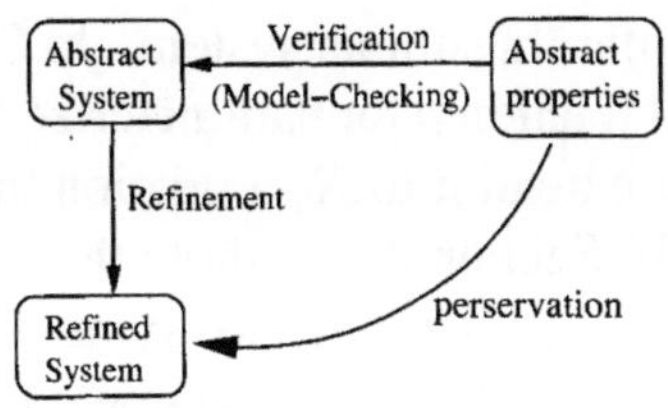


*Figure 8.1.* Specification and verification approach.

exploration and a verification of the *PLTL* properties by model-checking, as shown in Fig. 8.1.

We consider the class of parameterized systems composed of an arbitrary number of similar processes for which the parameter represents the number of processes. In this framework, there is a modelisation where symbolic states can be seen as regular languages [Bouajjani et al., 2000, Jonsson and Nilsson, 2000] and where the reachability problem can be solved using an acceleration method. On such an abstraction, it is possible to:

- verify safety properties of the system by model-checking;
- verify refinement by model exploration;
- verify properties expressed in Propositional Linear Time Logic ($PLTL$) about the behaviors of the system.

In this paper, the algorithmic verification of our refinement relation between finite state systems is extended to parameterized systems. For that, we define the semantics for this refinement relation we call $\rho$. But $\rho$ cannot be computed directly since we have to use acceleration techniques [Bouajjani et al., 2000, Jonsson and Nilsson, 2000] to compute a finite symbolic representation of a parameterized system. So, we define another refinement relation, $\mathcal{R}$, that can be computed using acceleration techniques. The main contribution of this paper is to ensure that if $\mathcal{R}$ is verified between two symbolic systems, then $\rho$ is also verified for the corresponding concrete systems. We give some conditions required for this to hold. We also provide an algorithmic verification of $\mathcal{R}$ based on fixed points computations. This algorithmic verification is not guaranteed to terminate because it relies on the constructions of the reachability graphs of the systems which are not guaranteed to terminate. The goal of this work is to take advantage of property preservation by the refinement. Indeed, when the refinement is verified, most of the $PLTL$ properties that hold on the symbolic abstract model are preserved for the refined one.

The paper is organized as follows. In Section 2, we give preliminary notions about transducers and regular languages to model parameterized systems and define our notion of symbolic transitions system. In Section 3, we define the semantics of the refinement relation $\rho$ for parameterized systems. As in general $\rho$ can not be computed, we extend it to $\mathcal{R}$, a relation that can be calculated by accelerations techniques. In Section 4, we show that, under some conditions, if the refinement relation $\mathcal{R}$ holds between two symbolic systems, then $\rho$ holds too. Moreover, we present how $\mathcal{R}$ can be computed in an iterative way. In Section 5, we give a short tool description, and illustrate its use on an example. We end in Section 6 by some concluding remarks, and directions for future work.

## 1. Symbolic Model

In this section, we define a symbolic model for transition systems based on regular languages and transducers.

$\Sigma^*$ denotes the set of finite words over $\Sigma$. The word $uv$ is the concatenation of $v$ after $u$; $\epsilon$ is the empty word. A language $L$ is a subset of $\Sigma^*$. As usual, $a \in \Sigma$ denotes the language $\{a\} \subseteq \Sigma^*$ and $\overline{L} = \Sigma^* \setminus L$.

We define a regular expression $E$ and its associated language $L(E)$ as follows:

- $\emptyset$ is a regular expression, and $L(\emptyset) = \emptyset$,
- $\epsilon$ is a regular expression, and $L(\epsilon) = \{\epsilon\}$,
- $a$ is a regular expression, and $L(a) = \{a\}$,
- $E_1 + E_2$ is a regular expression, and $L(E_1 + E_2) = L(E_1) \cup L(E_2)$,
- $E_1.E_2$ is a regular expression, and $L(E_1.E_2) = L(E_1).L(E_2)$,
- $E_1^*$ is a regular expression, and $L(E_1^*) = (L(E_1))^*$.

By the Kleene theorem, a language is regular iff there is a regular expression to which it is associated. From now on, we use $E$ and $L(E)$ indifferentely.

### 1.1 Representing States and Transitions

Let $\Sigma$ be a finite alphabet. In our approach, states of the system are words of a regular language $E \subseteq \Sigma^*$ and transitions are relations between these words. A state of a system (or global configuration) is a word $a_1 a_2 ... a_n$ over $\Sigma$, where $n$ is the number of processes and each $a_i$ is the state of the i-th process.

The transition relation is a relation between words of the same length. A transition labeled by an action $\alpha$ transforms (according to $\alpha$) a word into another word by changing the letters corresponding to the processes being transformed by $\alpha$. An action $\alpha$ can be written using the triple $\langle ww, (w \rightarrow r), a \rangle$. The second element of the triple describes the transformation made by the action using a word of pairs over $\Sigma \times \Sigma$. The first and last elements of the triple describe the context in which the action can be applied using words over $\Sigma$, i.e., states of the processes which are not concerned with the action.

EXAMPLE 1 *Consider an action transforming the most right process in state $w$ into a process in state $r$. Applying this action to the configuration $wwwa$ (case of a system composed of 4 processes) transforms it into the configuration $wwra$. If, now, the system is composed of an arbitrary number of processes, the regular expression $w^*a$ represents the global state of the system for which the last process is in state $a$ and every other process is in state $w$, this independently of the number of processes. Then, $w^*a$ characterizes a symbolic configuration (global state) of the parameterized system.*

We extend the transition relation between words to a symbolic transition relation between regular languages. Such an action can be represented by the triple $\langle \Sigma^*, (w \to r), \bar{w}^* \rangle$ where $\bar{w} = \Sigma \setminus \{w\}$. The first and the last elements of the triple are regular languages, called the *left* and the *right contexts* of the action. Their definition, according to [Abdulla et al., 1999], is as follows:

DEFINITION 2 *A left context $C_l$ is a regular language accepted by a deterministic finite state automaton having a single accepting state. A right context $C_r$ is a regular language such that its reverse language is a left context.*

An action $\alpha$ is represented by a triple $\langle C_l, T, C_r \rangle$ where $C_l$ is a left context, $T$ is a regular language over $\Sigma \times \Sigma$, and $C_r$ is a right context. For example, the symbolic action $\langle w^*, [(r \to a)(r \to r)^*(a \to w)], \epsilon \rangle$ indicates that:

- we first observe an arbitrary number of processes in state $w$, then,
- the first process being in state $r$ changes his state and becomes $a$ (in other words, it rewrites in $a$),
- the rightmost process must be in state $a$ and rewrites in $w$,
- all intermediate processes must be in state $r$ and do not change their state.

## 1.2 Symbolic Transition Systems

We now define the model using a symbolic transition system. In the formal framework of transition systems, a symbolic configuration is called a symbolic state.

DEFINITION 3 *A symbolic transition system is a quadruple $\langle \Sigma, E_0, Act, \to \rangle$ where $\Sigma$ is a finite alphabet, $E_0$ is a regular initial configuration on $\Sigma$, $Act$ is a set of (names of) symbolic actions, and $\to$ is a symbolic transition relation ($\to \subseteq \Sigma^* \times Act \times \Sigma^*$).*

The labeled transition relation $\rightarrow$ is defined by a set of triples $(E, \alpha, E')$ (written $E \xrightarrow{\alpha} E'$). It can be extended to a sequence of transitions in the standard way: $E'$ is reachable from $E$, written $E \rightarrow^* E'$, if there exists a sequence of symbolic states $E = E^0, E^1, ..., E^n = E'$ such that for each $i \geq 0$ we have $E^i \xrightarrow{\alpha_i} E^{i+1}$. We write $E \nrightarrow$ to express that no action can be applied from $E$. Let $Att(E_0) = \{E | E_0 \rightarrow^* E\}$ be the set of reachable states from $E_0$.

In Definition 4, $X$ is a variable such that its domain is included in $\Sigma^*$. The set $AP_\Sigma$ contains atomic propositions where each atomic proposition $e$ is either $X = L$ or $X \subseteq L$, where $L \subseteq \Sigma^*$. The set $SP_\Sigma$ contains propositions over $\Sigma$ where each proposition is defined by the following grammar: $p_1, p_2 := e | p_1 \vee p_2 | \neg p_1$.

DEFINITION 4 *We define a state proposition $p \in SP_\Sigma$ to be valid for a symbolic state E, written $E \models p$, as follows:*

- $E \models X = L$ *iff* $E = L$,
- $E \models X \subseteq L$ *iff* $E \subseteq L$,
- $E \models p_1 \vee p_2$ *iff* $E \models p_1$ *or* $E \models p_2$,
- $E \models \neg p$ *iff* $\neg(E \models p)$.

We call *invariant* a predicate $I$ which holds on each symbolic state. In other words, $I$ is an invariant of a transition system $TS$ over $\Sigma$ iff $I \in SP_\Sigma$ and $(\forall E.(E \in Att(E_0) \Rightarrow E \models I))$.

**Transducers for the transition relation.** A transition labeled by $\alpha$ ($\in Act$) can be represented by a finite state automaton over $\Sigma \times \Sigma$, called a transducer (see for example Fig. 8.2).

DEFINITION 5 *Let $\Sigma_1$ and $\Sigma_2$ be two alphabets. A transducer $\mathcal{T}$ over $\Sigma_2 \times \Sigma_1$ is a finite state automaton over $\Sigma_2 \times \Sigma_1$ accepting words of pairs $(a_i, b_i)$ for $1 \leq i \leq n$ coming from the words $(a_1 \dots a_n) \in \Sigma_2^*$ and $(b_1 \dots b_n) \in \Sigma_1^*$.*

Let $id(L)$ be the language recognized by a transducer accepting identical pairs of words of $L$. We write $id$ for $id(\Sigma)$. The transducer $\mathcal{T}_\alpha$ for the action $\alpha = \langle C_g, T, C_d \rangle$ accepts all words of pairs belonging to the language defined by $id(C_g).T.id(C_d)$. It expresses the relation $\xrightarrow{\alpha}$.

DEFINITION 6 *The symbolic transition relation $\xrightarrow{\alpha}$ induced by $\mathcal{T}_\alpha$ links all pairs of words $(a_1...a_n, b_1...b_n)$ such that words of pairs $(a_1, b_1)...(a_n, b_n)$ are accepted by $\mathcal{T}_\alpha$. We say that $\xrightarrow{\alpha}$ is a regular relation.*

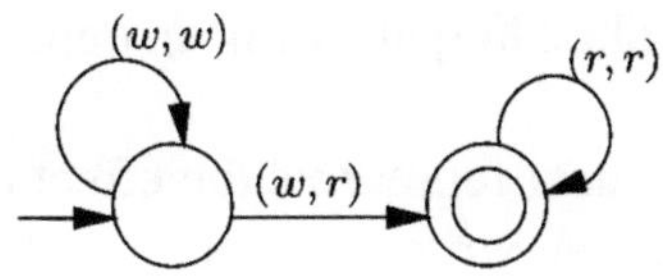


*Figure 8.2.* A transducer accepting all words of pairs of the form $(w,w)^*(w,r)(r,r)^*$

## 1.3 Calculating the Reachable Configuration for a Transition

Given a transducer $\mathcal{T}_\alpha$ and an automaton $\mathcal{A}$ accepting a symbolic configuration $E$, we can built a new automaton $\mathcal{A}'$ accepting all words of a symbolic configuration $E'$, such that $E \xrightarrow{\alpha} E'$ and resulting from the composition of $\mathcal{T}_\alpha$ and $\mathcal{A}$, written $\mathcal{T}_\alpha \circ \mathcal{A}$.

DEFINITION 7 *The composition $\mathcal{T}_\alpha \circ \mathcal{A}$ of a transducer $\mathcal{T}_\alpha = \langle \Sigma\times\Sigma, \mathcal{R}, r_0, \delta_T, F_{\mathcal{T}_\alpha}\rangle$ and an automaton $\mathcal{A} = \langle \Sigma, Q, q_0, \delta_{\mathcal{A}}, F_{\mathcal{A}}\rangle$ is the automaton $\langle \Sigma, Q \times \mathcal{R}, (q_0, r_0), \delta, F_{\mathcal{A}} \times F_{\mathcal{T}_\alpha}\rangle$ where $(q_2, r_2) \in \delta((q_1, r_1), b)$ if there exists an $a \in \Sigma$ such that $q_2 \in \delta_{\mathcal{A}}(q_1, a)$ et $r_2 \in \delta_{\mathcal{T}_\alpha}(r_1, (a, b))$.*

Using this composition, we can compute the next reachable state of the system. But this is not enough to have a finite representation of the behavior of the system because some actions can be executed again and again from an initial configuration. For instance, consider the symbolic state $w^*a$ and the action $\alpha$ transforming the last process in state $w$ into a process in state $r$. If we apply successively $\alpha$ from $w^*a$, we get an infinite sequence of symbolic states $w^*a \xrightarrow{\alpha} w^*ra \xrightarrow{\alpha} w^*rra \xrightarrow{\alpha} w^*rrra \xrightarrow{\alpha} ...$ since the transition can be applied again and again from the last computed state. To have a finite representation of the behavior of the system, we must compute in one step the reachable symbolic state corresponding to an arbitrary number of applications of the same action, i.e., compute the transitive closure of the action, called *acceleration.*

## 1.4 Acceleration

In [Abdulla et al., 1999], the authors give the sufficient conditions required to compute automatically the accelerated transducer corresponding to an action transforming only one process, i.e., where $T$ is limited to a pair of letters. In [Jonsson and Nilsson, 2000, Bouajjani et al., 2000], this is extended to actions transforming simultaneously many processes. Let $k$ be the maximal number of times a same symbol can be rewritten for the same action. If $k$ is bounded, an accelerated transducer having a state space in a size of $k$ exponential can be computed. When the construction cannot be calculated, there are widening techniques [Bouajjani et al., 2000] allowing an upper-approximation of the set of reachable states.

Each accelerated transducer represents the effect of an arbitrary number of applications of the action. We write $\alpha^*$ (resp. $\alpha^+$) the transitive and reflexive (resp. the transitive closure) of an action $\alpha$.

The application of $\alpha$ (resp. $\alpha^+$) to a symbolic configuration $E$ is written $\alpha(E)$ (resp. $\alpha^+(E)$). We also write $E(\alpha)$ (resp. $E(\alpha^+)$) the subset of $E$ from which the action $\alpha$ (resp. $\alpha^+$) can effectively be applied (source of the application). This subset is calculated by intersecting $E$ and the source language of $\alpha$ (resp. $\alpha^+$). For example, let $E = a^*\{b, c\}a^*$ be a symbolic configuration and $\alpha = \langle \Sigma^*, [b \rightarrow a], \Sigma^* \rangle$ be an action. We have $\alpha(E) = a^* + a^*ca^*$ and $E(\alpha) = E \cap \Sigma^* b \Sigma^* = a^*ba^*$.

Using these accelerated actions, we can compute, from the initial symbolic state of the system, the successive sets of reachable symbolic states of the system. If we get a set of configurations that has already been computed (this is detected by an inclusion test between regular languages), then the computation ends. Again, widening techniques [Bouajjani et al., 2000] can be used to help termination. This computation, which is useful to verify the refinement, can be viewed as the construction of a reachability graph where the transitions are labeled by $\alpha^*$ or $\alpha^+$ and where the states are symbolic configurations.

## 2. Refinement

The refinement definition between event systems has been defined in [Bellegarde et al., 2000] as a relation between transition systems. In this section, we extend this definition to our symbolic models. First, we give a definition of the refinement relation between symbolic transition systems, written $\rho$. Then, we extend $\rho$ to $\mathcal{R}$ by taking into account transition acceleration and show that under some conditions, $\mathcal{R}$ and $\rho$ are equal. Finally, we present an iterative computation to verify that $\mathcal{R}$ holds between two given systems.

### 2.1 Semantics for the Refinement Relation between Symbolic Models

Let $TS_1 = \langle \Sigma_1, E_{01}, Act_1, \rightarrow_1 \rangle$ and $TS_2 = \langle \Sigma_2, E_{02}, Act_2, \rightarrow_2 \rangle$ be two symbolic transition systems at two levels of refinement and $I_1$, $I_2$ be their respective invariants. As in [Bellegarde et al., 2000], we consider that a refinement introduces new actions, so we have $Act_1 \subseteq Act_2$. We also require that $\Sigma_1 \cap \Sigma_2 = \emptyset$, since refinement renames variables.

Our goal is to verify that $TS_2$ is a refinement of $TS_1$ by a relation $\rho$ between $TS_2$ and $TS_1$. For that, we first define the relation $\mu$, needed to link the states of the systems. Second, we present the refinement relation we call $\rho$. The relation $\mu$ is implemented using a deterministic transducer $\mathcal{T}_\mu$ over $\Sigma_2 \times \Sigma_1$ recognizing the pairs of words $(a_1 \ldots a_n, b_1 \ldots b_n)$ such that the words of pairs $(a_1, b_1), \ldots, (a_n, b_n)$ are accepted by $\mathcal{T}_\mu$. For instance, if $\Sigma_1 = \{a, b\}$ and $\Sigma_2 = \{x, y, z\}$, the configuration $x^* y x^* z x^*$ is linked to the configuration $a^* b a^*$ for the relation $\mu = \{(x, a), (y, b), (z, a)\}$. The transducer $\mathcal{T}_\mu$ accepts the language $((x, a) + (y, b) + (z, a))^*$.

To formalize $\mu$, we also need an invariant $I_{1,2}$ which is a formula of the form $\mathcal{T}(X_2) \subseteq X_1$ where $\mathcal{T}$ is a deterministic transducer over $\Sigma_2 \times \Sigma_1$. The relation $\mu$ depends on the validity of the conjunction of $I_1$, $I_2$ and $I_{1,2}$.

DEFINITION 8 *We define a proposition p to be valid for the pair of symbolic states $(E_2, E_1)$, written $(E_2, E_1) \models_g p$, as follows:*

- $(E_2, E_1) \models_g p$ *iff either* $p \in SP_{\Sigma_1}$ *and* $E_1 \models p$, *or* $p \in SP_{\Sigma_2}$ *and* $E_2 \models p$, *or* $p = \mathcal{T}(X_2) \subseteq X_1$ *and* $\mathcal{T}(E_2) \subseteq E_1$,
- $(E_2, E_1) \models_g p \vee q$ *iff* $(E_2, E_1) \models_g p$ *or* $(E_2, E_1) \models_g q$,
- $(E_2, E_1) \models_g \neg p$ *iff* $\neg((E_2, E_1) \models_g p)$.

A symbolic state $E_2$ in $TS_2$ is linked to a symbolic state $E_1$ in $TS_1$, written $E_2 \; \mu \; E_1$, iff $(E_2, E_1) \models_g I_1 \wedge I_2 \wedge I_{1,2}$. We also denote by $\mu(E)$ the language accepted by $\mathcal{T}_\mu(E)$. As in [Bellegarde et al., 2000], the refinement relation is defined as a restriction of $\mu$.

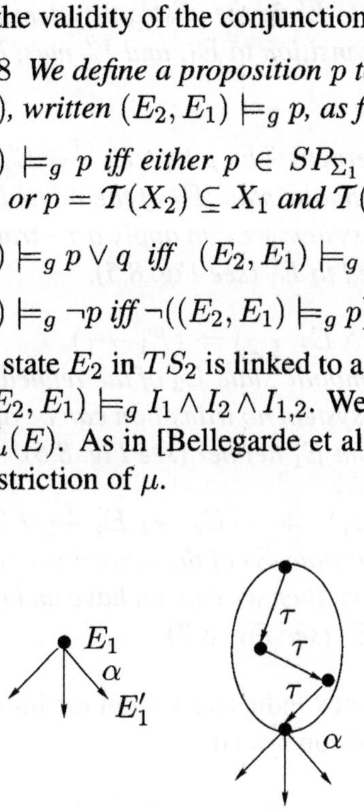


*Figure 8.3.* Silent transitions

**Definition of the refinement relation..** To describe the refinement relation, we keep the labels of $TS_2$ transitions belonging in $Act_1$. However, new transitions (those labeled in $Act_2 \setminus Act_1$) that are introduced by the refinement are considered as $\tau$-transitions, i.e. *silent* transitions (see Fig. 8.3), so $Act_2$ becomes $Act_1 \cup \{\tau\} = Act_{1\tau}$. Silent transitions are not allowed to take control of the system forever, so infinite paths are forbidden for silent transitions. We also require that transitions from $TS_2$ do not introduce new deadlocks. Consequently, vivacity and safety properties are preserved during refinement.

DEFINITION 9 *Let $\alpha$ be an action in $Act_1$. Let $TS_1 = \langle \Sigma_1, E_{10}, Act_1, \rightarrow_1 \rangle$ and $TS_2 = \langle \Sigma_2, E_{20}, Act_{1\tau}, \rightarrow_2 \rangle$ be two transition systems. The refinement relation $\rho \subseteq {\Sigma_2}^* \times {\Sigma_1}^*$ is defined as the greatest binary relation included in $\mu$ satisfying the following clauses :*

*1* **(strict transition refinement)** $(E_2 \ \rho \ E_1 \wedge E_2 \xrightarrow{\alpha}_2 E_2') \Rightarrow (\exists E_1'$ *s.t.* $E_1 \xrightarrow{\alpha}_1 E_1' \wedge E_2' \ \rho \ E_1')$
*This means that if from a symbolic state $E_2$ of the refined system linked to a state $E_1$ of the abstract system we can apply a transition leading to $E_2'$, then there must be a state $E_1'$ in the abstract system such that $E_1'$ is reachable by applying the transition to $E_1$, and $E_2'$ must be linked to $E_1'$ (see Fig. 8.4).*

*2* **(stuttering transition refinement)** $(E_2 \ \rho \ E_1 \wedge E_2 \xrightarrow{\tau}_2 E_2') \Rightarrow (E_2' \ \rho \ E_1)$
*This means that if from a symbolic state $E_2$ of the refined system, linked to a state $E_1$ of the abstract system we can apply a $\tau-$transition leading to $E_2'$, then $E_2'$ must be linked to $E_1$ (see Fig. 8.5).*

*3* **(non $\tau$-deadlock)** $(E_2 \ \rho E_1 \wedge E_2 \not\rightarrow_2) \Rightarrow (E_1 \not\rightarrow_1)$
*This means that if from a symbolic state $E_2$ of the refined system, linked to a state $E_1$ of the abstract system no transition can be applied, then no transition can be applied from $E_1$ neither (see Fig. 8.6).*

*4* **(non $\tau$-divergence)** $(E_2 \ \rho \ E_1) \Rightarrow \neg \, (E_2 \xrightarrow{\tau}_2 E_2' \xrightarrow{\tau}_2 E_2'' \xrightarrow{\tau}_2 \cdots)$
*This means that if a symbolic state $E_2$ of the refined system is linked to a state $E_1$ of the abstract system, then we can not have an infinite sequence of $\tau$-transition applying to $E_2$ (see Fig. 8.7).*

The refinement relation $\rho$ is defined inductively from the initial states of the refined and abstract symbolic transition systems.

DEFINITION 10 *Let $TS_1 = \langle \Sigma_1, E_{10}, Act_1, \rightarrow_1 \rangle$ and $TS_2 = \langle \Sigma_2, E_{20}, Act_{1\tau}, \rightarrow_2 \rangle$ be two transition systems. $TS_1$ is refined by $TS_2$, written $TS_1 \sqsupseteq TS_2$, if $E_{20} \rho E_{10}$.*

Since a symbolic state $E$ is a regular language, an action applies to a subset of $E$. Figure 8.4 and Figure 8.5 show that each configuration represented by $E_2(\alpha)$ is linked to a configuration of the abstract system by $\rho$. Moreover, $E_1(\alpha)$ equals to $\rho(E_2(\alpha))$. In Fig. 8.6, we label the respective parts of configurations $E_2$ and $E_1$, for which no action applies, with $E_2 \not\rightarrow_2$ and $E_1 \not\rightarrow_1$.

**Extending $\rho$ to $\mathcal{R}$ by acceleration.** Since a symbolic configuration represents an arbitrary number of configurations, a transition may be applied an arbitrary number of times, and the computation of the reachability graph of

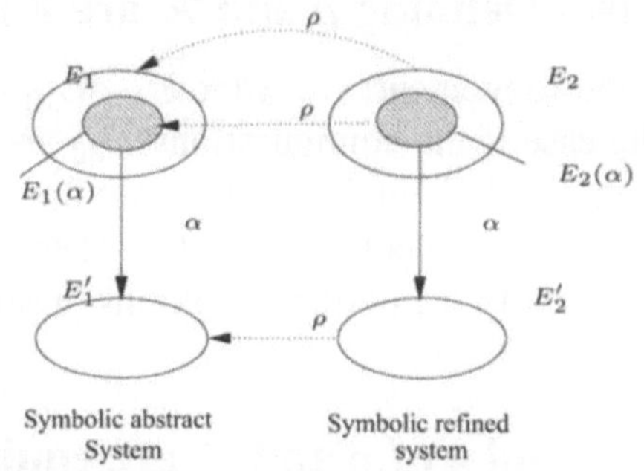


*Figure 8.4.* Strict transition

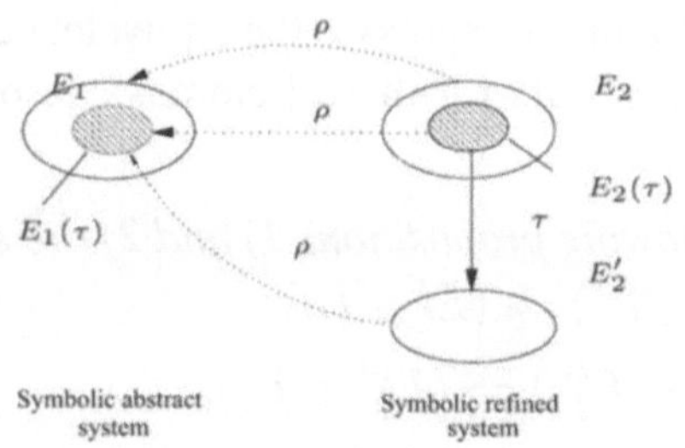


*Figure 8.5.* Stuttering transition

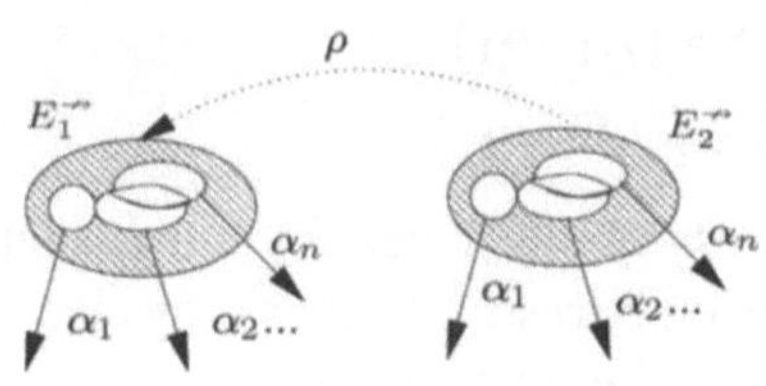


*Figure 8.6.* Non $\tau$-deadlock

the system does not terminate. To help termination, we have to use acceleration (see Section 1.4). We define another refinement relation, $\mathcal{R}$, using strict transition closure. Then, $\mathcal{R}$ is the greatest binary relation included in $\mu$ and satisfying clauses of Definition 9 where $\rightarrow$ is replaced by $\xrightarrow{+}$. The computation of $\mathcal{R}$ is more likely to be effective since acceleration helps termination. Using $\mathcal{R}$ instead of $\rho$ to verify refinement is possible iff $E_{20}\rho E_{10}$ is equivalent to $E_{20}\mathcal{R}E_{10}$. Unfortunately, in Section 4, we show that it is not the case. However, we can give and verify conditions that ensure this equivalence.

# 3. When Clauses Defining $\rho$ and $\mathcal{R}$ are Equivalent

In order to decide the refinement, we want $E_{20}\rho E_{10}$ and $E_{20}\mathcal{R}E_{10}$ to be equivalent. This is the case if the configurations $E_{20}$ and $E_{10}$ are equivalent for each clause defining $\rho$ and $\mathcal{R}$. In this section, we first show that Clauses 2, 3 and 4 of Definition 9 are equivalent for $\mathcal{R}$ and $\rho$. Then we show that it is not the case for Clause 1. Finally we define conditions, easily verified, ensuring the equivalence.

## 3.1 Clauses 2, 3 and 4 of $\rho$ and $\mathcal{R}$ are equivalent

Recall that $\mathcal{R} \subseteq \mu$ and $\rho \subseteq \mu$ and that $\mu$ is a relation linking states of transition systems. Then, since relations $\mathcal{R}$ and $\rho$ are defined recursively, we have to put $\mu$ in the right part of the implications of the clauses to verify their equivalence. The following three Lemmas expresses the equivalence for Clauses 2, 3 and 4 of relations $\rho$ and $\mathcal{R}$. Proofs for these Lemmas can be found in [Bellegarde et al., 2003].

LEMMA 11 *The following propositions 1) and 2) are equivalent:*
*1)* $(E_2\ \rho\ E_1 \wedge E_2 \xrightarrow{\tau}_2 E_2') \Rightarrow (E_2'\ \mu\ E_1)$
*2)* $(E_2\ \mathcal{R}\ E_1 \wedge E_2 \xrightarrow{\tau^+}_2 E_2'^+) \Rightarrow (E_2'^+\ \mu\ E_1)$.

This means that if two symbolic states $E_2$ and $E_1$ verify the stuttering transition clause for relation $\rho$ (resp. $\mathcal{R}$), then they also verify the stuttering transition clause for relation $\mathcal{R}$ (resp. $\rho$).

LEMMA 12 *The following propositions 1) and 2) are equivalent:*
*1)* $(E_2\ \rho\ E_1 \wedge E_2 \not\rightarrow_2) \ \Rightarrow (E_1 \not\rightarrow_1)$
*2)* $(E_2\ \mathcal{R}\ E_1 \wedge E_2 \overset{+}{\not\rightarrow}_2) \ \Rightarrow (E_1 \overset{+}{\not\rightarrow}_1)$.

In other words, if two symbolic states $E_2$ and $E_1$ verify the clause for lack of new deadloks for relation $\rho$ (resp. $\mathcal{R}$), then they also verify the clause for for lack of new deadlocks for relation $\mathcal{R}$ (resp. $\rho$).

LEMMA 13 *The following propositions 1) and 2) are equivalent:*
*1)* $(E_2\ \rho\ E_1) \ \Rightarrow \neg\ (E_2 \xrightarrow{\tau}_2 E_2' \xrightarrow{\tau}_2 \cdots)$
*2)* $(E_2\ \mathcal{R}\ E_1) \ \Rightarrow \neg\ (E_2 \xrightarrow{\tau^+}_2 E_2'^+ \xrightarrow{\tau^+}_2 \cdots)$.

In other words, if two symbolic states $E_2$ and $E_1$ verify the non $\tau$-divergence clause for relation $\rho$ (resp.$\mathcal{R}$), then they also verify the non $\tau$-divergence clause for relation $\mathcal{R}$ (resp. $\rho$).

## 3.2 Clause 1 of $\rho$ not equivalent to Clause 1 of $\mathcal{R}$

In this section, we explain why Clause 1 of $\rho$ is not equivalent to Clause 1 of $\mathcal{R}$.

LEMMA 14 *Clause 1 of $\rho$ implies Clause 1 of $\mathcal{R}$.*
$((E_2\,\rho\,E_1 \wedge E_2 \overset{\alpha}{\rightarrow}_2 E_2') \Rightarrow (\exists E_1' \text{ s.t. } E_1 \overset{\alpha}{\rightarrow}_1 E_1' \wedge E_2'\,\mu\,E_1')) \Rightarrow ((E_2\,\mathcal{R}\,E_1 \wedge E_2 \overset{\alpha^+}{\rightarrow}_2 E_2'^+) \Rightarrow (\exists E_1' \text{ s.t. } E_1 \overset{\alpha^+}{\rightarrow}_1 E_1'^+ \wedge E_2'^+\,\mu\,E_1'^+))$.

In other words, if two symbolic states $E_2$ and $E_1$ verify strict transition refinement clause for relation $\rho$, then they also verify strict transition refinement clause for relation $\mathcal{R}$.

LEMMA 15 *Clause 1 of $\mathcal{R}$ does not imply Clause 1 of $\rho$.*
$((E_2\,\mathcal{R}\,E_1 \wedge E_2 \overset{\alpha^+}{\rightarrow}_2 E_2'^+) \Rightarrow (\exists E_1'^+ \text{ s.t.} E_1 \overset{\alpha^+}{\rightarrow}_1 E_1'^+ \wedge E_2'^+\,\mu\,E_1'^+)) \not\Rightarrow ((E_2\,\rho\,E_1 \wedge E_2 \overset{\alpha}{\rightarrow}_2 E_2') \Rightarrow (\exists E_1' \text{ s.t. } E_1 \overset{\alpha}{\rightarrow}_1 E_1' \wedge E_2'\,\mu\,E_1'))$.

In other words, if two symbolic states $E_2$ and $E_1$ verify strict transition refinement clause for relation $\mathcal{R}$, then there is no guaranty that they also verify strict transition refinement clause for relation $\rho$.

*Proof idea.* Suppose Clause 1 is satisfied for $\mathcal{R}$. Then, a symbolic state $E_2$ is linked by $\mu$ to a symbolic abstract state $E_1$. From $E_2$, we can apply $\alpha^+$, leading to a $E_2'^+$. From $E_1$ we can also apply $\alpha^+$ and go to state $E_1'^+$. Moreover, $E_2'^+$ and $E_1'^+$ are linked by $\mu$. The refinement relation $\rho$ holds if all intermediate symbolic configurations reachable from $E_2$ and $E_1$ by applying successively $\alpha$ are also linked by $\mu$. But, this cannot be ensured since we loose the trace of the intermediate symbolic states. So, Clause 1 for $\rho$ is not guaranteed to be satisfied. This is illustrated by Example 16 in which we have Clause 1 verified for $\mathcal{R}$ but not for $\rho$.

EXAMPLE 16 *Let $\alpha_C$ and $\alpha_A$ be respectively an action of the refined system and an action af the abstract system such that:*

$$\mathcal{L}(\mathcal{T}_{\alpha_C}) = ((a_1,a_1)+(b_1,b_1))^*(b_1,c_1)(b_1,c_1)(c_1,c_1)^*$$
$$\mathcal{L}(\mathcal{T}_{\alpha_A}) = ((a,a)+(b,b))^*(b,c)(c,c)^*$$

*Then, using acceleration, we have:*

$$\mathcal{L}(\mathcal{T}_{\alpha_C^+}) =$$
$$((a_1,a_1)+(b_1,b_1))^*(b_1,c_1)(b_1,c_1)(b_1,c_1)^*(c_1,c_1)^*$$
$$\mathcal{L}(\mathcal{T}_{\alpha_A^+}) = ((a,a)+(b,b))^*(b,c)(b,c)^*(c,c)^*$$

*Let $\mu$ be the relation such that $\mathcal{L}(\mathcal{T}_\mu) = ((a_1,a)+(b_1,b)+(c_1,c))^*$ and $E_2$, $E_1$ be two symbolic states linked by $\mu$ such that $E_2 = (a_1+b_1)^*$ and $E_1 = (a+b)^*$.*

*If we apply $\alpha_C$ to $E_2$, we get the configuration $\alpha_C(E_2) = (a_1+b_1)^*c_1c_1$. If we apply $\alpha_A$ to $E_1$, we get the configuration $\alpha_A(E_1) = (a+b)^*c$. These two resulting configurations are not linked by $\mu$ because $\mu(\alpha_C(E_2)) = (a+b)^*cc$, which is not included in $(a+b)^*c$. Thus, Clause 1 for $\rho$ is not verified.*

*If we apply $\alpha_C^+$ to $E_2$, we get the configuration $\alpha_C^+(E_2) = (a_1+b_1)^*c_1c_1c_1^*$. If we apply $\alpha_A^+$ to $E_1$, we get the configuration $\alpha_A^+(E_1) = (a+b)^*cc^*$. These two resulting configurations are linked by $\mu$ because $\mu(\alpha_C^+(E_2)) = (a+b)^*ccc^*$, which is included in $(a+b)^*cc^*$. Thus, Clause 1 for $\mathcal{R}$ is verified.*

## 3.3 Ensuring Equivalence

In this section, we show that $E_{20}\rho E_{10}$ is equivalent to $E_{20}\mathcal{R}E_{10}$ under some sufficient conditions. In the preceding section, we saw that Clause 1 verified by $\mathcal{R}$ does not imply that Clause 1 is verified by $\rho$. But we show that the implication holds if each action of $TS_2$ *refines* its corresponding action in $TS_1$.

In our formalism, this means that each pair of words $(u, v)$ recognized by a transducer of the refined system is linked to a pair of words $(u', v')$ recognized by the corresponding transducer of the abstract system in the following way: $(u, u') \models I_1 \wedge I_2 \wedge I_{1,2}$, and $(v, v') \models I_1 \wedge I_2 \wedge I_{1,2}$, i.e., $\mu(u) = u'$ and $\mu(v) = v'$. To verify that an action of $TS_2$ *refines* its corresponding action in $ST_1$, we need the following notations and notions.

DEFINITION 17 *The synchronized product $\mathcal{A}^2$ of an automaton $\mathcal{A} = \langle\Sigma, Q, T, q_0\rangle$ with itself is the automaton $\langle\Sigma\times\Sigma, Q\times Q, T', (q_0, q_0)\rangle$ where $((q_1, q_2), (e_1, e_2), (q_1', q_2')) \in T'$ iff $(q_1, e_1, q_1') \in T$ and $(q_2, e_2, q_2') \in T$.*

Let $\mathcal{T}_\mu$ over $\Sigma_2\times\Sigma_1$ be a deterministic transducer (a deterministic automaton) corresponding to the relation $\mu$, and $\mathcal{T}_\mu{}^2$ be the synchronized product of $\mathcal{T}_\mu$ with itself. $\mathcal{T}_\mu{}^2$ is defined over $((\Sigma_2\times\Sigma_1)\times(\Sigma_2\times\Sigma_1))$.

Let $f$ be a function renaming labels of transitions for automata over pairs of pairs of letters like $\mathcal{T}_\mu{}^2$ such that $f : ((\Sigma_2\times\Sigma_1)\times(\Sigma_2\times\Sigma_1)) \to ((\Sigma_2\times\Sigma_2)\times(\Sigma_1\times\Sigma_1))$ and $f((e_1, e_2), (e_3, e_4)) = ((e_1, e_3), (e_2, e_4))$. If $\mathcal{T}_\mu{}^2 = \langle((\Sigma_2\times\Sigma_2)\times(\Sigma_1\times\Sigma_1)), Q\times Q, T', (q_0, q_0)\rangle$, then $\mathcal{T}_{\mu f}^2 = \langle f((\Sigma_2\times\Sigma_2)\times(\Sigma_1\times\Sigma_1)), Q\times Q, T', (q_0, q_0)\rangle$. The language accepted by $\mathcal{T}_{\mu f}^2$ is $\mathcal{L}(\mathcal{T}_{\mu f}^2) = \{((u, v), (u', v')) | (u, u') \in \mathcal{L}(\mathcal{T}_\mu) \wedge (v, v') \in \mathcal{L}(\mathcal{T}_\mu)\}$.

Consider the composition of the transducer $\mathcal{T}_\alpha$ for an action $\alpha$ of the refined system with our automaton $\mathcal{T}_{\mu f}^2$ according to Definition 7.

Clause 1 is verified by $\rho$ under the following hypothesis:

HYPOTHESIS 3.1 *1* $\forall i.((\alpha_{1i} \in Act_1 \wedge \alpha_{2i} \in Act_2) \Rightarrow \mathcal{L}(\mathcal{T}_{\alpha_{2i}} \circ \mathcal{T}_{\mu f}^2) \subseteq \mathcal{L}(\mathcal{T}_{\alpha_{1i}}))$.

2 $\forall(u, v).((u, v) \in \mathcal{L}(\mathcal{T}_{\alpha_{2i}}) \Rightarrow \exists(u', v').\ (u, u') \in \mathcal{L}(\mathcal{T}_\mu) \wedge (v, v') \in \mathcal{L}(\mathcal{T}_\mu))$

This condition means that for all pairs of words $(u, v)$ recognized by a transducer $\mathcal{T}_{\alpha_{2i}}$ corresponding to an action of the refined system, there exists a pair of words $(u', v')$ accepted by the transducer $\mathcal{T}_{\alpha_{1i}}$ corresponding to an action with the same label in the abstract system, such that $(u, u') \in \mathcal{T}_\mu$ and $(v, v') \in \mathcal{T}_\mu$, i.e, $\mu(u) = u'$ and $\mu(v) = v'$. Example 18 illustrates that Hypothesis 1 is necessary to verify the refinement.

EXAMPLE 18 *Consider Example 16. We have:*
$\mathcal{L}(\mathcal{T}_{\mu f}^2(\mathcal{T}_{\alpha_C})) = ((a, a) + (b, b))^*(b, c)(b, c)(c, c)^*$, *which is not included in*

$\mathcal{L}(T_{\alpha_A}) = ((a,a)+(b,b))^*(b,c)(c,c)^*$. *Consequently, Hypothesis 1 does not hold and the refinement is not verified.*

THEOREM 19 *Under Hypothesis 1, the strict refinement Clause for $\rho$ is satisfied.*

*Proof idea.* Let $\alpha_C$ be an action refining an action $\alpha_A$ of the abstract system according to Hypothesis 1. Given $E_2$ and $\alpha_C$, let $E_2'$ be $\{v \mid \exists u \in E_2 \wedge (u,v) \in \alpha_C\}$ such that $E_2'$ is not empty. Let $\tilde{u}, \tilde{v}$ be words such that $\tilde{u} \in E_2 \wedge (\tilde{u},\tilde{v}) \in \alpha_C$. Because of Hypothesis 1, the pair $(\mu(\tilde{u}), \mu(\tilde{v}))$ exists and belongs to $\alpha_A$. Thus, $\mu(\tilde{u}) \in E_1$, so $\mu(\tilde{v}) \in E_1'$ and $E_1'$ is not empty. By definition, $\mu(E_2') = \{\mu(v) | v \in E_2'\} = \{\mu(v) | \exists u \in E_2 \wedge (u,v) \in \alpha_C\}$. By Hypothesis 1, we also have $\{(u,v) | (u,v) \in \alpha_C\} \subseteq \{(u,v) | (\mu(u), \mu(v)) \in \alpha_A\}$. So, we can write $\mu(E_2') \subseteq \{\mu(v) | \exists u \in E_2 \wedge (\mu(u), \mu(v)) \in \alpha_A\}$. By definition, $\mu(E_2) = \{\mu(u) | u \in E_2\}$, and we can write $\mu(E_2') \subseteq \{\mu(v) | \exists x \in \mu(E_2) \wedge (x, \mu(v)) \in \alpha_A\}$. We also have $\mu(E_2) \subseteq E_1$. Consequently, we can write $\mu(E_2') \subseteq \{\mu(v) | \exists x \in E_1 \wedge (x, \mu(v)) \in \alpha_A\}$. Therefore, $E_1' = \{u | \exists x \in E_1 \wedge (x,u) \in \alpha_A\}$. So, $\mu(E_2') \subseteq E_1'$ and we are done.

THEOREM 20 *Under Hypothesis 1, $E_{20}\mathcal{R}E_{10}$ is equivalent to $E_{20}\rho E_{10}$.*

*Proof.* From Theorem 19 and Lemma 14, Clause 1 of $\rho$ is equivalent to Clause 1 of $\mathcal{R}$. Since Lemma 11, 12, 13 ensure the equivalence between the other clauses, we have $E_{20}\mathcal{R}E_{10} \Leftrightarrow E_{20}\rho E_{10}$.

Hypothesis 1 can be verified according to Definitions 7 and 17 before the algorithmic verification of the refinement relation $\mathcal{R}$, which is described in the next section.

## 3.4 Iterative Computation of a $\tau$-simulation

In [Bellegarde et al., 2000], we show that the refinement relation is a special case of $\tau$-simulation. It is well-known that a (bi)simulation can be computed in an iterative way when one of the systems is finitely branching. In our framework, the transition systems are finitely branching since the sets *Act* and $\Sigma$ are finite sets.

Consequently, the relation $\mathcal{R}$ can be calculated by a width-first exploration of the set of reachable symbolic configurations. This exploration uses successive approximations of the least fixed points calculation for the set of reachable configurations. To give a semi-effective procedure to decide refinement, we must be able to perform completely the simultaneous exploration of the two systems symbolic reachable configurations. This exploration is done from the refined system reachable states. For that, it is enough to answer the following questions:

1 How to glue symbolic configurations? We must decide whether the current configuration $E_2$ of $TS_2$ is linked by $\mu$ to the current configuration $E_1$ of $TS_1$ or not. This is effective since we can decide whether the validity $(E_2, E_1) \models_g I_1 \wedge I_2 \wedge I_{1,2}$.

2 How to compute the actions accelerations for the abstract and refined systems to calculate the simulation? This is semi-effective because the computation of the accelerations may not terminate.

3 How to decide strict transition refinement? This is effective by composing the transducers for the actions of the refined system with $T_f^2$ and checking that whether the language of the resulting automaton is included in the language of the transducer of the corresponding action of the abstract system or not.

4 How to decide stuttering transition refinement? For that,

   (a) we have to compute the $\tau$-transitions, i.e, the new transitions accelerations.

   (b) we must construct the symbolic reachable configurations using the $\tau$-transitions from a configuration $E_2$ already linked to $E_1$. Remark that this needs a width-first exploration of the set of symbolic configurations reachable by the $\tau$-transitions and, consequently, successive approximations of a least fixed point computation.

5 How to decide lack of $\tau$-deadlocks? For that, we detect deadlocks from each symbolic reachable configuration of the refined model during the joint exploration (see Fig. 8.6).

6 How to decide non $\tau$-divergence? A sufficient condition for that is the absence of cycles of $\tau$-transitions in the reachability graph. Detection of such cycles, called $\tau$-cycles, is done using a semi-effective procedure, detailed below.

We have a *potential cycle* when, in a sequence of $\tau$-transitions, a symbolic configuration $E'$ has a non empty intersection with one of its preceeding configurations. Let $E = E_0, E_1, \ldots, E_n = E'$ be the symbolic configurations of the sequence of $\tau$-transitions from $E$ to $E'$ such that $E \cap E' \neq \emptyset$. Let $S_0, S_1, \ldots, S_n$ be the source parts of each $\tau$-application of the potential cycle (see Fig. 8.7). We have the three following cases.

- Case 1. A potential cycle corresponds to a $\tau$-cycle (then it is a fixed point of the sequence of $\tau$-transitions) if $S_n = S_0$ (see Example 21).
- Case 2. A potential cycle does not correspond to any $\tau$-cycle if $S_n \cap S_0 = \emptyset$ (see Example 22) .

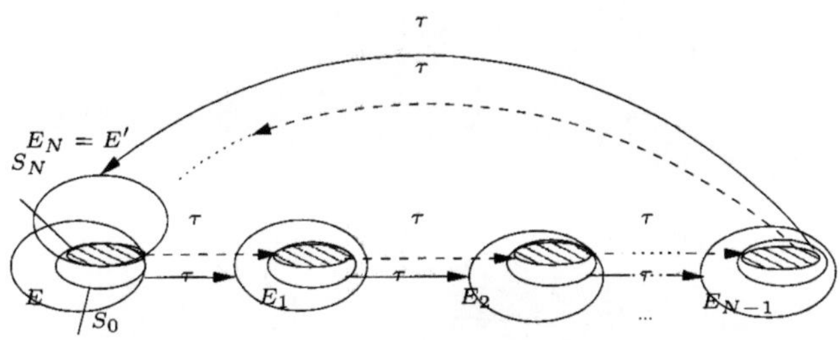


*Figure 8.7.* Non $\tau$-divergence

- Case 3. Otherwise, we cannot conclude and we must apply the sequence of $\tau$-transitions again from the configuration $S_n \cap S_0$ (see Example 23) .

EXAMPLE 21 *Let $T_1$ and $T_2$ be two transducers for $\tau$-transitions and $E_0$ be an initial state. We have:*

$$\mathcal{L}(T_1) = (a \to b)id(\Sigma^*)$$
$$\mathcal{L}(T_2) = (b \to a)id(\Sigma^*)$$
$$E_0 = aa^*.$$

*Then, by applying successively $T_1$ and $T_2$ from $E_0$, we get the following sequence: $E_0 = aa^*, E_1 = ba^*, E_2 = aa^*$. Here, the source parts of the $\tau$-transitions are equal to the computed configurations. So, we have $E_0 = S_0 = aa^*$, $E_1 = S_1 = ba^*$, and $E_2 = S_2 = aa^*$. We have a potential cycle because $E_0 \cap E_2$ is not empty. Moreover, we have $S_0 = S_2$. Consequently the potential cycle corresponds to a $\tau$-cycle.*

EXAMPLE 22 *Consider $T_1$ and $T_2$, two transducers for $\tau$-transitions, and an initial state $E_0$ such that:*

$$\mathcal{L}(T_1) = id(w+r)^*(a \to w)id(w+r)^*(r \to a)id(w+r)^*$$
$$\mathcal{L}(T_2) = id(w^*a)(r \to w)((r \to w + id(w))^*$$
$$E_0 = (w+r)^*a(w+r)^*.$$

*Then, by applying successively $T_1$ and $T_2$ from $E_0$, we get the following sequence $E_0 = (w+r)^*a(w+r)^*, E_1 = (w+r)^*w(w+r)^*a(w+r)^*, E_2 = ww^*aw^*$. There is a potential cycle because $E_0 \cap E_2$ is not empty. $E_0 \cap E_2 = ww^*aw^*$. We want to know if this potential cycle is a $\tau$-cycle. For that we compute the source parts of the transducers for the $\tau$-transitions. We get $S_0 = (w+r)^*a(w+r)^*r(w+r)^*$, $S_1 = w^*w(r+w)^*$ and $S_2 = \emptyset$. So, the potential cycle is not a $\tau$-cycle.*

EXAMPLE 23 *Consider $T_1$ and $T_2$, two transducers for $\tau$-transitions, and an initial state $E_0$ such that:*

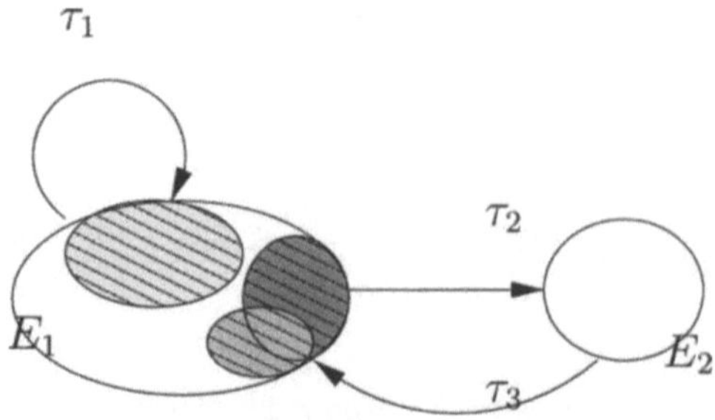


*Figure 8.8.* Nested cycles

$$\mathcal{L}(\mathcal{T}_1) = id(a^*a)(b \to c)id(b)^*$$
$$\mathcal{L}(\mathcal{T}_2) = id(\Sigma)^*(c \to a)id(b)^*$$
$$E_0 = a^*ab^*.$$

*Then, by applying successively $\mathcal{T}_1$ and $\mathcal{T}_2$ from $E_0$, we get the following sequence: $E_0 = a^*ab^*, E_1 = a^*acb^*, E_2 = a^*aab^*$. As $E_0 \cap E_2$ is not empty, we have a potential cycle. We compute the source parts of the transducers, and we get $S_0 = a^*abb^*$, $S_1 = E_1 = a^*acb^*$, and $S_2 = a^*aabb^*$. As $S_2$ is not empty, we cannot conclude. We have to apply the transducers again from $S_0 \cap S_2 = a^*aabb^*$. Then, we generate a growing sequence of configurations like $S_2 = a^*aaabb^*$, then $S_2 = a^*aaaabb^*$, etc, and we cannot conclude. However, it is easy to see that the cause of the loop is the arbitrary number of processes. Since this number is finite, we can trust that there is no real $\tau$-cycle for this example.*

REMARK 24 *When there is a nested loop (see Fig. 8.8) in the reachable symbolic graph, there are infinitely many potential cycles. So, it is impossible to look for $\tau$-cycles using the semi-algorithm above. In such a case, we cannot conclude. A way out is to find judicious and effective hypotheses such that if a finite number of potential cycles are not $\tau$-cycles then all the potential cycles are also not $\tau$-cycles.*

*For a simple example, in the graph in Fig. 8.8, we have $(E_1(\tau_1) \cup \tau_1(E_1)) \cap (E_1(\tau_2) \cup \tau_3(E_2)) = \emptyset$. This is enough to guarantee that the potential loops $C_1 = E_1, E_1$ with $\tau_1$ and $C_2 = E_1, E_2, E_1$ with $\tau_2$ and $\tau_3$ cannot generate infinitely many potential nested loops so that there are only $C_1$ and $C_2$ to check for $\tau$-divergence.*

*Another way to bypass the problem is to put restrictions to the refinement so that no nested potential loops are possible. For example, a (certainly too strong) restriction is to introduce new actions one by one at each refinement step.*

The refinement relation provides a formal framework for verifying parameterized systems, since when the refinement holds, most of the properties that have been verified on the abstract model, are preserved by the refined one (see [Bellegarde et al., 2002] for more details).

# 4. Tool and Example

## 4.1 Tools

We have implemented a tool, using $LASH$ [1] libraries to verify the refinement relation for parameterized systems, and the tool $RMC$ [2] to compute transducers acceleration. Presently our tool generates symbolic reachability graphs, verifies refinement between two symbolic systems and also verifies safety properties for these systems. When the refinement is not verified, our tool finds the origin of the error. For example, it takes 13 seconds to verify automatically the refinement between two levels of refinement of the PID system (see [Dick and Faivre, 1993] for a specification).

We illustrate a computation of refinement on the example of an elevator parameterized by the number of its floors.

## 4.2 Abstract level

At this level, we do not observe the nature of a call (in front of the door or inside the cabin). Moreover, a call can come from anywhere at any time. When stopped at a floor $i$, the cabin moves up when called from a floor $j > i$ above, and it moves down when a call is made at a floor $k < i$ below. The effect of a call from the floor $j$ above (or $k$ below) puts the cabin at the floor $i$ (or $k$). So doing, the elevator might not answer the call at the floor $i$, if any, because of the unknown nature of this call. When the cabin moves, it keeps its direction (up or down) up until the elevator answers all the calls above it. Initially, the cabin is in a stop position at the first floor, and there is no call to answer yet.

Therefore, a floor can be in one of the following states:

- 0 (no call at this floor), or
- $A$ (call at this floor), or
- $L$ (the cabin stopped at the floor), or
- $M$ (the cabin is moving up from the floor), or
- $D$ (the cabin is moving down from the floor).

| *Name* | $C_l$ | $T$ | $C_r$ |
|---|---|---|---|
| *Call* | $\Sigma^*$ | $(O \rightarrow A)$ | $\Sigma^*$ |
| *CallAbove* | $(O+A)^*$ | $(L \rightarrow M)$ | $(O+A)^*A$ $(O+A)^*$ |
| *CallBelow* | $(O+A)^*A$ $(O+A)^*$ | $(L \rightarrow D)$ | $(O+A)^*$ |
| *Up* | $(O+A)^*$ | $((M \rightarrow O)+(M \rightarrow A))$ $id(O+A)^*(A \rightarrow M)$ | $(O+A)^*$ |
| *Down* | $(O+A)^*$ | $(A \rightarrow D)id(O+A)^*$ $((D \rightarrow O)+(D \rightarrow A))$ | $(O+A)^*$ |
| *StopUp* | $(O+A)^*$ | $(M \rightarrow L)$ | $O^*$ |
| *StopDown* | $O^*$ | $(D \rightarrow L)$ | $(O+A)^*$ |

*Figure 8.9.* **Seven abstract actions**

| *Name* | *Acceleration* |
|---|---|
| $Call^+$ | $id(\Sigma)^*(O \rightarrow A)(id(\Sigma)+(O \rightarrow A))^*$ |
| *CallAbove* | $id(O+A)^*(L \rightarrow M)id(O+A)^*id(A)id(O+A)^*$ |
| *CallBelow* | $id(O+A)^*id(A)id(O+A)^*(L \rightarrow D)id(O+A)^*$ |
| $Up^+$ | $id(O+A)^*((M \rightarrow O)+(M \rightarrow A))id(O+A)^*$ $((id(O+A)+(A \rightarrow O))^*$ $(A \rightarrow M)$ |
| $Down^+$ | $id(O+A)^*(A \rightarrow D)$ $(id(O+A)+(A \rightarrow O))^*id(O+A)^*$ $(D \rightarrow O)+(D \rightarrow A))$ |
| *StopUp* | $id(O+A)^*(M \rightarrow L)id(O)^*$ |
| *StopDown* | $id(O)^*(D \rightarrow L)id(O+A)^*$ |

*Figure 8.10.* **Accelerations of the abstract actions**

The elevator abstract symbolic transition system $STS_1$ is $(\Sigma_1, E_{0_1}, Act_1, \rightarrow_1)$ where

$$\Sigma_1 \stackrel{\text{def}}{=} \{O, A, L, M, D\}$$
$$E_{0_1} \stackrel{\text{def}}{=} LO^*$$
$$Act_1 \stackrel{\text{def}}{=} \{Call, CallAbove, CallBelow, Up, Down, StopUp, StopDown\},$$

and $\rightarrow_1$ is a result of the computation of a symbolic reachability graph as shown in Fig. 8.11. The seven actions are described in Fig.8.9.

Before computing the reachable state space, we show the result of computing the acceleration of the actions obtained by using the tool $RMC$. The result is in Fig. 8.10.

The computation of the reachable state space by our tool gives 9 reachable regular configurations. It ends since the two last configurations are included in the union of the already computed configurations. Figure 8.11 shows the reachability graph. Its nodes are labeled by the reachable regular configurations.

| Origin | Regular Configuration Origin | Action | Extremity |
|---|---|---|---|
| $E_0 \stackrel{def}{=}$ | $L O^*$ | $Call^+$ | $E_1$ |
| $E_1 \stackrel{def}{=}$ | $L O^* A (A+O)^*$ | $Call^+$<br>$CallAbove$ | $\subseteq E_1$<br>$E_2$ |
| $E_2 \stackrel{def}{=}$ | $M O^* A (A+O)^*$ | $Call^+$<br>$Up^+$ | $\subseteq E_2$<br>$E_3$ |
| $E_3 \stackrel{def}{=}$ | $(A+O)(A+O)^* M (A+O)^*$ | $Call^+$<br>$Up^+$<br>$StopUp$ | $\subseteq E_3$<br>$\subseteq E_3$<br>$E_4$ |
| $E_4 \stackrel{def}{=}$ | $(A+O)(A+O)^* L O^*$ | $Call^+$<br>$CallBelow$ | $E_5$<br>$E_6$ |
| $E_5 \stackrel{def}{=}$ | $(A (A+O)^* L (A+O)^*)$<br>$+(O O^* A (A+O)^* L (A+O)^*)$<br>$+(OO^* L O^* A (A++)^*)$ | $Call^+$<br>$CallAbove$<br>$CallBelow$ | $\subseteq E_5$<br>$\subseteq E_3$<br>$E_7$ |
| $E_6 \stackrel{def}{=}$ | $O^* A (A+O)^* D O^*$ | $Call^+$<br>$Down^+$ | $\subseteq E_7$<br>$E_8$ |
| $E_7 \stackrel{def}{=}$ | $O^* A (A+O)^* D (A+O)^*$ | $Call^+$<br>$Down^+$ | $\subseteq E_7$<br>$\subseteq E_8$ |
| $E_8 \stackrel{def}{=}$ | $(A+O)^* D (A+O)(A+O)^*$ | $Call^+$<br>$Down^+$<br>$StopDown$ | $\subseteq E_8$<br>$\subseteq E_8$<br>$E_9$ |
| $E_9 \stackrel{def}{=}$ | $O^* L (A+O)(A+O)^*$ | $Call^+$<br>$CallAbove$ | $E_{10}$<br>$E_{11}$ |
| $E_{10} \stackrel{def}{=}$ | $(O^* A (A+O)^* L (A+O)(A+O)^*)$<br>$+(O^* L O^* A (A+O)^*$ | | |
| $E_{11} \stackrel{def}{=}$ | $O^* M O^* A (A+O)^*$ | | |

*Figure 8.11.* Abstract reachability graph

The reachability graph allows us to verify progress properties (see [Manna and Pnueli, 1992]) like:

- if the elevator is moving up and there exist calls from above the cabin then it will keep moving up until there are no calls from above the cabin, and its dual property,
- if the elevator is moving down and if there exists calls from below the cabin then it will keep moving down until there are no calls from below the cabin,
- the cabin will never stay forever in a stop state.

## 4.3 Refined Level

The refinement observes if the calls are coming either from the cabin, or from a floor, or from both. When the cabin is not at the floor $i$, either a call from the cabin or a call from the floor $i$ can be made. So, at the floor $i$, the calls may accumulate by action $ManyCalls$. When the cabin is at the floor $i$ a call from the cabin has no effect. However a call from the floor, included in the call from the floor $i$, is observed. More precisely

- The cabin is seen as answering a call from the cabin at the floor $i$ when it quits the floor $i$.
- However, when the cabin quits the floor $i$, it is not possible to see whether the call from the floor $i$ is answered or not.
- Moreover, when a cabin stops at the floor $i$ where there is a call, this calls remains unanswered.

The elevator must always answer in priority the closest call from the cabin. Notice how this refinement level introduces the next refinement level. At that level it would be important to determine the calls from the floor with an intention to go up or to go down by not always answering a –not yet distinguished– call from a floor when the cabin is moving (up or down).

Therefore, a floor can be in one of the following states:

- 0 (no call, and no cabin at this floor), or
- $A_1$ (call from the cabin, and no cabin at this floor), or
- $A_2$ (call from the floor, and no cabin at this floor), or
- $A_3$ (call from both the floor and the cabin, and no cabin at this floor), or
- $L$ (the cabin is stopped at the floor, and there is no call from the floor), or
- $L_2$ (the cabin is stopped at the floor, and a call at the floor is not answered), or
- $M$ (the cabin is moving up from the floor, and there is no call at this floor), or
- $M_2$ (the cabin is moving up from the floor, and a call at this floor is not answered), or
- $D$ (the cabin is moving down from the floor, and there is no call at this floor), or
- $D_2$ (the cabin is moving down from the floor, and a call at this floor is not answered).

The elevator's refined symbolic transition system $STS_2$ is $(\Sigma_2, E_{0_2}, Act_2, \rightarrow_2)$ where

$$\Sigma_2 \stackrel{\text{def}}{=} \{O, A_1, A_2, A_3, L, L_2, M, D, D_2\}$$
$$E_{0_2} \stackrel{\text{def}}{=} LO^*$$
$$Act_2 \stackrel{\text{def}}{=} \{Call, ManyCalls, CallAbove, CallBelow, Up, Down, StopUp, StopDown\},$$

and $\rightarrow_2$ is the result of a computation of a symbolic reachability graph as shown in Fig. 8.14. The eight actions are presented in Fig. 8.12.

The accelerations of the actions are shown in Fig. 8.13.

Figure 8.14 shows the beginning of the computation of the reachability graph which ends into 252 reachable regular configurations.

## 4.4 Refinement Verification

The transducer $\mathcal{T}_\mu$ corresponding to the relation $\mu$ recognizes the following language:

$$((O,O) + (A_1, A) + (A_2, A) + (A_3, A) + (L, L) + (L_2, L_2) + (M, M) + (M_2, M) + (D, D) + (D_2, D))^*$$

The verification of Clauses 1, 2 and 3 of the refinement has been performed with success in 57 seconds using our tools. However, the verification of Clause 4 fails. Indeed, this verification loops while deciding the absence of cycles of $\tau$-transitions in the reachability graph. Potential cycles are due to the new action $ManyCalls$ which is silent for the verification. For example, consider the regular configuration $E_4 \stackrel{\text{def}}{=} (L\,(O + A_1 + A_2 + A_3)^*\, A_3\, (O + A_1 + A_2 + A_3)^*) + (L_2\, O^*\, (A_1 + A_2 + A_3)\, (O + A_1 + A_2 + A_3)^*)$, target of a transition labeled by $ManyCalls$. The action $ManyCalls$ applied to $E_4$ gives as result the configuration $E'_4 \stackrel{\text{def}}{=} (L\,(O{+}A_1{+}A_2{+}A_3)^*\, A_3\, (O{+}A_1{+}A_2{+}A_3)^* A_3 (O + A_1 + A_2 + A_3)^*) + (L_2\, O^*\, (A_1 + A_2 + A_3)\, (O + A_1 + A_2 + A_3)^*\, A_3\, (O + A_1 + A_2 + A_3))$ included in $E_4$. Therefore, a potential cycle is detected. This kind of potential cycles happens each time the action $ManyCalls$ applies. The detection of the $\tau$-cycle loops like in Example 23. So, we cannot conclude with our tool. However, a good look at the trace of the computation allows seeing that the cause of the loop is the arbitrary number of floors. So, any instance is a refinement.

# 5. Conclusions and Future Works

In this paper, we show that the refinement relation that we have defined using acceleration can be semi-computed. This allows us to verify the refinement relation for an interesting class of parameterized systems, where the parameter is

| *Name* | $C_l$ | $T$ | $C_r$ |
|---|---|---|---|
| *Call* | $\Sigma^*$ | $((O \rightarrow A_1) + (O \rightarrow A_2))$ | $\Sigma^*$ |
| *CallAbove* | $(O + A_1 + A_2 + A_3)^*$ | $((L \rightarrow M) + (L_2 \rightarrow M_2)$<br>$id(O)^* id(A_1 + A_2 + A_3)$ | $(O + A_1 + A_2 + A_3)^*$ |
| *CallBelow* | $O^*(A_1 + A_2 + A_3)$<br>$(O + A_1 + A_2 + A_3)^*$ | $((L \rightarrow D) + (L_2 \rightarrow D_2))$ | $(O + A_1 + A_2 + A_3)^*$ |
| *Up* | $(O + A_1 + A_2 + A_3)^*$ | $((M \rightarrow O) + (M_2 \rightarrow O) + (M_2 \rightarrow A_2))$<br>$(id(A_2 + O)^*((A_3 \rightarrow M) + (A3 \rightarrow M_2)$<br>$+(A1 \rightarrow M) + (A2 \rightarrow M) + (A2 \rightarrow M_2))$ | $(O + A_1 + A_2 + A_3)^*$ |
| *Down* | $(O + A_1 + A_2 + A_3)^*$ | $((A_1 \rightarrow D) + (A_2 \rightarrow D2) + (A_2 \rightarrow D)$<br>$+(A_3 \rightarrow D) + (A_3 \rightarrow D_2))id(A_2 + O)^*$<br>$((D \rightarrow O) + (D_2 \rightarrow O) + (D_2 \rightarrow A_2))$ | $(O + A_1 + A_2 + A_3)^*$ |
| *StopUp* | $(O + A_1 + A_2 + A_3)^*$ | $((M \rightarrow L) + (M_2 \rightarrow D_2))$ | $O^*$ |
| *StopDown* | $O^*$ | $((D \rightarrow L) + (D_2 \rightarrow L_2))$ | $(O + A_1 + A_2 + A_3)^*$ |
| *ManyCalls* | $\Sigma^*$ | $((A_2 \rightarrow A_3) + (A_1 \rightarrow A_3)$<br>$+(L \rightarrow L_2) + (M \rightarrow M_2) + (D \rightarrow D_2))$ | $\Sigma^*$ |

*Figure 8.12.* **Eight refined actions**

| Name | Accelerations |
|---|---|
| $Call^+$ | $id(\Sigma)^* ((M \rightarrow M_2) + (L \rightarrow L_2) + (D \rightarrow D_2) + (O \rightarrow A_2) + (O \rightarrow A_3) + (A_2 \rightarrow A_3) + (O \rightarrow A) + (A \rightarrow A_3)) (id(\Sigma) + (D \rightarrow D_2) + (O \rightarrow A_2) + (O \rightarrow A_3) + (A_2 \rightarrow A_3) + (O \rightarrow A) + (A \rightarrow A_3))*$ |
| $CallAbove$ | $id(O + A_3 + A_2 + A)^*((L \rightarrow M) + (L_2 \rightarrow M_2))\, id(O)^*.id(A_3 + A_2 + A)$ $id((O + A_3 + A_2 + A)^*$ |
| $CallBelow$ | $id(O)^*.id(A_3 + A_2 + A)id((O + A_3 + A_2 + A)^* ((L \rightarrow D) + (L_2 \rightarrow D_2))$ $id(O + A_3 + A_2 + A)^*$ |
| $Up^+$ | $id(O + A_3 + A_2 + A)^* ((M_2 \rightarrow O) + (M_2 \rightarrow A_2)(M \rightarrow O))\, id(O + A_2)$ $((A_3 \rightarrow O) + (A_3 \rightarrow A_2) + (A_2 \rightarrow O) + (A_2 \rightarrow A_2) + (A \rightarrow O))\, id(O + A_2)^*$ $((A_3 \rightarrow M_2) + (A_2 \rightarrow M_2) + (A_3 \rightarrow M) + (A_2 \rightarrow M) + (A \rightarrow M))$ $id(O + A_3 + A_2 + A)^*$ |
| $Down^+$ | $id(A_3 + A + O + A_2)^* ((A_3 \rightarrow D_2) + (A_3 \rightarrow D) + (A \rightarrow D) + (A_2 \rightarrow D_2) + (A_2 \rightarrow D)) ((A_3 \rightarrow O) + (A \rightarrow O) + (O \rightarrow O) + (A_3 \rightarrow A_2) + (A_2 \rightarrow O) + (A_2 \rightarrow A_2))^* ((D_2 \rightarrow O) + (D \rightarrow O) + (D_2 \rightarrow A_2))\, id(A_3 + A + O + A_2)^*$ |
| $StopUp$ | $id(O + A_1 + A_2 + A_3)^* ((M \rightarrow L) + (M_2 \rightarrow L_2))id(O)^*$ |
| $StopDown$ | $id(O)^* ((D_2 \rightarrow A_2) + (D \rightarrow L))\, id(O + A_1 + A_2 + A_3)^*$ |
| $ManyCalls$ | $id(\Sigma^*)$ $((A_2 \rightarrow A_3) + (A_1 \rightarrow A_3) + (L \rightarrow L_2) + (M \rightarrow M_2) + (D \rightarrow D_2))$ $(id(\Sigma) + (A_1 \rightarrow A_3) + (A_2 \rightarrow A_3) + (L \rightarrow L_2) + (M \rightarrow M_2) + (D \rightarrow D_2))^*$ |

*Figure 8.13.* Acceleration of the refined actions

| Ori. | Regular Configuration Origin | Action | Extr. |
|---|---|---|---|
| $E_0$ | $L\,O^*$ | $Call^+$<br>$ManyCalls^+$ | $E_1$<br>$E_2$ |
| $E_1$ | $L\,O^*\,(A_1+A_2)(A_1+A_2+O)^*$ | $Call^+$<br>$CallAbove$<br>$ManyCalls^+$ | $\subseteq E_1$<br>$E_3$<br>$E_4$ |
| $E_2$ | $L_2\,O^*$ | $Call^+$ | $\subseteq E_4$ |
| $E_3$ | $M\,O^*(A_1+A_2)(A_2+A_2+O)^*$ | $Call^+$<br>$Up^+$<br>$ManyCalls^+$ | $\subseteq E_3$<br>$E_5$<br>$E_6$ |
| $E_4$ | $(L_2\,O^*(A_1+A_2+A_3+)(A_1+A_2+A_3+O)^*)$<br>$+(L\,(A_1+A_2+O)^*\,A_3\,(A_1+A_2+A_3+O)^*$ | $Call^+$<br>$CallAbove$<br>$ManyCalls^+$ | $\subseteq E_4$<br>$\subseteq E_6$<br>$\subseteq E_4$ |
| $E_5$ | $O\,(A_2+O)(A_2+O)^*\,(A_2+O)(M+M_2)$<br>$(A_1+A_2+O)^*$ | $Call^+$<br>$Up^+$<br>$StopUp$<br>$ManyCalls^+$ | $E_7$<br>$\subseteq E_5$<br>$E_8$<br>$E_9$ |
| $E_6$ | $(M_2\,O^*(A_1+A_2+A_3+)(A_1+A_2+A_3+O)^*)+$<br>$(M\,(A_1+A_2+O)^*\,A_3\,(A_1+A_2+A_3+O)^*$ | $Call^+$<br>$Up^+$<br>$ManyCalls^+$ | $\subseteq E_6$<br>$E_{10}$<br>$\subseteq E_6$ |
| $E_7$ | $(A+A_2)(A_1+A_2+O)(A_1+A_2+O)^*$<br>$(A_1+A_2+O)(M+M_2)(A_1+A_2+O)^*$ | $Call^+$<br>$Up^+$<br>$StopUp$<br>$ManyCalls^+$ | $\subseteq E_7$<br>$E_{11}$<br>$E_{12}$<br>$E_{13}$ |
| $E_8$ | $O\,(A_2+O)(A_2+O)^*\,(A_2+O)(L_2+L)\,O^*$ | $Call^+$<br>$CallBelow$<br>$ManyCalls^+$ | $E_{14}$<br>$E_{15}$<br>$E_{16}$ |
| $E_9$ | $OA_3(M+M_2)+(A_3A_2+A_3+O)(A_2+A_3+O)$<br>$(A_2+A_3+O)^*(A_2+A_3+O)(M+M_2)$<br>$(A_1+A_2+A_3+O)^*$ | $Call^+$<br>$Up^+$<br>$StopUp$<br>$ManyCalls^+$ | $\subseteq E_{13}$<br>$E_{17}$<br>$\subseteq E_{16}$<br>$\subseteq E_9$ |
| $E_{10}$ | $(A_2+O)(A_2+O)(A_2+O)^*(A_2+O)$<br>$(M+M_2)(A_1+A_2+A_3+O)^*$ | $Call^+$<br>$Up^+$<br>$StopUp$<br>$ManyCalls^+$ | $E_{18}$<br>$\subseteq E_{10}$<br>$E_{19}$<br>$E_{20}$ |
| ... | ... | ... | |

*Figure 8.14.* Refined reachability graph

an arbitrary number of similar processes. The refinement relation is computed on the symbolic models of the systems. These symbolic models must also be computed. For that, we represent states and actions of the system by regular languages and transducers and compute the accelerations of the actions (it is possible when the actions satisfy the conditions given in [Abdulla and Jonsson, 1999, Abdulla et al., 1999]). We generate a reachability graph where the states are regular languages and transitions are accelerated actions by using an effective construction of the set of reachable states. In this paper, we also provide semi-algorithms

- to compute the reachability graph of the system, and
- to detect the lack of new $\tau$-cycles.

The reachability graph is used to verify the refinement clauses. Although the detection of the $\tau$-cycles cannot be done in many cases with the current version of the semi-algorithms presented in this paper (see Example 23 and Section 4), we are looking for a well-ordering to replace the equality allowing us to conclude more often.

We must complete our study by adding the verification of $PLTL$ properties. Then, we will be in measure to evaluate the contribution of our verification approach for parameterized systems.

## Notes

1. For the Liège Automata-based Symbolic Handler (LASH), see the url *http://www.montefiore.ulg.ac.be/boigelot/research/lash/*.
2. For the Regular Model Checking (RMC), see the url *http://www.regularmodelchecking.com/*

## References

[Abdulla and Jonsson, 1999] Abdulla, P. and Jonsson, B. (1999). On the existence of network invariants for verifying parametrized systems. In Olderog and Steffen, editors, *Correct System Design – Recent Insights and Advances*, volume 1710 of *LNCS*, pages 180–197. Springer-Verlag.

[Abdulla et al., 1999] Abdulla, P. A., Bouajjani, A., Jonsson, B., and Nilsson, M. (1999). Handling global conditions in parametrized system verification. In *Proc. 11th Int. Conf. on Computer Aided Verification CAV'99*, volume 1633 of *LNCS*, pages 134–145. Springer Verlag.

[Abrial, 1996] Abrial, J.-R. (1996). *The B Book.* Cambridge University Press - ISBN 0521-496195.

[Annichini et al., 2000] Annichini, A., Asarin, E., and Bouajjani, A. (2000). Symbolic techniques for parametric reasoning. In *Proc. Int. Conf. on Computer Aided Verification, CAV'00, volume 1855 of Lecture Notes in Computer Science*, pages 419–434. Springer-Verlag.

[Apt and Kozen, 1986] Apt, K. R. and Kozen, D. C. (1986). Limits for automatic verification of finite-state concurrent systems. *Information Processing Letters*, 22(6):307–309.

[Arons et al., 2001] Arons, T., Pnueli, A., Ruah, S., Xu, Y., and Zuck, L. (2001). Parameterized verification with automatically computed inductive assertions. In Berry, G., H.Comon, and Finkel, A., editors, *Proc. Int. Conf.*

*CAV'01, volume 2102 of Lecture Notes in Computer Science*, pages 221–234. Springer Verlag.

[Bellegarde et al., 2002] Bellegarde, F., Charlet, C., and Kouchnarenko, O. (2002). Raffiner pour vérifier des systèmes paramétrés. *Technique et Science Informatiques*, 8(21):1121–1149.

[Bellegarde et al., 2003] Bellegarde, F., Charlet, C., and Kouchnarenko, O. (2003). Using acceleration to compute parameterized system refinement. Technical report, INRIA RR 4716.

[Bellegarde et al., 2000] Bellegarde, F., Julliand, J., and Kouchnarenko, O. (2000). Ready-simulation is not ready to express a modular refinement relation. In *Proc. Int. Conf. on Fondamental Aspects of Software Engineering, FASE'2000*, volume 1783 of *LNCS*, pages 266–283. Springer-Verlag.

[Bouajjani et al., 2000] Bouajjani, A., Jonsson, B., Nilsson, M., and Touili, T. (2000). Regular model checking. In Emerson, E. A. and Sistla, A. P., editors, *Proc. Int. Conf. on Computer Aided Verification, CAV'00, volume 1855 of Lecture Notes in Computer Science*, pages 403–418. Springer-Verlag.

[Butler, 2000] Butler, M. J. (2000). csp2B: A practical approach to combining CSP and B. *Formal Aspects of Computing*, 12:182–198.

[Clarke et al., 1995] Clarke, E. M., Grumberg, O., and Jha, S. (1995). Verifying parameterized networks using abstraction and regular languages. In *Proc. of the 6th International Conference on Concurrency Theory (CONCUR'95), Philadelphia, PA*, pages 395–407.

[Comon and Jurski, 1998] Comon, H. and Jurski, Y. (1998). Multiple counters automata, safety analysis and presburger arithmetic. In *Proc. Int. Conf. CAV'98, volume 1427 of LNCS*, Springer-Verlag.

[Dick and Faivre, 1993] Dick, J. and Faivre, A. (1993). Automating the generation and sequencing of test cases from model-based specifications. *FME'93: Industrial-Strength Formal Methods*, LNCS 670 Springer-Verlag:268–284.

[Emerson and Kahlon, 2000] Emerson, E. and Kahlon, V. (2000). Reducing model checking of the many to the few. In *In 17th International Conference on Automated Deduction (CADE-17)*, volume 14, pages 236–255.

[Emerson and Namjoshi, 1995] Emerson, E. A. and Namjoshi, K. S. (1995). Reasoning about rings. In *Proc. 22nd ACM SIGPLAN-SIGACT Int. Symp. on Principles of Programming Languages (POPL'95)*, pages 85–94, San Francisco, California. ACM Press.

[Emerson and Namjoshi, 1996] Emerson, E. A. and Namjoshi, K. S. (1996). Automatic verification of parameterized synchronous systems. In *Proc. Int. Conf. CAV'98*, volume 1102 of *LNCS*.

[Jonsson and Nilsson, 2000] Jonsson, B. and Nilsson, M. (2000). Transitive closures of regular relations for verifying infinite-state systems. In *Proc.*

*6th Int. Conf. on Tools and Algorithms for the Construction and Analysis of Systems (TACAS'2000)*, LNCS, pages 220–234. Springer-Verlag.

[Kesten et al., 1997] Kesten, Y., Maler, O., Marcus, M., Pnueli, A., and Shahar, E. (1997). Symbolic model checking with rich assertional languages. In *In O. Grumberg, editor, Proceedings of the 9th International Conference on Computer Aided Verification (CAV 1997)*, LNCS, pages 424–435. Springer Verlag.

[Lamport, 1999] Lamport, Leslie (1999). Specifying concurrent systems with tla +. In *Calculational System Design*, pages 183–247, Amsterdam. In Manfred Broy and Ralf Steinbruggen editors.

[Maidl, 2001] Maidl, M. (2001). A unifying model checking approach for safety properties of parameterized systems. In Berry, G., H.Comon, and Finkel, A., editors, *Proc. Int. Conf. CAV'01, volume 2102 of Lecture Notes in Computer Science*, pages 311–323. Springer Verlag.

[Manna and Pnueli, 1992] Manna, Z. and Pnueli, A. (1992). *The Temporal Logic of Reactive and Concurrent Systems: Specification.* Springer-Verlag - ISBN 0-387-97664-7.

[Pnueli and Shahar, 2000] Pnueli, A. and Shahar, E. (2000). Liveness and acceleration in parametrized verification. In *Computer Aided Verification, CAV'00*, volume 1855 of *LNCS*, pages 328–343.

[Sistla and Gyuris, 1999] Sistla, A. P. and Gyuris, V. (1999). Parameterized verification of linear networks using automata as invariants. *Formal Aspects of Computing*, 11(4):402–425.

III

# TYPE THEORETIC MODELS AND METHODS FOR SYSTEM DESIGN

# Chapter 9

# BEHAVIORAL TYPE INFERENCE

## *Part I – Algebraic theory for behavioral type inference*

Jean-Pierre Talpin and Paul Le Guernic

*INRIA, project Espresso*
*IRISA, Campus de Beaulieu*
*F-35042 Rennes Cedex, France*
{Jean-Pierre.Talpin,Paul.LeGuernic}@irisa.fr

**Abstract** The design productivity gap has been recognized as one of the major threats to the continued growth of embedded systems design productivity. Ad-hoc system-level design methodologies, that lift modeling to higher levels of abstraction, and the concept of intellectual property (IP), that promotes reuse of existing components, are essential steps to manage design complexity. However, the issue of compositional correctness arises with these steps. Given components from different manufacturers, designed with heterogeneous models, at different levels of abstraction, assembling them in a correct-by-construction manner is a difficult challenge. We address it by proposing a process algebraic model to support system design with a formal model of computation that serves as a behavioral type system to capture the behavior of system components at the interface level. The proposed algebra is conceptually minimal, equipped with a formal semantics defined in a synchronous model of computation, and supports a scalable notion and a flexible degree of abstraction. We demonstrate its benefits by considering the type-based synthesis of latency-insensitive protocols. We show that the synthesis of component wrappers can be optimized by the behavioral information carried by interface type descriptions and yield minimized stalls and maximized throughput.

## Introduction

The design productivity gap has been recognized by the semiconductor industry as one of the major threats to the continued growth of embedded systems design productivity. System level design methodologies that lift design methods at higher-level of abstraction are essential to manage design complexity.



*R. Gupta et al. (eds.), Formal Methods and Models for System Design*, 213-243.

A number of advances in high-level modeling and validation have been proposed over the past decade in the aim of improving the level of abstraction in system design. Most of these enable greater reuse of existing intellectual property (IP) blocks.

Design correctness is an important step in this move. Given the complexity of system-level designs, it is important that the composition of system-level IP blocks be guaranteed correct. However, *a posteriori* validation of component compositions is a difficult problem. Techniques are needed that ensure design correctness as part of the design process itself. To address this issue, methodological precepts have been developed that separately focus on design reuse and correctness by construction.

Both reuse and elevation of abstraction critically depend on guaranteed design correctness. To improve the state of the art in component composition from existing component libraries, we specifically seek to address the following issue: given a high level architectural description (e.g., a virtual architecture) and a library of implemented components, i.e. the elements of an execution platform, how can one automate the selection of virtual components from the library, and automatically ensure composability ?

Our approach is based on a high-level modeling and specification methodology that ensures compositional correctness through an algebra capturing behavioral aspects of component interfaces. It is presented via a minimalist algebra, called the iSTS (implicit and synchronous transition systems), that is akin to Pnueli's synchronous transition systems (STS, [19]) and Dijkstra's guarded commands [9]. This minimalist formalism allows to specify the state transitions, synchronization relations and scheduling constraints implied by a given system, in the presence of multi-clocked synchrony (i.e. polychrony).

Our approach consists of using this formalism as a type system to describe the behavior of system components and allow for global model transformations to be performed on the system based on behavioral type information. This approach builds upon previous work on scalable design exploration using refinement/abstraction-based design methodologies [21], implemented in the Polychrony workbench [23]. It is both conceptually minimal and equipped with a formal semantics defined in a multi-clocked synchronous model of computation [15].

After a detailed and informal exposition of the iSTS and of its model of computation, sections 1 and 2, we exemplify its use as a behavioral type system in system design by the definition of model transformations, section 9.6, and formal methodologies, section 3, applied to the characterization of a formal design methodology for protocol synthesis based on a relation of finite flow equivalence, section 4.6.

# 1. Rationale

We start with a detailed and informal outline of an algebraic formalism which we call the iSTS (implicit synchronous transition systems). The key notions put to work in this notation are essentially borrowed to Pnueli's STS [19] and Dijkstra's guarded command language [9]. In the iSTS, a process consists of simultaneous propositions that manipulate signals. A signal is an infinite flow of values that is sampled by a discrete series of instants or reactions. An event corresponds to the value carried by a signal during a particular reaction or instant. The main features of the iSTS notation are put together in the example of figure 9.1, that describes the behavior of a counter modulo 2, noted $P$ ($\stackrel{\text{def}}{=}$ means "is defined by"), through a set of simultaneous *propositions*, labeled from (1) to (3).

$$P \stackrel{\text{def}}{=} \begin{array}{lrcll} & & & \neg s^0 & (1) \\ | & \hat{x} & \Rightarrow & s' = \neg s & (2) \\ | & s & \Rightarrow & \hat{y} & (3) \end{array}$$

*Figure 9.1.* Specification of a counter modulo 2

- The proposition (1) is an invariant. It says that the initial value of the signal $s$, denoted by $s^0$, is false. This is specified by the proposition $\neg s^0$ (equivalent to $s^0 = 0$).
- The proposition (2) is a guarded command. It says that if the signal $x$ is present during a given reaction then the value of $s$ is toggled.
  - The leftmost part of the sentence, the proposition $\hat{x}$, is a condition or a guard. It denotes the *clock* of $x$. It is true iff the signal $x$ is present during the current reaction.
  - The rightmost part of the sentence, the proposition $s' = \neg s$, is a transition. The term $s'$ refers to the next value of the signal $s$. The proposition $s' = \neg s$ says that the next value of $s$ is the negation of the current value of $s$.
- The proposition (3) is another guarded command. It says that if $s$ is true then $y$ is present.

Notice that, in proposition (3), the guard expects the signal $s$ to hold the value true but that its action does not actually specify the value of the signal $y$: it simply requires it to be present. Proposition (3) is hence an *abstraction*: a proposition that partially describes the properties of the system under consideration without implying a particular implementation.

To implement a *function* or a *system*, this proposition needs to be compositionally *refined* by another, saying which value $y$ should hold when present. To this end, one may for instance compose the counter $P$ of figure 9.1 with the proposition $Q$:

$$Q \stackrel{\text{def}}{=} (y' = y + 1 \,|\, y^0 = 0)$$

Notice that the iteration specified by the proposition $Q$ is not guarded. This means that it is an invariant which describes the successive values of the signal $y$ in time but not the particular time samples at which the signal $y$ should occur. The composition of $Q$ with $P$ has the effect of synchronizing the signal $y$ in $Q$ to the *clock* $s$ in $P$. This composition is called a refinement: the system obeys the specification denoted by the initial proposition $P$ and is constrained to further satisfy the additional requirements of $Q$.

$$P \text{ is an abstraction of } P\,|\,Q \qquad P\,|\,Q \text{ is a refinement of } P$$

The notions introduced so far hold necessary and sufficient ingredients to specify the behavior of multi-clocked synchronous systems. Pnueli's original STS notation features two additional notions which, in retrospect, are essentially geared towards verification by model-checking.

- One feature is choice $P \vee Q$. For instance, the STS $a = 1 \vee b = 1$ allows to non-deterministically have $a$ or $b$ present with the value true at all time. Non-determinism can equally be modelled using guarded commands and a partially defined signal $s$ whose scope is lexically restricted to the desired proposition; In the iSTS, we write:

  $$R \stackrel{\text{def}}{=} (s \Rightarrow a \,|\, \neg s \Rightarrow b)/s$$

  to mean the non-deterministic proposition of choosing $a$ or $b$ upon the value of an internal signal $s$, whose calculation is left unspecified. Here, the notation $P/s$ means that the scope of $s$ is local to the process $P$. Again, notice that the proposition $R$ may best be understood as the *abstraction* of an executable specification, e.g., one that specifies the actual value of $s$ in time (for instance, the toggle $s$ of the counting process $P$).

  $$R \text{ is an abstraction of } \left(s \Rightarrow a \,|\, \neg s \Rightarrow b \,|\, \left(\neg s^0 \,|\, s' = \neg s\right)\right)/s$$

- Another feature of the STS is explicit absence $\bot$. The proposition $x = \bot$ explicitly specifies that $x$ does not hold a value in the context in which it is considered. For instance, the STS:

  $$(a = 1 \wedge x \neq \bot) \vee (x = \bot \wedge b = 1)$$

means that either $a$ is true and $x$ is present or that $b$ is true and $x$ is absent. In the iSTS, this notion is implicit. It can for instance be specified by a refinement of $R$ with the invariant "$x$ is present iff $b$ is true":

$$R \,|\, b = \hat{x}$$

In the aim of moving back and forth from abstraction to refinement, one last essential feature of the iSTS is the notion of scheduling specification. A scheduling specification is designed to imply an order of execution to otherwise purely logical propositions. Whereas a transition, e.g. $s' = \neg s$, implicitly means that the next value of $s$ is computed using the present value of $s$, a proposition, e.g. $y = x$, just means that $x$ and $y$ are equal. It does not specify any order of execution. An order of execution can be imposed to this proposition by its refinement with a scheduling constraint, noted $y \rightarrow x$. Then,

$$x = y \text{ is an abstraction of } x = y \,|\, y \rightarrow x$$

where $y \rightarrow x$ informally means that "$x$ cannot happen before $y$". This additional requirement refines the time scale, from one in which $x$ and $y$ happen simultaneously, to a more precise one in which one observes that $x$ cannot happen before $y$. We will adopt the following syntax, borrowed to the Signal language, and write $x := y$ for an assignment of $y$ to $x$:

$$(x = y \,|\, y \rightarrow x) \text{ is an abstraction of } x := y$$

**Formal syntax.** The formal syntax of the iSTS is defined by the inductive grammar $P$ in table 9.1. A process $P$ manipulates boolean values noted $v \in \{0, 1\}$ and signals noted $x, y, z$. A location $l$ refers to the initial value $x^0$, the present value $x$ and the next value $x'$ of a signal. A reference $r$ stands for either a value $v$ or a signal $x$.

A clock expression $e$ is a proposition on boolean values. When true, a clock $e$ defines a particular period in time. The clocks 0 and 1 denote events that never/always happen. The clock $x = r$ denotes the proposition: "$x$ is present and holds the value $r$". Particular instances are:

- the clock $\hat{x} \stackrel{\text{def}}{=} (x = x)$ which means that "$x$ is present",
- the clock $x \stackrel{\text{def}}{=} (x = 1)$ which means that "$x$ is true",
- the clock $\neg x \stackrel{\text{def}}{=} (x = 0)$ which means that "$x$ is false".

Clocks are propositions combined using the logical combinators of conjunction $e \wedge f$, to mean that both $e$ and $f$ hold, disjunction $e \vee f$, to mean that either $e$ or $f$ holds, and symmetric difference $e \setminus f$, to mean that $e$ holds and not $f$.

A process $P$ consists of the simultaneous composition of elementary propositions. 1 is the process that does nothing. The proposition $l = r$ means that "$l$ holds the value $r$". In the introductory example, we wrote $x' = \neg x$ to mean the proposition $x' = \neg x =^{\text{def}} x = 1 \Rightarrow x' = 0 \,|\, x = 0 \Rightarrow x' = 1$.

The proposition $x \to l$ means that "$l$ cannot happen before $x$". The process $e \Rightarrow P$ is a guarded command. It means: "if $e$ is present then $P$ holds". Processes are combined using synchronous composition $P \,|\, Q$ to denote the simultaneity of the propositions $P$ and $Q$. Restricting a signal name $x$ to the lexical scope of a process $P$ is written $P/x$. We refer to the *free variables* vars$(P)$ of a process $P$, table 9.A.1, as the set of signal names that occur free in the lexical scope of $P$.

*Table 9.1.* Formal syntax of iSTS algebra

$$\begin{array}{rr}
\text{(reference)} & r ::= x \,|\, v \\
\text{(location)} & l ::= x^0 \,|\, x \,|\, x' \\
\text{(clock)} & e, f ::= 0 \,|\, x = r \,|\, e \wedge f \,|\, e \vee f \,|\, e \setminus f \,|\, 1 \\
\text{(process)} & P, Q ::= 1 \,|\, l = r \,|\, x \to l \,|\, e \Rightarrow P \,|\, (P \,|\, Q) \,|\, P/x
\end{array}$$

**Notational conventions.** In the formal presentation of the iSTS, we restrict ourself to a subset of the elementary propositions in the grammar of table 9.1, which we call atoms $a$:

$$\text{(atoms)} \quad a, b ::= x^0 = v \,|\, l = y \,|\, x \to l \; s.t. \, l ::= x \,|\, x'$$

Other propositions as well as additional syntactic shortcuts, used in the examples, can be defined by using this restricted subset as follows.

$$\begin{array}{ll}
l = v \stackrel{\text{def}}{=} (l = x \,|\, x^0 = v \,|\, x' = x)/x \text{ iff } x \neq l \neq x' & \hat{x} \stackrel{\text{def}}{=} (x = x) \\
l := x \stackrel{\text{def}}{=} (l = x \,|\, x \to l) & l \stackrel{\text{def}}{=} (l = 1) \\
\hat{x} = \hat{y} \stackrel{\text{def}}{=} (\hat{x} \Rightarrow \hat{y} \,|\, \hat{y} \Rightarrow \hat{x}) & \neg l \stackrel{\text{def}}{=} (l = 0)
\end{array}$$

## 2. A polychronous model of computation

To deal with the apparent heterogeneity of synchrony and asynchrony in GALS architectures, designers usually consider stratified models, such as CSP (communicating sequential processes) or Kahn networks (communicating dataflow functions).

By contrast, polychrony (multi-clocked synchrony) establishes a continuum from synchrony to asynchrony: modeling, design, transformation, verification issues are captured within the same model and hence independently of spatial and temporal considerations implied by a synchronous (local) or to an asynchronous (global) viewpoint. The definition of uniform methodologies for the formal design of GALS architectures has been the subject of recent and detailed studies in [15] and [21].

In this section, we cast polychrony in the context of the iSTS algebra to further explore abstraction and refinement relations between system-level models in a way geared towards the aim of compositional system design.

## 2.1 Domain of polychrony

The polychronous model of computation, proposed in [15], consists of a unique *domain* of traces, that does not differentiate synchrony from asynchrony, and semi-lattice structures, that render synchrony and asynchrony using specific timing equivalence relations.

We consider a partially-ordered set $(\mathcal{T}, \leq, 0)$ of tags. A tag $t \in \mathcal{T}$ denotes a symbolic instant or a eriod in time. We note $C \in \mathcal{C}$ a *chain* of $\mathcal{T}$. Events, signals, behaviors and processes are defined starting from tags as follows:

DEFINITION 9.1 (POLYCHRONY)

- *An* event $e \in \mathcal{E} = \mathcal{T} \times \mathcal{V}$ *is the pair of a value and a tag.*
- *A* signal $s \in \mathcal{S} = C \rightarrow \mathcal{V}$ *is a function from a* chain *to a set of values.*
- *A* behavior $b \in \mathcal{B}$ *is a function from names* $x \in \mathcal{X}$ *to signals* $s \in \mathcal{S}$.
- *A* process $p \in \mathcal{P}$ *is a set of behaviors that have the same domain.*

Figure 9.2 depicts a behavior $b$ in the polychronous domain $\mathcal{P}$.

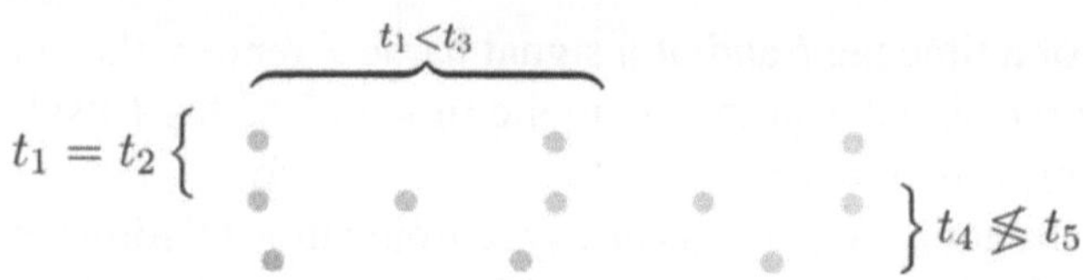


*Figure 9.2.* A behavior in the polychronous model of computation

Tags $t_1$ and $t_2$ (top and middle green signals) are equal, meaning that the events they time are synchronous. Tag $t_1$ precedes $t_3$, written $t_1 < t_3$, to mean the scheduling relation that causally relates them in time. The blue signal at the bottom has no tag comparable to either of the other green ones, e.g. $t_4 \lessgtr\!\!\!\!/\ t_5$. It denotes a signal belonging to a different clock domain.

**Notations.** In the remainder, we write $\text{tags}(s)$ and $\text{tags}(b) = \cup_{x \in \text{vars}(b)} \text{tags}(b(x))$ for the tags of a signal $s$ and of a behavior $b$, $b|_X$ for the projection of a behavior $b$ on $X \subset \mathcal{X}$ and $b/X = b|_{\text{vars}(b) \setminus X}$ for its complementary, $\text{vars}(b)$ and $\text{vars}(p)$ for the domains of $b$ and $p$.

**Synchronous composition.** $p \,|\, q$ is defined by the union of all behaviors $b$ (from $p$) and $c$ (from $q$) which are synchronous: all signals along the interface

$I = \text{vars}(p) \cap \text{vars}(q)$ between $p$ and $q$ carry the same values at the same time tags.

$$p|q = \{b \cup c \mid (b,c) \in p \times q, I = \text{vars}(p) \cap \text{vars}(q), b|_I = c|_I\}$$

## 2.2 Scheduling structure of polychrony

To render the scheduling relations between events occurring at the same time tag $t$, we refine the domain of polychrony with scheduling relations. A scheduling $t_x \to t'_y$ means that the event along the signal named $y$ at $t'$ may not happen before $x$ at $t$.

Figure 9.3 depicts three scheduling relations superimposed to the signals $x$ and $y$ of figure 9.2. The scheduling relation $t_x \to t_y$ denotes the observation that the event occurring along $x$ at $t$ precedes the event along $y$.

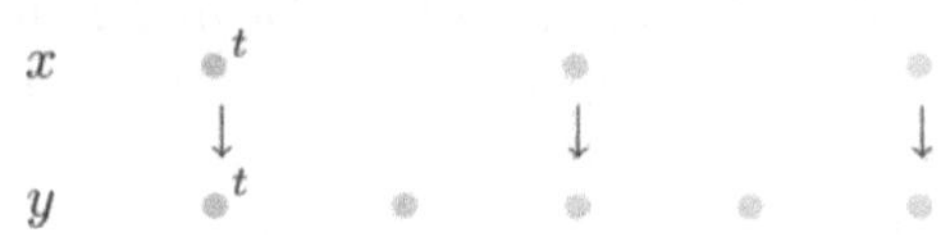


*Figure 9.3.* Scheduling relations between simultaneous events

The pair $t_x$ of a time tag $t$ and of a signal name $x$ renders the very date $d$ of an event along the signal $x$ at the symbolic time $t$. The tag $t$ itself represents the period during which multiple events take place to form a reaction: the tag $t$ corresponds is the equivalence class of a synchronization relation between dates $d$, as in the synchronous structures [18]. The domain of dates $\mathcal{D} = \mathcal{T} \times \mathcal{X}$ of a given behavior $b$ is subject to a pre-order relation $\to^b$ that denotes scheduling and contains causality $<$. When no ambiguity is possible on the identity of $b$ in a scheduling constraint $x \to^b y$, we write it $x \to y$.

$$\begin{aligned} \forall b \in \mathcal{B}, \forall x \in \text{vars}(b), \forall t, t' \in \text{tags}(b(x)),\ & t < t' \Rightarrow t_x \to^b t'_x \\ & t_x \to^b t'_x \Rightarrow \neg(t' < t) \end{aligned}$$

## 2.3 Synchronous structure of polychrony

In the previous section, we gave a structural definition of a domain of traces for capturing the possible behavior of processes in the iSTS algebra. Building upon this domain, we define the semi-lattice structure which relationally denotes synchronous behaviors in this domain.

The intuition behind this relation is depicted figure 9.4. It is to consider a signal as an elastic with ordered marks on it (tags). If the elastic is stretched, marks remain in the same relative and partial order but have more space (time) between each other.

The same holds for a set of elastics: a behavior. If elastics are equally stretched, the order between marks is unchanged. In the figure 9.4, the time scale of $x$ and $y$ change but the partial timing and scheduling relations are preserved. Stretching is a partial-order relation which defines clock equivalence (definition 9.2).

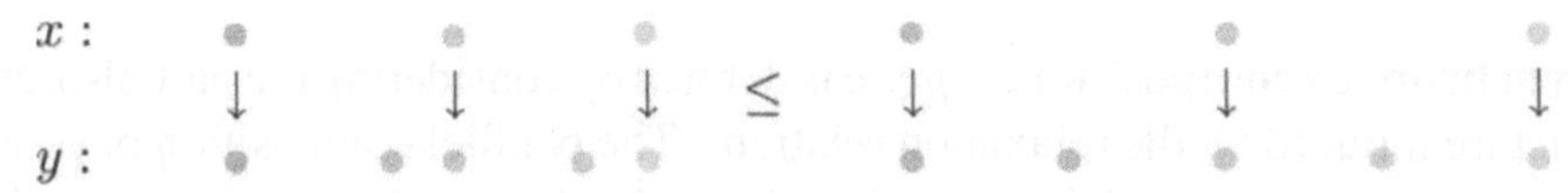


*Figure 9.4.* Relating synchronous behaviors by stretching.

DEFINITION 9.2 (CLOCK EQUIVALENCE)
*A behavior c is a* stretching *of b, written* $b \leq c$*, iff* $\text{vars}(b) = \text{vars}(c)$ *and there exists a bijection f on T which satisfies*

$$\left| \begin{array}{l} \forall t, t' \in \text{tags}(b), t \leq f(t) \wedge (t < t' \Leftrightarrow f(t) < f(t')) \\ \forall x, y \in \text{vars}(b), \forall t \in \text{tags}(b(x)), \forall t' \in \text{tags}(b(y)), t_x \rightarrow^b t'_y \Leftrightarrow f(t)_x \rightarrow^c f(t')_y \\ \forall x \in \text{vars}(b), \text{tags}(c(x)) = f(\text{tags}(b(x))) \wedge \forall t \in \text{tags}(b(x)), b(x)(t) = c(x)(f(t)) \end{array} \right.$$

*b and c are* clock-equivalent*, written* $b \sim c$*, iff there exists a behavior d s.t.* $d \leq b$ *and* $d \leq c$*.*

## 2.4 Asynchronous structure of polychrony

The asynchronous structure of polychrony is modeled by weakening the clock-equivalence relation to allow for comparing behaviors w.r.t. the sequences of values signals hold regardless of the time at which they hold these values. The *relaxation* relation allows to individually stretch the signals of a behavior in a way preserving scheduling constraints. Relaxation is a partial-order relation that defines the flow-equivalence relation: two behaviors are flow-equivalent iff their signals hold the same values in the same order.

Figure 9.5 depicts two asynchronously equivalent behaviors related by relaxation. The first event along $x$ has been shifted as the effect of delaying its transmission using e.g. a FIFO buffer.

*Figure 9.5.* Relating asynchronous behaviors by relaxation.

Relaxation is a partial-order relation which defines flow-equivalence (definition 9.3).

DEFINITION 9.3 (FLOW EQUIVALENCE)
*A behavior c is a* relaxation *of b, written* $b \sqsubseteq c$, *iff* $\mathrm{vars}(b) = \mathrm{vars}(c)$ *and, for all* $x \in \mathrm{vars}(b)$, $b_{|x} \leq c_{|x}$. *the behaviors b and c are* flow-equivalent, *written* $b \approx c$, *iff there exists a behavior d s.t.* $d \sqsubseteq b$ *and* $d \sqsubseteq c$.

**Asynchronous composition.** $p \parallel q$ is defined by considering the partial-order structure induced by the relaxation relation . The parallel composition of $p$ and $q$ consists of behaviors $d$ that are relaxations of behaviors $b$ and $c$ from $p$ and $q$ along shared signals $I = \mathrm{vars}(p) \cap \mathrm{vars}(q)$ and that are stretching of $b$ and $c$ along the independent signals of $p$ and $q$.

$$p \parallel q = \left\{ d \in \mathcal{B}|_{\mathrm{vars}(p) \cup \mathrm{vars}(q)} \;\middle|\; \exists (b, c) \in p \times q, \begin{array}{c} d/I \geq b/I \wedge d/I \geq c/I \\ \wedge\, b|_I \sqsubseteq d|_I \;\; \wedge\, d|_I \sqsupseteq c|_I \end{array} \right\}$$

## 2.5 From synchronous to asynchronous structures

The original definition of the polychronous model of computation [15] is designed to render the synchronous hypothesis as implemented in the multi-clocked data-flow notation Signal and relate it to asynchronous architectures using communication with unbounded delay. In an embedded architecture, however, the flow of a signal usually slides from another as the result of introducing finite delays using, e.g., a finite FIFO buffer or a *relay station*. To account for such a behavior, we refine the flow-equivalence of [15] by a series of (reflexive-anti-symmetric) relations $\sqsubseteq_N$ (for $N > 0$) which yields the (series of) reflexive-symmetric flow relations $\approx_N$ to identify processes of same flows up to a flow-preserving first-in-first-out (FIFO) buffer of size $N$.

DEFINITION 9.4 (FINITE RELAXATION)
*A behavior c is a* 0-relaxation *of b iff* $b \leq c$. *A behavior c is a* 1-relaxation *of b, written* $b \sqsubseteq_1 c$, *iff* $\mathrm{vars}(b) = \mathrm{vars}(c)$ *and there exists* $d \geq b$ *s.t. for all* $x \in \mathrm{vars}(d)$, $d(x) = (t_i, v_i)_{i \geq 0}$, $c(x) = (t'_i, v_i)_{i \geq 0}$ *and for all* $i \geq 0$, $t_i \leq t'_i < t_{i+1}$. *For all* $N > 0$, *c is a* N-relaxation *of b, written* $b \sqsubseteq_N c$, *iff there exists d s.t.* $b \sqsubseteq_1 d \sqsubseteq_{N-1} c$

Notice that the union $\cup_{N \geq 0} \sqsubseteq_N$ of the finite relaxation relations is the asynchronous partial-order of relaxation $\sqsubseteq$.

PROPERTY 1 (TRANSITIVITY)
$b \sim b'$ *implies* $b \approx_0 b'$,
$b \approx_M b' \approx_N b''$ *implies* $b \approx_{M+N} b''$,
$b \approx_M b'$ *implies* $b \approx b'$.

The series of relations $(\approx_N)_{N\geq 0}$ yields the largest equivalence classes of behaviors that can be modeled in the polychronous model of computation without asynchrony. It consists of behaviors equal up to finite-flow equivalence $\approx^*$ and corresponds to behaviors equal up to timing deformations performed by a finite protocol.

DEFINITION 9.5 (FINITE FLOW-EQUIVALENCE)
*$b$ and $c$ are finitely flow-equivalent, written $b \approx^* c$, iff there exists $N \geq 0$ st. $b \sqsubseteq_N c$.*

## 2.6 Denotational semantics of the iSTS algebra

The detailed presentation and extension of the polychronous model of computation allows to give a denotational model to the iSTS notation introduced in section 1. This model consists of relating a proposition $P$ to the set of behaviors $p$ it denotes.

**Meaning of clocks.** Let us start with the denotation of a clock expression $e$ (table 9.2). The meaning $[\![e]\!]_b$ of a clock $e$ is defined relatively to a given behavior $b$ and consists of the set of tags satisfied by the proposition $e$ in the behavior $b$.

In table 9.2, the meaning of the clock $x = v$ (resp. $x = y$) in $b$ is the set of tags $t \in \text{tags}(b(x))$ (resp. $t \in \text{tags}(b(x)) \cap \text{tags}(b(y))$) such that $b(x)(t) = v$ (resp. $b(x)(t = b(y)(t))$. In particular, $[\![\hat{x}]\!]_b = \text{tags}(b(x))$. The meaning of a conjunction $e \wedge f$ (resp. disjunction $e \vee f$ and difference $e \setminus f$) is the intersection (resp. union and difference) of the meaning of $e$ and $f$. Clock 0 has no tags.

*Table 9.2.* Denotational semantics of clock expressions

$$
\begin{array}{l}
[\![x = v]\!]_b = \{t \in \text{tags}(b(x)) \mid b(x)(t) = v\} \qquad [\![0]\!]_b = \emptyset \\
[\![x = y]\!]_b = \{t \in \text{tags}(b(x)) \cap \text{tags}(b(y)) \mid b(x)(t) = b(y)(t)\} \\
[\![e \wedge f]\!]_b = [\![e]\!]_b \cap [\![f]\!]_b \\
[\![e \vee f]\!]_b = [\![e]\!]_b \cup [\![f]\!]_b \\
[\![e \setminus f]\!]_b = b[\![e]\!]_b \setminus [\![f]\!]_b \\
[\![1]\!]_b = \text{tags}(b)
\end{array}
$$

**Meaning of propositions.** The denotation of a clock expression by a set of tags yields the denotational semantics of propositions $P$, written $[\![P]\!]$, table 9.3. The meaning $[\![P]\!]^e$ of a proposition $P$ is defined with respect to a clock expression $e$. Where this information is absent, we assume $[\![P]\!] = [\![P]\!]^1$ to mean that $P$ is an invariant (and is hence independent of a particular clock).

The meaning of an initialization $x^0 = v$ , written $[\![x^0 = v]\!]^e$, consists of all behaviors defined on $x$, written $b \in \mathcal{B}|_x$ such that the initial value of the signal $b(x)$ equals $v$. Notice that it is independent from the clock expression $e$ provided by the context. In table 9.3, we write:

- $\mathcal{B}|_X$ for the set of all behaviors of domain $X$
- $\min(C)$ for the minimum of the chain of tags $C$
- $\text{succ}_t(C)$ for the immediate successor of $t$ in the chain $C$
- $\text{vars}(P)$ and $\text{vars}(e)$ for the set of free signal names of $P$ and $e$.

The meaning of a proposition $x = y$ at the clock $e$ consists of all behaviors $b$ defined on $\text{vars}(e) \cup \{x, y\}$ such that all tags $t \in [\![e]\!]_b$ at the clock $e$ belong to $b(x)$ and $b(y)$ and are associated with the same value. A scheduling specification $y \to x$ at the clock $e$ denotes the set of behaviors $b$ defined on $\text{vars}(e) \cup \{x, y\}$ which, for all tags $t \in [\![e]\!]_b$, requires $x$ to preceed $y$: if $t$ is in $b(x)$ then it is necessarily in $b(y)$ and satisfies $t_y \to^b t_x$. The propositions $x' = y$ and $y \to x'$ is interpreted similarly by considering the tag $t'$ that is the successor of $t$ in the chain $C$ of $x$.

*Table 9.3.* Denotational semantics of propositions

$$
\begin{aligned}
[\![x^0 = v]\!]^e &= \{b \in \mathcal{B}|_x \mid b(x)(\min(\text{tags}(b(x)))) = v\} \\
[\![x = y]\!]^e &= \{b \in \mathcal{B}|_{\text{vars(e)} \cup \{x,y\}} \mid \forall t \in [\![e]\!]_b, \\
&\qquad t \in \text{tags}(b(x)) \wedge t \in \text{tags}(b(y)) \wedge b(x)(t) = b(y)(t)\} \\
[\![y \to x]\!]^e &= \{b \in \mathcal{B}|_{\text{vars(e)} \cup \{x,y\}} \mid \forall t \in [\![e]\!]_b, \\
&\qquad t \in \text{tags}(b(x)) \Rightarrow t \in \text{tags}(b(y)) \wedge t_y \to^b t_x\} \\
[\![x' = y]\!]^e &= \{b \in \mathcal{B}|_{\text{vars(e)} \cup \{x,y\}} \mid \forall t \in [\![e]\!]_b, \\
&\qquad t \in C = \text{tags}(b(x)) \wedge t \in \text{tags}(b(y)) \wedge b(x)(\text{succ}_t(C)) = b(y)(t)\} \\
[\![y \to x']\!]^e &= \{b \in \mathcal{B}|_{\text{vars(e)} \cup \{x,y\}} \mid \forall t \in [\![e]\!]_b, \\
&\qquad t \in C = \text{tags}(b(x)) \Rightarrow t \in \text{tags}(b(y)) \wedge t_y \to^b (\text{succ}_t(C))_x\}
\end{aligned}
$$

The behavior of a guarded command $f \Rightarrow P$ at the clock $e$ is equal to the behavior of $P$ at the clock $e \wedge f$. The meaning of a restriction $P/x$ consists of the behaviors $c$ of which a behavior $b/x$ from $P$ are a stretching of. The behavior of $P \,|\, Q$ consists the synchronous composition of the behaviors of $P$ and $Q$.

*Table 9.4.* Denotational semantics of propositions

$$
[\![f \Rightarrow P]\!]^e = [\![P]\!]^{e \wedge f} \quad [\![P \,|\, Q]\!]^e = [\![P]\!]^e \,|\, [\![Q]\!]^e \quad [\![P/x]\!]^e = \{c \leq b/x \mid b \in [\![P]\!]^e\}
$$

## 2.7 Formal properties of the denotational semantics

In the previous section, we gave a definition of the iSTS notation in the polychronous model of computation. To assess the compositionally of this notation, it is important to define the extent of this definition, to analyse its stability to syntactic transformations, to characterize equivalence and closure properties pertaining on the structure and on the context of a given iSTS process, in order to give a proper assessment on its compositionally.

**Structural equivalence.** As an example, the common process algebraic properties of structural equivalence are directly implied by the definition of table 9.3. We write $P[y/x]$ for the substitution of $x$ by $y$ in $P$.

$$
\begin{array}{rl}
P\,|\,1{=}P & P/x{=}(P[y/x])/y \\
P\,|\,Q{=}Q\,|\,P & (P/x)/y{=}(P/y)/x \\
P\,|\,(Q\,|\,R){=}(P\,|\,Q)\,|\,R & P\,|\,(Q/x){=}(P\,|\,Q)/x \quad x \notin \mathrm{vars}(P) \\
(e \vee f) \Rightarrow P{=}e \Rightarrow P\,|\,f \Rightarrow P & e \Rightarrow P/x{=}(e \Rightarrow P)/x \quad x \notin \mathrm{vars}(e)
\end{array}
$$

**Congruence.** Property 2 defines a remarkable syntactic property of iSTS with respect to the semi-lattice of clock equivalence. It means that, if clock equivalence holds between two processes $P$ and $Q$, then it also holds in any context formed by "pluging" either of the processes $P$ or $Q$ to a term $K[]$ of the form

$$K[] ::= [] \,|\, (P\,|\,K[]) \,|\, (e \Rightarrow K[]) \,|\, (K[]/x)$$

From the point of view of compositional design, this property is important. It allows one to rely on the behavior expressed once and for all by a process $P$ to represent a class of processes equivalent by clock equivalence and in any possible context $K$, the *"write once run anywhere"* slogan.

PROPERTY 2 (CONGRUENCE)
*If $P \sim Q$ then $K[P] \sim K[Q]$.*

**Closure.** Property 3 states that the semantics of the iSTS is closed by clock-equivalence. It means that, whenever a process $P$ has a given behavior $b \in [\![P]\!]$, then it also has any behavior $c$ that is a stretch of $b$. This is an important assessment on the expressive capability of the iSTS: the property of closure implies that, whenever the intent of a given proposition is to model a particular scenario, a particular invariant of a system, then its implicit extent is the class of all behaviors that are clock-equivalent to this scenario.

PROPERTY 3 (CLOSURE)
*If $b \in [\![P]\!]$ and $b \leq c$ then $c \in [\![P]\!]$.*

**Expressibility.** Property 4 extends this result to finite flow equivalence with the aim of showing that the semantics of iSTS is expressive enough to model systems composed of synchronous processes communicating via reliable and bounded protocols, and hence, encompass the specification of GALS architectures. Property 4 can also be seen as a justification (*a posteriori*) on the lake of an operator of parallel composition in the iSTS, that would model an unreliable because unbounded communication medium.

Property 4 says that, whenever a given process or system has a given behavior $b$, and whenever $c$ is a finite relaxation of $b$, then another process $Q$ can be found such that $c$ is a behavior of $P \,|\, Q$, demonstrating the closure of the iSTS for finitely flow equivalent behaviors.

PROPERTY 4 (EXPRESSIBILITY)
*If* $b \in [\![P]\!]$ *and* $b \sqsubseteq^* c$ *then there exists* $Q$ *s.t.* $c \in [\![P \,|\, Q]\!]$.

Finding a process $Q$ satisfying property 4 amounts to considering the finite FIFO buffer protocols that are directly implied by the definition of the relation of relaxation. The model of a 1-place FIFO buffer from $x$ to $y$, noted $\mathsf{fifo}_1^{x,y}$ is defined in the table 9.5 (it is divided in four small chunks to ease comprehension). It uses two local signals $r$ and $s$. The initial value of the state variable $s$ is set to true (proposition $s^0 = 1$). The state variable $s$ and the register are synchronized (proposition $\hat{s} = \hat{r}$). When $s$ is true, the FIFO awaits the presence of the input signal $x$ (proposition $s = \hat{x}$) and defines the next value of the register $r$ by the value of $x$ (proposition $s \Rightarrow r' := x$). When $s$ is false, it synchronizes with output signal $y$ (proposition $\neg s = \hat{y}$), sends the current value of the register $r$ (proposition $\neg s \Rightarrow y := r$) and sustains the value of $r$ (proposition $r' := r$). If either a value is loaded or stored then the status of $s$ changes (proposition $\hat{s} \Rightarrow s' = \neg s$).

*Table 9.5.* Model of FIFO buffers in the iSTS algebra

$$\mathsf{fifo}_1^{x,y} \stackrel{\text{def}}{=} \left[ \left( \begin{array}{r} s^0 = 1 \\ \hat{s} = \hat{r} \\ s = \hat{x} \\ s \Rightarrow r' := x \end{array} \right) \,\Big|\, \left( \begin{array}{r} \neg s = \hat{y} \\ \neg s \Rightarrow y := r \\ \neg s \Rightarrow r' := r \\ \hat{s} \Rightarrow s' := \neg s \end{array} \right) \right] / rs$$

Compared to the input signal $x$, one easily observe that the output signal $y$ of the protocol $\mathsf{fifo}_1^{x,y}$ has all tags shifted by at most one place (the shift cannot pass the next input, by definition of table 9.5).

$$\forall x, y \in \mathcal{X}, \forall b \in [\![\mathsf{fifo}_1^{x,y}]\!],\ z \mapsto b(x) \sqsubseteq_1 z \mapsto b(y)$$

This development yields the result expected in property 4: if $b \in [\![P]\!]$ and $b \sqsubseteq^* c$ then there exists $N \geq 0$ and $Q =^{\text{def}} \mathsf{fifo}_N$ s.t. $c \in [\![P \,|\, Q]\!]$

$$\begin{array}{c} b(x) \\ \sqcap \\ b(y) \end{array}$$

Starting from the protocol $\mathsf{fifo}_1^{x,y}$, finite FIFO buffers of length $N > 1$ and of width $M > 1$ can easily be defined by induction on $N$ and $M$.

$$\forall M, N > 1, \qquad \mathsf{fifo}_N^{x,y} \stackrel{\text{def}}{=} \left(\mathsf{fifo}_1^{x,z} \,|\, \mathsf{fifo}_{N-1}^{z,y}\right) / z$$

$$\mathsf{fifo}_N^{x_{1..m}, y_{1..m}} \stackrel{\text{def}}{=} \left(\mathsf{fifo}_N^{x_{1..m-1}, y_{1..m-1}} \,|\, \mathsf{fifo}^{x_m, y_m}\right)$$

To attach a finite FIFO buffer to a given process $P$, we define the template or generic process $\mathsf{fifo}_N\langle P\rangle$ to redirect the input signals $x_{1..m} = \text{in}(P)$ and the output signals $x_{m+1..n} = \text{vars}(P) \setminus \text{in}(P)$ of $P$ to an input buffer $\mathsf{fifo}_N^{x_{1..m}, y_{1..m}}$ and an output buffer $\mathsf{fifo}_N^{y_{m+1..n}, x_{m+1..n}}$. This is done by substituting the free variables $x_{1..n}$ by fresh signal names $y_{1..n}$. The definition of the function $\text{in}(P)$ is given in appendix 9.A.

$$\begin{array}{rcl} \mathsf{fifo}_N\langle P\rangle & \stackrel{\text{def}}{=} & \left(\mathsf{fifo}_n^{x_{1..m}, y_{1..m}} \,|\, (P[y_i/x_i]_{i=1}^n) \,|\, \mathsf{fifo}_N^{y_{m+1..n}, x_{m+1..n}}\right) / y_{1..n} \\ \text{where} & & \{x_{1..m}\} = \text{in}(P) \end{array}$$

## 3. From processes to behavioral types

In the terminology of [10], the process $P$ specified in the grammar 9.1 defines the *dynamic or statefull interface* of a given system: it defines a transition system that specifies the invariants of its evolution in time. The abstraction of a process $P$ by its *static or stateless interface* $\hat{P}$ is defined by the type inference system $e \vdash P : Q$ of table 9.6. To a process $P$ in a context of clock $e$, it associates the clock and scheduling relations $Q$ that form a static abstraction of $P$.

*Table 9.6.* Abstraction of propositions by behavioral types

$$\vdash 1 : 1 \qquad \vdash x^0 = v : 1 \qquad e \vdash x \to y : e \Rightarrow x \to y$$

$$e \vdash x = y : e \Rightarrow x = y \qquad e \vdash x' = y : e \Rightarrow \hat{x} = \hat{y} \,|\, e \Rightarrow \hat{x} \,|\, e \Rightarrow \hat{y}$$

$$\frac{e \wedge f \vdash P : Q}{e \vdash f \Rightarrow P : Q} \qquad \frac{e \vdash P : Q}{e \vdash P/x : Q/x} \qquad \frac{e \vdash P_1 : Q_1 \quad e \vdash P_2 : Q_2}{e \vdash P_1 \,|\, P_2 : Q_1 \,|\, Q_2}$$

The *static* interface $\hat{P}$ of a process $P$ consists of the abstraction of propositions $P$ by a set of clock and scheduling relations $Q$, defined by the type inference relation $e \vdash P : Q$ of table 9.6, and of the closure of scheduling relations by transitivity, i.e. $e \Rightarrow x \rightarrow y \,|\, f \Rightarrow y \rightarrow z$ implies $(e \wedge f) \Rightarrow x \rightarrow z$; and by distributivity, i.e. $e \Rightarrow x \rightarrow y \,|\, f \Rightarrow x \rightarrow y$ implies $(e \vee f) \Rightarrow x \rightarrow y$. The correctness of the inference system is stated by a denotational containment relation.

PROPERTY 5 (SOUNDNESS)
*If* $e \vdash P : Q$ *then* $[\![P]\!]^e \subseteq [\![Q]\!]^e$.
$[\![P]\!] \subseteq [\![\hat{P}]\!]$.

### 3.1 Scalability

Just as the theory of interfaces automata [10], the iSTS algebra allows for a scalable level of abstraction to be automatically obtained starting from the inferred type of a given module (table 9.6) and its hierarchical representation (example 2). The iSTS algebra shares with interface automata the possibility to define *static* interfaces (boolean relations) and *dynamic* interfaces (a transition system). It further allows to relate a given process $P$ to abstract types $Q$ in many ways:

1/ a transition $e \Rightarrow x' = u$ is abstracted by clock relations,

2/ a signal $x$ is abstracted from $P/x$ by any type $Q$ equivalent to $P$ not referencing $x$,

More generally, checking a user-specified type $Q$ consistent with the inferred type $\hat{P}$ of a component amounts to the satisfaction of the containment relation $[\![Q]\!] \subseteq [\![\hat{P}]\!]$ (i.e. the denotation of $Q$ contains the denotation of $\hat{P}$). If $P$ is a *static interface* which amounts to the satisfaction of a boolean equation. Similarly, checking that a *dynamic interface* $Q$ is an abstraction of a process $P$ amounts to verifying that $Q$ simulates $P$ by model-checking.

EXAMPLE 1 *Let us refine the specification of the counter modulo 2 defined figure 1. The process P (left) constrains the event $y$ to occur upon two successive inputs $x$. Its state $s$ is initially false. It is synchronized to the input signal $x$. The next value of $s$ is defined by the negation of its current value. When $s$ is true then $y$ is present. From the type inference relation of table 9.6, we obtain the static interface of $\hat{P}$ (right).*

$$P \stackrel{\text{def}}{=} \left( \begin{array}{c} s^0 = 0 \\ \hat{x} = \hat{s} \\ s' := \neg s \\ s = \hat{y} \\ y := s \end{array} \right) / s \qquad \hat{P} = \left( \begin{array}{c} \hat{x} = \hat{s} \\ s \rightarrow s' \\ s = \hat{y} \\ s = y \\ s \rightarrow y \end{array} \right) / s$$

*Let $Q = (\hat{x} > \hat{y})$ be the type declared in place of $\hat{P}$ to define a static interface type for some component implementing the proposition $P$. It just requires the* *clock $\hat{y}$ to be a sampling of $\hat{x}$. Checking that the declared type $Q$ is compatible with the inferred type $\hat{P}$ amounts to checking that $\hat{P}$ implies $Q$ to meet the required denotational containment relation: $[\![\hat{P}]\!] \subseteq [\![Q]\!]$. This is implemented by observing that $(\hat{x} = \hat{s} \mid \hat{y} = s)/s$ implies $\hat{x} > \hat{y}$.*

## 3.2 Hierarchization

Behavioral type inference allows to associate a system of clock equations $\hat{P}$ with an iSTS process $P$. In order to perform type-based manipulation of programs: merging several processes into one component or, conversely, splitting one into several distributed components, *hierarchization* [2] is the key concept put to work. Hierarchization consists of determining a canonical representation of the partial order between the signal clocks of a process $P$. This data-structure is the medium which we use to perform the verification, transformation and exploration of system design models.

EXAMPLE 2 *As an example, consider the type $P$ of the counter modulo 2 from example 1 (table 9.7, left). The abstraction of $P$ (table 9.7, middle left) consists of a set of clock and scheduling relations.*

*Hierarchization (table 9.7, middle right), consists of building a partial order relations between clocks starting from $\hat{P}$. The clock equivalence class at the top of the tree comprises $\hat{x}$ and $\hat{s}$. It is associated to the scheduling relation $\hat{x} \to s$. The left-most subtree corresponds to the clock $s = 0$ and the right-most sub-tree to the clocks $s = 1$ and $\hat{y}$.*

*Table 9.7.* From behavioral types to sequential C code

| (process) | (type) | (hierarchy) | (code) |
|---|---|---|---|
| $s^0 = 0$ | $\hat{x} = \hat{s}$ | | bool s = false; |
| $\hat{x} = \hat{s}$ | $s \to s'$ | $\hat{x}, \hat{s}$ | while true { |
| $s' := \neg s$ | $s = \hat{y}$ | / \ | if x then if s then { y = true; |
| $s = \hat{y}$ | $y = s$ | $\neg s$ $\quad$ $\hat{y}, s$ | s = false; } |
| $y := s$ | $s \to y$ | | else { s = true; }} |

*The C program (table 9.7, right) demonstrates the capability to build a control-flow graph starting from the system of clock equations and scheduling relations implied by a process $P$.*

Definition 9.6 formalizes the algorithm introduced in [2] for the construction of a hierarchy starting from a set of boolean equations. From the example of figure 9.7, we observe that both free and bound signals of a proposition $P$ are

involved by hierarchization. Therefore, definition 9.6 considers the Skolemization $\overline{P}$ of $P$. By induction on $P$:
- $\overline{a/X} = a/X$ and $\overline{P/x} = Q/X \cup \{x\}$ iff $\overline{P} = Q/X$
- $\overline{e \Rightarrow P} = (e \Rightarrow Q)/X$ iff $\overline{P} = Q/X$ and $\text{vars}(e) \cap X = \emptyset$
- $\overline{P_1 \,|\, Q_1} = (P_2 \,|\, Q_2)/XY$ iff $\overline{P}_1 = P_2/X$ and $\overline{Q}_1 = Q_2/Y$

Notice that $[\![\overline{P}]\!] = [\![P]\!]$. Definition 9.6 uses the static abstraction $\hat{Q}$ of $\overline{P} = Q/X$ to find clock equivalence classes and partially relate them. Rule 1 defines equivalence class for synchronous signals and constructs elementary partial order relations: the clock $(x = v)$ is smaller than $\hat{x}$. Rule 2 defines the insertion of a partial order of maximum $z$ below a clock $x$. The insertion algorithm, introduced in [2], yields a canonical representation of the corresponding partial order by observing that there exists a unique minimum clock below $x$ such that rule 2 holds. In definition 9.6, we write $P \models Q$, e.g. $\hat{Q} \models \hat{x} = \hat{y}$, iff the proposition $P$ implies the proposition $Q$, e.g. the static interface $\hat{Q}$ implies the clock equation $\hat{x} = \hat{y}$.

DEFINITION 9.6 (HIERARCHIZATION)

*The* clock relation $\preceq_P$ *of a process* $P$ *is the partial order relation between elementary clocks* $(x = r)$*, written* $h$*, and defined by application of rules* $(1-2)$ *with* $\overline{P} = Q/X$.

1. *if* $\hat{Q} \models \hat{x} = h$ *or* $\hat{Q} \models (x = v) = h$ *then* $\hat{x} \preceq h$.
2. *if* $\hat{x} \preceq h_1$, $\hat{x} \preceq h_2$ *and* $\hat{Q} \models h = h_1 \star h_2$ *for any* $\star \in \{\wedge, \vee, \setminus\}$ *then* $\hat{x} \preceq h$.

$h_1$ *and* $h_2$ *are clock-equivalent, written* $h_1 \mathrel{\prec\!\!\succ} h_2$*, iff* $h_1 \preceq h_2$ *and* $h_2 \preceq h_1$.

A process whose clock relation is hierarchical, in the sense of definition 9.7, meets a necessary criterion for being implemented by a sequential program.

DEFINITION 9.7 (HIERARCHIZATION)

*A process* $P$ *is* hierarchical *iff* $\preceq_P$ *has a minimum class* $\min \preceq_P$

## 3.3 Clock synthesis

In order to lift a larger class of processes to the rank of hierarchical processes, the Polychrony workbench implements additional program transformations. These transformations consist of refinements that promote the clock equivalence classes of an initial process $P$ until a minimum $\min \preceq_Q$ is found in the final process $Q$.

One refinement consists of promoting a bound signal in an oversampling process. A process $P/x$ is oversampling iff $P$ is hierarchical but its minimun is $\hat{x} \in \min \preceq_P$. In that case, one may promote $x$ by refining $P$ as

$$Q \stackrel{\text{def}}{=} (P \,|\, \hat{y} = \hat{x})/x \ s.t. \ y \notin \text{vars}(P)$$

which, by definition of the initial process $P$, satisfies $\hat{y} \in \min \preceq_Q$ and is hence hierarchical.

Another possible refinement is concerned with the simulation of non-deterministic processes $P$, whose clock relation usually consists of a set $\{\preceq_i \mid 0 < i \leq n\}$ of independent clock relations: for all $0 < i, j \leq n$, for all $x_i \in \min \preceq_i$ and $x_j \in \min \preceq_j$, $x_i \leftrightarrow\!\!\!\!\diamond\; x_j$ iff $i = j$. In that case, one may define a sequence of signals $y_i$ in order to simulate the behavior of the process $P$ using a hierarchical process $Q$ such that $\min \preceq_Q = \{y_{1..n}\}$ defined by:

$$Q \stackrel{\text{def}}{=} \left(P \mid \left(\textstyle\prod_{0<i\leq n} y_i = \hat{x}_i\right) \mid \left(\textstyle\prod_{0<i<n} \hat{y}_i = \hat{y}_{i+1}\right)\right)$$

## 3.4 Transformation of hierarchical processes

On the path towards a sequential implementation of a hierarchical process, an important step is the structural transformation of a process $P$ into a model
- from which formal design properties can easily be checked
- from which target code can easily and structurally be generated

We outline such transformations implemented in the Polychrony workbench [23], They consist of representing a process $P$ either into a disjunctive form, noted $\mathcal{D}_P$, or into a hierarchical form, noted $\mathcal{H}_P$. Table 9.8 examplifies such transformations operated on the counting process.

*Table 9.8.* Disjunctive and hierarchical representations of a process

| (process) | (disjunctive form) | (clock relation) | (hierarchical form) |
|---|---|---|---|
| $s^0 = 0$<br>$\hat{x} = \hat{s}$<br>$s' := \neg s$<br>$s = \hat{y}$<br>$y := s$ | $s^0 = 0$<br>$\hat{x} \Rightarrow \hat{s}$<br>$\hat{s} \Rightarrow \hat{x}$<br>$\hat{s} \Rightarrow s' := \neg s$<br>$s \Rightarrow \hat{y}$<br>$\hat{y} \Rightarrow s$<br>$s \Rightarrow y := s$ | $\hat{x}, \hat{s}$<br>/ \<br>$\neg s$    $s, \hat{y}$ | $s^0 = 0$<br>$\hat{x} = \hat{s}$<br>$\hat{x} \Rightarrow \begin{pmatrix} s' := \neg s \\ \hat{y} = s \\ s \Rightarrow y := s \\ \neg s \Rightarrow 1 \end{pmatrix}$ |

The disjunctive form $\mathcal{D}_P$ of the process $P$ consists of its representation as the composition of elementary guards $g \Rightarrow a$ consisting of a conjunctive clock proposition $g$ and an atomic action $a$.

$$\begin{array}{ll} D ::= G \mid D/x & \text{(disjunction)} \\ G ::= g \Rightarrow a \mid (G \mid G) & \text{(clause)} \\ g ::= 0 \mid 1 \mid (x = r) \mid g \wedge g \mid g \setminus g & \text{(guard)} \end{array}$$

The hierarchical form $\mathcal{H}_P$ of the process $P$ consists of representing it according to the tree-like structure of its associated clock relation $\preceq_P$. One of the benefits of this representation is to make the control-flow of the process explicit to facilitate code generation.

$$\begin{array}{lll} H ::= a \,|\, e \Rightarrow F & \text{(hierarchy)} \\ F ::= H \,|\, (H \,|\, F) & \text{(forest)} \end{array}$$

In table 9.8, one easily observes the interplay between the disjunctive and hierarchical forms of a process. The process $\mathcal{D}_P$ is easily obtained from the initial specification $P$ by a structural decomposition of its propositions.

Then, using the information carried by the clock relation $\preceq_P$, one iteratively picks the elementary guarded commands $g \Rightarrow a$ of $\mathcal{D}_P$ that match the successive clocks of $\preceq_P$ to form the hierarchical decomposition of $P$.

The disjunctive form of a process $P$ is defined by application of the function $\mathcal{D}_P =^{\text{def}} \mathcal{D}[\![P]\!]^1$ in table 9.6. We observe that $[\![\mathcal{D}_P]\!] = [\![P]\!]$ by construction. It is used, e.g., in appendix 9.A to easily identify the input and output signals of a process, or, e.g., in definition 9.12 to define conflict-free transitions.

$$\begin{array}{rlrl} \mathcal{D}[\![l = r]\!]^g = g \Rightarrow l = r & & \mathcal{D}[\![f \Rightarrow P]\!]^e = \mathcal{D}[\![P]\!]^{e \wedge f} \\ \mathcal{D}[\![x \rightarrow l]\!]^g = g \wedge \hat{x} \Rightarrow x \rightarrow l & & \mathcal{D}[\![P/x]\!]^e = (\mathcal{D}[\![P]\!]^e)/x \\ \mathcal{D}[\![a]\!]^{e \vee g} = \mathcal{D}[\![a]\!]^e \,|\, \mathcal{D}[\![a]\!]^g & & \mathcal{D}[\![P \,|\, Q]\!]^e = \mathcal{D}[\![P]\!]^e \,|\, \mathcal{D}[\![Q]\!]^e \end{array}$$

*Figure 9.6.* Disjunctive normal form $\mathcal{D}_P$ of a process $P$

The hierarchical form of a process $P$ is defined by application of the function $\mathcal{H}_P =^{\text{def}} \mathcal{H}[\![P]\!]^1_{\preceq}$ in table 9.7. It takes an initial clock 1 and the clock relation $\preceq_P$ as parameters. This relation is iteratively decomposed using the function top $\preceq$, which returns the clock equivalence class $H$ on top of $\preceq$, and the function bot $\preceq$, which returns the set of clock relations under $H$ in $\preceq$.

$$\begin{array}{l} \mathcal{H}[\![\prod_{i=1}^n g_i \Rightarrow a_i]\!]^g_{\preceq} = \prod_{i \in I} a_i \,|\, \mathsf{sync}^H \,|\, h \Rightarrow F \\ \quad \text{where } h \in H = \text{top} \preceq \\ \qquad\qquad I = \{0 < i \leq n \,|\, \hat{P} \models g \wedge h = g_i \wedge a_i \notin H\} \\ \qquad\qquad F = \prod_{\preceq_j \in \text{bot} \preceq} \mathcal{H}[\![\prod_{i \notin I} g_i \Rightarrow a_i]\!]^{g \wedge h}_{\preceq_j} \end{array}$$

*Figure 9.7.* hierarchical form $\mathcal{H}_P$ of a process $P$

The algorithm $\mathcal{H}$ iteratively constructs a subset $I$ of the indexed clauses $g_i \Rightarrow a_i$ of the disjunctive form $\mathcal{D}_P$. It selects propositions $g_i \Rightarrow a_i$ whose guard $g_i$ matches $g$, the clock on the path to the present hierarchy $\preceq$, and $h$, the clock of the present hierarchy $\preceq$. It discards those propositions $g_i \Rightarrow a_i$ whose actions $a_i$ are redundant with a clock expression of $H$. The iteration continues breadth-first to define the forest corresponding to bot $\preceq$ using the remaining propositions $g_i \Rightarrow a_i, i \notin I$. We write $\text{sync}^{h_1,h_2} =^{\text{def}} h_1 = h_2$ to synchronize the clock of the equivalence class $H$ of top $\preceq$.

Again, one easily observe that, by construction, we have $[\![\mathcal{H}_P]\!] = [\![P]\!]$ thanks to the semantics information provided by the clock relation $\preceq$. The hierarchical representation of $P$ allows one to easily identify the block of actions associated to every clock equivalence class of $P$. In the Polychrony workbench, it is used to directly generate sequential code of the form of if-then-else statements. It can also be used to construct a hierarchical abstraction of the process $P$ that privileges the exposure of behaviors at the top of the hierarchy and abstract behaviors occurring at the bottom of the hierarchy.

## 4. Formal analysis of processes

A hierarchical process satisfies necessary conditions for the property of *endochrony* to hold. A process is said endochronous (definition 9.8) iff, given an external (asynchronous) stimulation of its inputs, it reconstructs a unique synchronous behavior (up to clock-equivalence).

The behavior of an endochronous process $P$ is depicted in figure 9.8. The process accepts asynchronous yet flow-equivalent inputs $x$ and $y$ (left). Upon request from the process, inputs are fed from an input buffer in a clock equivalent manner (middle) so as to produce the same outputs in the same order at clock-equivalent rates (right).

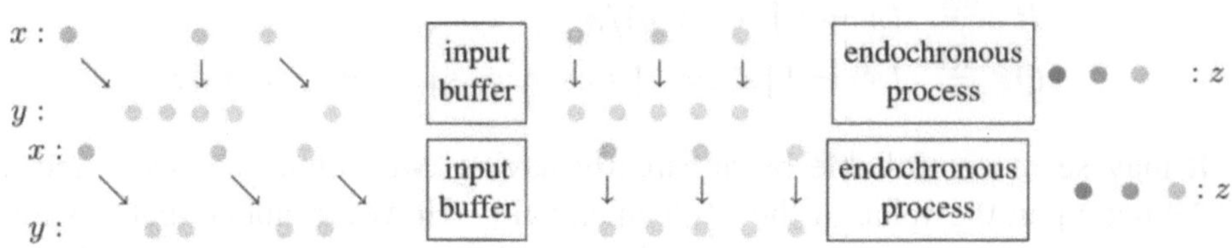


*Figure 9.8.* Endochrony: from flow-equivalent inputs to clock-equivalent outputs

In other words, endochrony denotes the class of processes that are stallable or insensitive to internal and external propagation delays: patient processes, in the sense of [8].

In the Polychrony workbench [23], GALS architectures are modeled starting from the property of endochrony and the criterion of *endo-isochrony* [15]: two endochronous processes $P$ and $Q$ of interface $I = \text{vars}(P) \cap \text{vars}(Q)$ are endo-isochronous iff $(P|_I) \,|\, (Q|_I)$ is endochronous. This criterion ensures that the refinement of the specification $P \,|\, Q$ by a distributed implementation $P \parallel Q$

is semantics-preserving. Although restrictive, it is directly amenable to static verification by checking $P$, $Q$ and $(P|_I)\,|\,(Q|_I)$ endochronous.

DEFINITION 9.8 (ENDOCHRONY [15])
*$P$ is endochronous iff* $\forall b, c \in [\![P]\!]$, $(b|_{\mathrm{in}(P)}) \approx (c|_{\mathrm{in}(P)}) \Rightarrow b \sim c$.

## 4.1 Formal analysis of hierarchical processes

Starting from the algorithm specified in definition 9.6 and the class of hierarchical processes isolated in definition 9.7, we seek towards processes which satisfies the property of endochrony in the context of the iSTS algebra by introducing the notion of controllability, definition 9.13. A correct definition of controllability requires a precise partition of the signals of a process $P$ into input signals $\mathrm{in}(P)$, output signals $\mathrm{out}(P)$ and state variables $\mathrm{def}(P)$. This partition is defined in appendix 9.A, table 9.A.2. A signal $x \in \mathrm{out}(P)$ (resp. $x \in \mathrm{def}(P)$) is an output (resp. state) of $P$ iff, whenever its presence is implied by a clock $e$, then there exists another clock $f$ implied by $e$ whose action is $x := y$ (resp. $x' := y$) for some reference $y$. The disjoint sets $\mathrm{def}(P)$ and $\mathrm{out}(P)$ are complementary of the input signals $\mathrm{in}(P)$ in $\mathrm{vars}(P)$.

## 4.2 Definability

A controllable process needs to correctly define its local and bound signals in order to meet the property of determinism, and hence endochrony. For instance, let us reconsider (left) the example of the specification $R$, section 1.

$$
\begin{aligned}
R &\stackrel{\mathrm{def}}{=} (s \Rightarrow x \,|\, \neg s \Rightarrow y)/s \\
Q &\stackrel{\mathrm{def}}{=} (s^0 = 1 \,|\, s' := z \,|\, s \Rightarrow x := s \,|\, \neg s \Rightarrow y := s)/s
\end{aligned}
$$

It may seem controllable by apparently having two output signals $x$ and $y$ defined upon the value of the boolean signal $s$. However, notice that $s$ is left undefined. Hence the behavior of $R$ is non-deterministic, since the choice of the value of $s$ is free. Had $s$ been associated to some external input $z$, as in $Q$ (right), one could have concluded that it is indeed controllable. In fact, notice that $Q$ is a refinement of $R$: $[\![Q]\!] \subseteq [\![R]\!]$.

DEFINITION 9.9 (DEFINABILITY)
*A process $P$ is* well-defined *iff all bound variables of $P$ are either output signals or state variables. By induction on the structure of $P$,*

- *an atomic proposition $a$ is well-defined.*
- *a guarded command $e \Rightarrow P$ is well defined iff $P$ is well-defined*
- *a composition $P\,|\,Q$ is well-defined iff $P$ and $Q$ are well-defined*
- *$P/x$ is well-defined iff $P$ is well-defined and* $x \in \mathrm{def}(P) \cup \mathrm{out}(P)$.

## 4.3 Causality

Causality is another possible origin of non-determinism. Let us consider the example of a clock proposition $\hat{x} =^{\text{def}} (x = x)$. It does not define an output signal, it just says that the signal $x$ is present. It can be regarded as an abstraction of the proposition $x := y =^{\text{def}} (x = y \,|\, y \to x)$, which defines it as an output signal and assigns the value of the input $y$ to it. What happens next, if we additionally write $y = x$ or $y := x$ ? In the first case, one can still regard the process:

$$x := y \,|\, y = x$$

as a redundant yet deterministic specification: $x$ takes the value of $y$. This is not the case in the latter specification, which tries to mutually define the output signals $x$ and $y$:

$$x := y \,|\, y := x$$

It is a non-deterministic specification, as both true and false are possible values of $x$ and $y$ every time this is needed. Thanks to the definition of $x := y$ by $x = y \,|\, y \to x$, the analysis of causal specifications relies on the information provided by the scheduling relation of a process $P$: a cyclic definition of output signals yields a proposition $e \Rightarrow x \to x$ at some clock $e$ in the transitive closure of the scheduling relation implied by $\hat{P}$. Hence the definition 9.10.

DEFINITION 9.10 (CAUSALITY)
*A process $P$ is* non-causal *iff $\hat{P} \models e \Rightarrow x \to x$ implies $\hat{P} \models e = 0$.*

## 4.4 Determinism

The isolation of well-defined and non-causal processes allows to touch an important intermediate result. When a process both defines its bound variables from the value of free variables and does not define its outputs in a cyclic manner, it is determinism in the sense of definition 9.11.

DEFINITION 9.11 (DETERMINISM [15])
*$P$ is deterministic iff $\forall b, c \in [\![P]\!]$, $(b|_{\text{in}(P)}) = (c|_{\text{in}(P)}) \Rightarrow b = c$.*

The relation to well-defined and non-causal processes is stated in property 6.

PROPERTY 6 (DETERMINISM)
*If $P$ is well-defined and non-causal then $P$ is deterministic*

One may be surprised by the absence of a notion of conflict-freedom in the criterion 9.11. For instance, consider the process $P =^{\text{def}} e \Rightarrow x = r \,|\, f \Rightarrow x = s$, one may wonder why the additional invariant $Q =^{\text{def}} e \wedge s \Rightarrow s = r$ is not required to check $P$ deterministic. A careful attention to the denotational semantics of $P$ shows that $[\![P]\!] = [\![P \,|\, Q]\!]$ because synchronous composition meaning filters conflicting behaviors. Checking this property would nonetheless be mandatory, should we consider to specify a correct sequential code generation scheme for iSTS. In the Polychrony workbench, the resolution of conflicting specifications is performed by the synthesis of the corresponding proof obligations $Q$, which are called *clock constraints*.

DEFINITION 9.12 (CONFLICT-FREEDOM)
*A process $P$ is* conflict-free *iff* $\mathcal{D}_P \stackrel{\text{def}}{=} (\prod_{i=1}^{m} g_i \Rightarrow a_i) / x_{1..n}$ *and for all* $0 < i, j \leq m$, $a_i =^{\text{def}} (l := x)$ *and* $a_j =^{\text{def}} (l := y)$ *implies* $\hat{P} \models g_i \wedge g_j \Rightarrow x = y$.

## 4.5 Controllability

If the definition 9.11 is compared to the behavior depicted in figure 9.9, one easily notices that a deterministic process still misses some important capabilities to meet the requirements of endochrony. A deterministic process must have a way to initiate a reaction upon the arrival of a pre-determined input signal (e.g. $x$ in figure 9.8), to check whether it needs further inputs (e.g. should it load $y$ from the input buffer or not) to finally produce an output (e.g. produce $z$ when the input $y$ is absent).

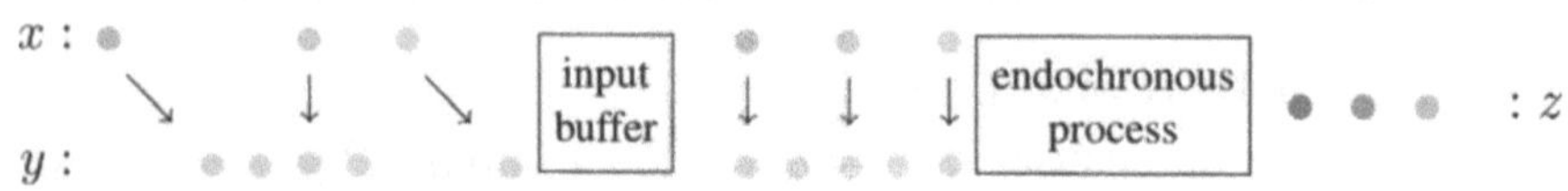


*Figure 9.9.* Endochrony as the reconstruction of clock-equivalence classes

The notion of hierarchy allows to fill this remaining gap. A hierarchical process $P$ has a pre-determined input tick, the minimum min $\preceq_P$ of its clock relation. Upon the presence of the tick (e.g. $x$ in figure 9.9), it is able to decide whether an additional input is needed or not (e.g. $\hat{x} \prec \hat{y}$ in figure 9.9), it is able to decide whether it should produce outputs (e.g. $\hat{x} \prec \hat{z}$ in figure 9.9). However, controllability requires this decision to be made starting from the input signals of a process (and not its output signals or state variables).

DEFINITION 9.13 (CONTROLLABILITY)
*A process $P$ is* controllable *iff $P$ is well-defined, non-causal, hierarchical and for all $y \in$* vars$(P)$*, there exists $x \in$* in$(P)$ *such that $x \preceq_P y$.*

The definition of controllability bridges the remaining gap from a deterministic specification to an endochronous program. It is stated by property 7. Property 7 generalizes the proposition of [15] to the context of partially defined signals of the iSTS algebra.

PROPERTY 7 (CONTROLLABILITY)
*If $P$ is controllable then $P$ is endochronous.*

## 4.6 A design methodology based on finite flow-equivalence

Building upon the iSTS notation and its implementation in the Polychrony workbench, we seek towards a formal methodology to characterize the invariants of protocol synthesis and propose the use of global model transformations to synthesize specialized and optimized component wrappers using the behavioral type information provided by iSTS processes.

To this end, we apply the iSTS algebra to the characterization of a correct-by-construction methodology for the compositional design of architectures starting from elementary endochronous processes. We focus on the definition of a refinement methodology and semantics preservation criterion that naturally proceed from the polychronous model of computation and define the spectrum of bounded protocol synthesis in system design.

The property of *flow-invariance* [15] proceeds from that of endochrony to ensure that the refinement of a synchronous specification $P \,|\, Q$ by its distributed implementation $P \parallel Q$ preserves flows: the refinement from $P$ to $Q$ (s.t. $I \subset \text{in}(P) = \text{in}(Q)$) is *flow-invariant* iff for all $b \in [\![P]\!]$, for all $c \in [\![Q]\!]$, $b|_I \approx c|_I$ implies $b \approx c$. Refinement-based design with the Polychrony workbench and using the criterion of flow-invariance has been studied in a number of related works on system design [15, 16, 21].

In the present study, we chose to restrict ourselves to the scope of the polychronous model of computation of iSTS excluding unbounded asynchrony. We therefore define the following criteria, which recast endochrony and flow-invariance to the context of the largest congruence expressible in the polychronous model of computation: finite flow-equivalence $\approx^*$.

We say that a process $P$ is finitely flow-preserving iff given finitely flow-equivalent inputs, it can only produce behaviors that are finitely flow equivalent.

DEFINITION 9.14 (FINITE FLOW-PRESERVATION)
$P$ *is* finitely flow-preserving *with* $I \subset \text{in}(P)$ *iff*

$$\forall b, c \in [\![P]\!],\ (b|_I) \approx^* (c|_I) \Rightarrow b \approx^* c$$

Example of finitely flow-preserving processes are endochronous processes. An endochronous process which receives finitely flow equivalent inputs produces clock-equivalent outputs. It hence forms a restricted sub-class of finitely-flow preserving processes. Furthermore, notice that flow-preservation is stable to the introduction a wrapper of $P$ consisting of a finite FIFO buffering protocol.

PROPERTY 8 (FINITE FLOW-PRESERVATION)
*1. If $P$ is endochronous then $P$ is finitely-flow preserving.*

2. *If $P$ is finitely flow-preserving then* $\mathsf{fifo}_N\langle P\rangle$ *is finitely flow-preserving.*

A refinement-based design methodology based on the property of finite flow-preservation consists of characterizing sufficient invariants for a given model transformation to preserve flows.

DEFINITION 9.15 (FINITE FLOW-INVARIANCE)
*The transformation of $P$ into $Q$ such that $I \subset \mathrm{in}(P) = \mathrm{in}(Q)$ is* finitely flow-invariant *iff*

$$\forall b \in [\![P]\!],\ \forall c \in [\![Q]\!],\ (b|_I) {\approx}^* (c|_I) \Rightarrow b \approx^* c$$

The property of finite flow-invariance is a very general methodological criterion. For instance, it can be applied to the characterization of correctness criteria for model transformations such as protocol insertion or desynchronization.

Let $P$ and $Q$ be two finitely flow-preserving processes and $R$ a protocol to link $P$ and $Q$, such as a finite FIFO buffer $\mathsf{fifo}_N$, or a double hand-shake protocol [21], or a relay station [8], or a loosely time-triggered architecture [5].

DEFINITION 9.16 (FLOW-PRESERVING PROTOCOL)
*The process $R$ is a flow-preserving protocol iff there exists $n > 0$ such that inputs* $\mathrm{in}(R) = \{x_{1..n}\}$ *are finitely flow-equivalent to outputs* $\mathrm{out}(R) = \{y_{1..n}\}$

$$\forall b \in [\![R]\!], b|_{x_{1..n}} \approx^* (b|_{y_{1..n}}[x_i/y_i]_{0<i\leq n})$$

The wrapper $R\langle P\rangle$ of a process $P$ with a protocol $R$ is defined by redirecting the signals of $P$ to $R$ using substitutions to install the connexions depicted in figure 9.10.

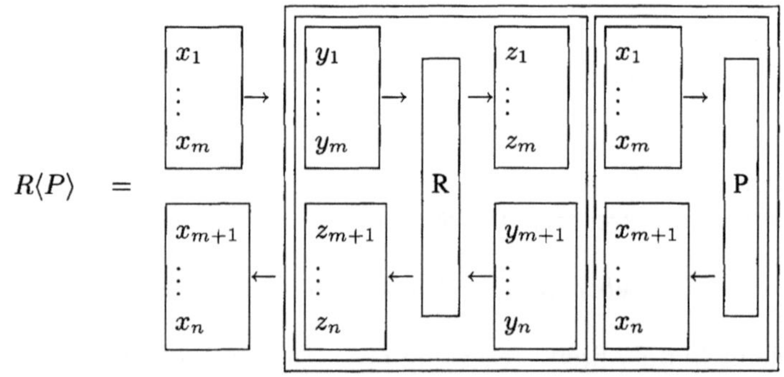


*Figure 9.10.* Connexions between a process $P$ and a protocol $R$ in the wrapper $R\langle P\rangle$

DEFINITION 9.17 (WRAPPER)
*Let $P$ be a process such that* $\mathrm{in}(P) = \{x_{1..m}\}$ *and* $\mathrm{out}(P) = \{x_{m+1..n}\}$. *Let $R$ be a flow-preserving protocol such that* $\mathrm{in}(R) = \{y_{1..n}\}$ *and* $\mathrm{out}(R) =$

$\{z_{1..n}\}$. *The wrapper of* $P$ *with* $R$ *is the template process noted* $R\langle P\rangle$ *and defined by:*

$$R\langle P\rangle \stackrel{\text{def}}{=} (((R[x_i/z_i]_{m<i\leq n})\,[x_i/y_i]_{0<i\leq m})\mid(P[y_i/x_i]_{m<i\leq n})\,[z_i/x_i]_{0<i\leq m})$$

A sufficient condition for the insertion of a protocol between two synchronous processes $P$ and $Q$ to finitely preserve flow is to guaranty that $P|_I \mid Q|_I$ is finitely flow preserving for $I = \text{vars}(P) \cap \text{vars}(Q)$, meaning that all communications between $P$ and $Q$ via shared signals $x \in I$ should be flow preserving and that $P$ and $Q$ may otherwise evolve independently. This condition is stated by property 9.

PROPERTY 9 (PROTOCOL INSERTION)
*If* $R$ *is a flow-preserving protocol and* $P$ *is finitely flow-preserving then* $R\langle P\rangle$ *is finitely flow-preserving. If* $R$ *is a flow-preserving protocol and* $P$, $Q$, $P|_I \mid Q|_I$ *are finitely flow-preserving then* $R\langle P\rangle \mid R\langle Q\rangle$ *is finitely preserving (*$I = \text{vars}(P) \cap \text{vars}(Q)$*).*

## 5. Related works

In [16], we consider the mirror approach. It consists of building an asynchronous model of the system, by considering synchronous processes connected with unbounded FIFO channels. This approach allows to equally have compositionally given by design. Then, we consider the refinement of this model with an implementation of FIFO channels with finite FIFO buffers, whose size is estimated in order to meet the expected property of semantics preservation. This methodology has the advantage to yield directly to the issue of estimating minimal buffer sizes to guaranty preservation of the specification's meaning.

Synchronous programming being a computational model which is popular in hardware design, and desynchronization being a technique to convert that computational model into a more general, globally asynchronous and locally synchronous computational model, suitable for system-on-chip design, one may naturally consider investigating further the links between these two models understood as Ptolemy domains [7] and study the refinement-based design of GALS architectures starting from polychronous specifications captured from heterogeneous elementary components.

The aim of capturing both synchrony and asynchrony in a unifying model of computation is shared by several approaches: interaction categories of Abramsky et al. in [1], communicating sequential processes of Hoare [11], Kahn networks [13], latency insensitive protocols of Carloni et al. [8], the heterogeneous systems of Benveniste et al. [4]. These models partition systems into synchronous islands (i.e. predefined *pearls* or IPs) and asynchronous networks. For instance, the heterogeneous model of [4] defines a tag-less model of asynchrony and a densely tagged model of synchrony (where, actually, stuttering equivalence is algebraically equivalent to the clock equivalence relation of the polychronous model of computation)

Synchrony and asynchrony are not partitioned in the present model. Both are captured within the same partially ordered trace structure, and related by the clock and finite-flow equivalence relations $\sim$ and $\approx^*$. The iSTS algebra carry over this unified model to capture modeling, transformation and verification of embedded systems from the highest levels of requirements specification down to its clock-accurate implementation as a Kahn network or a GALS architecture.

In the theory of latency-insensitive protocols [8], the status of an architecture components is rendered by the notion of *pearls* ("intellectual property under a shell") and supposed to satisfy a generic invariant of *patience* (robustness to external delays). Our approach consists of partly revealing the structure of the *pearl* by a behavioral type and of giving a multi-clocked specifications to the *shell* (the wrapper). The compilation of the behavioral type and of the wrapper provides a *controller* of the IP, that still satisfies the expected invariant of patience yet in a way that is cautious of the internal behavior of the IP, as revealed by its behavioral type. This results in the synthesis of an optimized protocol allowing for the correct-by-construction architecture construction with minimized of local latency and a maximized global throughput.

## 6. Conclusions and future works

Our approach is based on a high-level modeling and specification methodology that ensures compositional correctness through an algebra capturing behavioral aspects of component interfaces. This minimalist algebra, called iSTS, specifies the state transitions, synchronization relations and scheduling constraints implied by a given system, in the presence of multiple-clocked synchrony. Our approach consists of using the iSTS algebra as a type system to describe the behavior of system components and allow for global model transformations to be performed on the system based on behavioral type information. This approach builds upon previous work on scalable design exploration using refinement/abstraction-based design methodologies [21], implemented in the Polychrony workbench [23]. It is both conceptually minimal and equipped with a formal semantics defined in a synchronous model of computation [15] and implemented by the Polychrony workbench [23].

## References

[1] ABRAMSKY, S., GAY, S. J., NAGARAJAN, R. Interaction categories and the foundations of typed concurrent programming. In *Deductive Program Design: Proceedings of the 1994 Marktoberdorf International Summer School.* NATO ASI Series F, Springer-Verlag, 1996.

[2] AMAGBEGNON, T. P., BESNARD, L., LE GUERNIC, P. "Implementation of the data-flow synchronous language SIGNAL". In *Conference on Programming Language Design and Implementation.* ACM Press, 1995.

[3] Hoe, J., Arvind. "Synthesis of Operation-Centric Hardware Descriptions". *Proceedings of International Conference on Computer Aided Design.* IEEE Press, November 2000.

[4] BENVENISTE, A., CASPI, P., CARLONI, L. P., SANGIOVANNI-VINCENTELLI, A. L. "Heterogeneous Reactive Systems Modeling and Correct-by-Construction Deployment". In *Embedded Software Conference.* Lecture Notes in Computer Science, Springer Verlag, October 2003.

[5] BENVENISTE, A., CASPI, P., LE GUERNIC, P., MARCHAND, H., TALPIN, J.-P., TRIPAKIS, S. "A protocol for loosely time-triggered architectures". In *Embedded Software Conference.* Lecture Notes in Computer Science, Springer Verlag, October 2002.

[6] BERRY, G., GONTHIER, G. "The ESTEREL synchronous programming language: design, semantics, implementation". In *Science of Computer Programming*, v. 19, 1992.

[7] J.T. BUCK, S. HA, E.A. LEE AND D.G. MESSERSCHMITT. Ptolemy: A Framework for Simulating and Prototyping Heterogeneous Systems. In *International Journal of Computer Simulation, special issue on "Simulation Software Development"*v. 4, pp. 155-182. Ablex, April 1994.

[8] CARLONI, L. P., MCMILLAN, K. L., SANGIOVANNI-VINCENTELLI, A. L. "Latency-Insensitive Protocols". In *Proceedings of the 11th. International Conference on Computer-Aided Verification.* Lecture notes in computer science v. 1633. Springer Verlag, July 1999.

[9] E. DIJKSTRA "A Discipline of Programming". Prentice Hall, 1976.

[10] De Alfaro, L., Henzinger, T. A. "Interface theories for component-based design". *International Workshop on Embedded Software.* Lecture Notes in Computer Science v. 2211. Springer-Verlag, 2001.

[11] HOARE, C. *Communicating sequential processes.* Prentice Hall, 1985.

[12] Jeffords, R. and Heitmeyer, C. "A Strategy for Efficiently Verifying Requirements Specifications Using Composition and Invariants". *Sympo-*

*sium on the Foundations of Software Engineering*. ACM Press, September 2003

[13] KAHN, G. The semantics of a simple language for parallel programming. In IFIP *Congress*. North Holland, 1974.

[14] LEE, E. A., SANGIOVANNI-VINCENTELLI, A. "A framework for comparing models of computation". In *IEEE transactions on computer-aided design*, v. 17, n. 12. IEEE Press, December 1998.

[15] LE GUERNIC, P., TALPIN, J.-P., LE LANN, J.-L. Polychrony for system design. In *Journal of Circuits, Systems and Computers. Special Issue on Application-Specific Hardware Design*. World Scientific, 2002.

[16] MOUSAVI, M., R., LE GUERNIC, P., TALPIN, J.-P., SHUKLA, S., BASTEN, T. Modeling and validation of asynchronous systems in synchronous frameworks. In *Digital Automation and Test Europe*. IEEE Press, February 2004.

[17] NOWAK, D., BEAUVAIS, J.-R., TALPIN, J.-P. "Co-inductive axiomatization of a synchronous language". In *International Conference on Theorem Proving in Higher-Order Logics*. Lecture Notes in Computer Science, Springer Verlag, October 1998.

[18] NOWAK, D., TALPIN, J.-P., LE GUERNIC, P. "Synchronous structures". In *International Conference on Concurrency Theory*. Lecture Notes in Computer Science, Springer Verlag, August 1999.

[19] PNUELI, A., SHANKAR, N., SINGERMAN, E. Fair synchronous transition systems and their liveness proofs. *International School and Symposium on Formal Techniques in Real-time and Fault-tolerant Systems*. Lecture Notes in Computer Science v. 1468. Springer Verlag, 1998.

[20] S. K. RAJAMANI AND J. REHOF, "A BEHAVIORAL MODULE SYSTEM FOR THE $\pi$-CALCULUS". *Static Analysis Symposium*. Lectures Notes in Computer Science. Springer Verlag, July 2001.

[21] J.-P. TALPIN, P. LE GUERNIC, S. K. SHUKLA, R. GUPTA, AND F. DOUCET. "Polychrony for formal refinement-checking in a system-level design methodology". *Application of Concurrency to System Design*. IEEE Press, June 2003.

[22] J.-P. TALPIN, D. BERNER, S. SHUKLA, P. LE GUERNIC, A. GAMATIÉ, R. GUPTA "Behavioral type inference for compositional system design". In *Formal Methods and Models for System Design (this volume)*. Kluwer Academic Publishers, June 2004.

[23] The Polychrony workbench. `http://www.irisa.fr/espresso/Polychrony`, February 2004.

# Appendix: Naming conventions

We refer to the *free variables* $\mathrm{vars}(P)$ of a process $P$, table 9.A.1, as the set of signal names that occur free in the lexical scope of $P$.

$$\begin{aligned}
\mathrm{vars}(0) &= \emptyset \\
\mathrm{vars}(1) &= \emptyset \\
\mathrm{vars}(x^0 = v) &= \{x\} \\
\mathrm{vars}(x = v) &= \{x\} \\
\mathrm{vars}(x' = v) &= \{x\} \\
\mathrm{vars}(x = y) &= \{x, y\} \\
\mathrm{vars}(x' = y) &= \{x, y\}
\end{aligned}
\qquad
\begin{aligned}
\mathrm{vars}(x \rightarrow y) &= \{x, y\} \\
\mathrm{vars}(x \rightarrow y') &= \{x, y\} \\
\mathrm{vars}(e \wedge f) &= \mathrm{vars}(e) \cup \mathrm{vars}(f) \\
\mathrm{vars}(e \vee f) &= \mathrm{vars}(e) \cup \mathrm{vars}(f) \\
\mathrm{vars}(e \setminus f) &= \mathrm{vars}(e) \cup \mathrm{vars}(f) \\
\mathrm{vars}(e \Rightarrow P) &= \mathrm{vars}(e) \cup \mathrm{vars}(P) \\
\mathrm{vars}(P \,|\, Q) &= \mathrm{vars}(P) \cup \mathrm{vars}(Q) \\
\mathrm{vars}(P/x) &= \mathrm{vars}(P) \setminus \{x\}
\end{aligned}$$

*Figure 9.A.1.* Free variables $\mathrm{vars}(P)$ of a process $P$

The defined state and output signals $\mathrm{def}(P)$ and $\mathrm{out}(P)$ of a process $P$ are defined by the relation of figure 9.A.2 starting from the disjunctive form of $P$ (figure 9.6). A signal $x \in \mathrm{out}(P)$ (resp. $x \in \mathrm{def}(P)$) is an output of $P$ iff, whenever its presence is implied by a guard $g$, then there exists a guard $h$ implied by $g$ whose action is $x := r$ (resp. $x' := r$ ) for some $r$. The sets $\mathrm{def}(P)$ and $\mathrm{out}(P)$ define the variables $x \in \mathrm{vars}(P)$ that are the input signals $\mathrm{in}(P)$ of $P$. In the definition of the generic function $\mathrm{locs}(P)$, we overload $\hat{x}$ to $\hat{l}$ and assume that $\hat{x'} =^{\mathrm{def}} \hat{x}$.

$$\begin{aligned}
x \in \mathrm{def}(P) &\Leftrightarrow x' \in \mathrm{locs}(\mathcal{D}_P) \wedge \exists 0 < i \leq n, g_i \stackrel{\mathrm{def}}{=} 1 \wedge a_i \stackrel{\mathrm{def}}{=} x^0 = v \\
x \in \mathrm{out}(P) &\Leftrightarrow x \in \mathrm{locs}(\mathcal{D}_P) \wedge x \notin \mathrm{def}(P) \\
\mathrm{in}(P) &= (\mathrm{vars}(P) \setminus \mathrm{def}(P)) \setminus \mathrm{out}(P)
\end{aligned}$$

$$\begin{aligned}
&l \in \mathrm{locs}\left(\left(\textstyle\prod_{i=1}^{n} g_i \Rightarrow a_i\right) / x_{1..m}\right) \Leftrightarrow \\
&\qquad \forall 0 < i \leq n\, s.t.\, \hat{P} \models g_i \Rightarrow \hat{l}, \exists 0 < j \leq n, \hat{P} \models g_i \Rightarrow g_j \wedge a_j \stackrel{\mathrm{def}}{=} l := r
\end{aligned}$$

*Figure 9.A.2.* Input and output signals of a process

Chapter 1

# BEHAVIORAL TYPE INFERENCE

## *Part II – Behavioral type inference for system design*

Jean-Pierre Talpin,[1] David Berner,[1] Sandeep Kumar Shukla,[2] Paul Le Guernic,[1] Abdoulaye Gamatié,[1] and Rajesh Gupta[3]

[1] *INRIA, project Espresso*
*IRISA, Campus de Beaulieu*
*F-35042 Rennes Cedex, France*
FirstName.LastName@irisa.fr

[2] *Department of Electrical and Computer Engineering*
*Virginia Tech*
*Blacksburg, Virginia, USA*
shukla@vt.edu

[3] *Department of Computer Science and Engineering*
*University of California et San Diego*
*La Jolla, California, USA*
gupta@cs.ucsd.edu

**Abstract** The design productivity gap has been recognized by the semiconductor industry as one of the major threats to the continued growth of system-on-chips and embedded systems. Ad-hoc system-level design methodologies, that lifts modeling to higher levels of abstraction, and the concept of intellectual property (IP), that promotes reuse of existing components, are essential steps to manage design complexity. However, the issue of compositional correctness arises with these steps. Given components from different manufacturers, designed with heterogeneous models, at different levels of abstraction, assembling them in a correct-by-construction manner is a difficult challenge. We address this challenge by proposing a process algebraic model to support system design with a formal model of computation and serve as a behavioral type system to capture the behavior of system components at the interface level. The proposed algebra is conceptually minimal, equipped with a formal semantics defined in a synchronous model of computation, and supports a scalable notion and a flexible degree of abstraction.



R. Gupta et al. (eds.), *Formal Methods and Models for System Design*, 245-282.

We demonstrate its benefits by considering the type-based synthesis of latency-insensitive protocols. We show that the synthesis of component wrappers can be optimized by the behavioral information carried by interface type descriptions and yields minimized stalls and maximized throughput.

## 1. Introduction

The design productivity gap has been recognized by the semiconductor industry as one of the major threats to the continued growth of complex system-chips and their applications. System level design methodology that moves from RTL-based design entry into design methods at higher-level of abstraction is essential to manage design complexity. A number of advances in high-level modeling and validation have been proposed over the past decade in an attempt to improve the level of abstraction in system design. Most of these enable greater reuse of existing intellectual property (IP) blocks.

Design correctness is an important part of this move. Given the complexity of system-level designs, it is important that the composition of system-level IP blocks be guaranteed correct. However, *a posteriori* validation of component compositions is a difficult problem. Techniques are needed that ensure design correctness as a part of the design process itself. To address this issue, methodological precepts have been developed that separately focus on design reuse and correctness by construction.

Both reuse and elevation of abstraction at design entry critically depend on guaranteed design correctness. To enable design reuse and raise the level of design entry there is industry momentum towards standardization efforts such as the Virtual Socket Interface Alliance (VSIA) and Accellera. These efforts target standards in specification of the interfaces, test data and properties that should be verified in design of components for system-chips.

At the same time with the advent of high-level design specification languages from SystemC, SpecC to Superlog, a promise is held out for a successful move towards designing at the system level through adoption of a dominant high-level specification language standard. While useful, standardization alone, whether of design specification languages or of design properties, will not solve the fundamental problem of design correctness. Techniques are needed to ensure correctness through the composition process itself.

To improve the state of the art in component composition from existing IP libraries, we specifically seek to address the following issue: given a high level architectural description (e.g., a virtual architecture ) and a library of implemented components, how can one automate the selection of implementation of virtual components from the library, and automatically ensure composability of data and behavior?

Our approach is based on a high-level modeling and specification methodology that ensures compositional correctness through a type theory capturing behavioral aspects of component interfaces. The proposed system builds upon previous work on scalable design exploration using refinement/abstraction-based design methodologies [15], implemented in the Polychrony workbench [17], and on a layered Component Composition Environment, Balboa [5], which allows specification of SystemC components with mixed levels of structural and behavioral details and high degree of concurrency and timing requirements.

Our behavioral type system consists of a minimalist formalism, called the iSTS (implicit synchronous transition systems), that is akin to Pnueli's synchronous transition systems (STS, [12]) and Dijkstra's guarded commands [3]. The iSTS is used as a type system to describe the behavior of system components and allows for global model transformations to be performed on the system based on behavioral type information. It is equipped with a formal semantics defined in a multi-clocked synchronous model of computation [7] and implemented by the Polychrony workbench [17].

## 1.1 Roadmap

We put principles of type theory to work for the definition of a behavioral type inference system targetting the high-level design language standard SystemC.

After an introduction to the core of SystemC under consideration, Section 3, we present our behavioral type inference method, Section 3.2. Being defined for a static single-assignment (SSA) intermediate representation of programs, the proposed inference technique is generic and language-independent. It is generalized to the structuring elements of the SystemC class system, Section 1.9. This extension yields the formulation of a SystemC design methodology that reduces compositional design correctness verification to the validation of synthesized proof obligations.

Applications of the technique to the compositional assembly of SystemC modules and classes and to the verification of design correctness properties are outlined, Section 3.4. Its implementation using the Polychrony workbench is described, Section 3.9.

## 1.2 Rationale

To allow for an easier grasp on the proposed behavioral type inference technique, we outline the analysis of a small fragment of a SystemC program, Figure 1.1, the construction of its dynamic behavioral type, Figure 1.2, and the inference of its static abstraction, Figure 1.3. Then we elaborate the notion of proof obligations synthesis by giving a brief outline of the design correctness issues which can be modeled and checked in the framework of our type system.

**Static single assignment.** Figure 1.1 (left) depicts a simple C code fragment consisting of an iterative program that counts the number of bits set to one in the variable idata. While idata is not equal to zero, it adds its right-most bit to an output count variable ocount and shifts it right in order to process the next bit. In the static single-assignment (SSA) representation of the program Figure 1.1 (right), all variables (idata and ocount) are read and written once per cycle.

Label L2 is the entry point of the SSA block that represents the while loop. The first instruction loads the input variable idata into the register T1. The second instruction stores the result of its comparison with 0 in the register T0. If T0 is false, control is passed to the block L3.

Otherwise, the next instruction is executed: the variable ocount is loaded into T2, the last bit of T1 is loaded into T3, the sum of T2 and T3 assigned to ocount and the right-shift of T1 assigned to idata. The block terminates with an unconditional branch back to label L2. L2.

| (C source) | (SSA code) |
|---|---|
| ```while (idata != 0)``` | L2:T1 = idata; |
| { | T0 = T1 == 0; |
| ocount = ocount + (idata & 1); | if T0 then goto L3; |
| idata = idata >> 1; | T2 = ocount; |
| } | T3 = T1 & 1; |
| | ocount = T2 + T3; |
| | idata = T1 >> 1; |
| | goto L2; |

*Figure 1.1.* Translation of a source program into static single assignment form

**Behavioral types.** Although verbose, the SSA intermediate representation of an imperative program can present an otherwise arbitrarily obfuscating C program in a form that can be easily manipulated by an automatic program analyzer. Let us focus on the block L2 in the example of Figure 1.2, left.

The behavioral type of the block L2 consists of the simultaneous composition of logical propositions that form a synchronous transition system. Each proposition is associated with one instruction: it specifies its *invariants*. In particular, it tells when the instruction is executed, what it computes, when it passes control to the next statement, when it branches to another block.

On the first line, for instance, we associate the instruction T1 = idata of block label L2 to the proposition $x_{\mathsf{L2}} \Rightarrow \mathsf{T1} = \mathsf{idata}$. In this proposition, the new variable $x_{\mathsf{L2}}$ is a boolean that is true iff the label L2 is being executed. So, the proposition says that if the label L2 is being executed, then T1 is always equal to idata. If not, then another proposition may hold. In our case, all subsequent propositions are conditioned by $x_{\mathsf{L2}}$ meaning that they hold when L2 is being executed.

The extent of a proposition is for the duration of a reaction. A reaction can last arbitrarily long period of time provided that it is finite and that every variable or register changes its value at most once during this period. For instance, consider the instruction if T0 then L3. It is likely that label L3 will, just as L2, perform some operation on the input idata. Therefore, its execution is delayed until after the current reaction. We refer to $x'_{\mathsf{L3}}$ as the next value of the state variable $x_{\mathsf{L3}}$, to indicate that it will be active during the next reaction. Hence, the proposition $x_{\mathsf{L2}} \Rightarrow \mathsf{T0} \Rightarrow x'_{\mathsf{L3}}$ says that control will be passed to L3 at the next reaction when control is presently at L2 and when T0 is true. The instructions that follow this test are conditioned by the negative $\neg$T0, this means: "in the block L2 and not in its branch to L3".

| (SSA code) | (behavioral type) |
|---|---|
| L2:T1 = idata; | $x_{\mathsf{L2}} \Rightarrow$ T1 := idata |
| T0 = T1 == 0; | T0 := (T1 = 0) |
| if T0 then goto L3; | T0 $\Rightarrow x'_{\mathsf{L3}}$ |
| T2 = ocount; | $\neg$T0$\Rightarrow$T2 := ocount |
| T3 = T1 & 1; | T3 := T1&1 |
| ocount = T2 + T3; | ocount′ := T2 + T3 |
| idata = T1 >> 1; | idata′ := T1 >> 1 |
| goto L2; | $x'_{\mathsf{L2}}$ |

*Figure 1.2.* Behavioral type of the SSA program

**Static abstraction.** We have seen that every instruction of an SSA program could be associated with a proposition to render its control-flow and data-flow behaviors. This representation provides a formal and expressive way to model, analyze, optimize and verify the behavior of ordinary SystemC programs.

To ease both optimization and verification of such programs based on that representation, we abstract it over its control flow, characterized by boolean relations between *clocks*, and its data flow, characterized by scheduling relations between *signals*.

Let us first define some terminology. A clock $\hat{x}$ is associated with a signal $x$. The signal $x$ corresponds to the flow of the successive values of a variable, sampled by the discrete periods of time that we call reactions. The clock of $\hat{x}$ denotes that set of periods or instants.

Figure 1.3, all operations on integers and bits reported in the behavioral type on the left have been abstracted by boolean relations between clocks, middle, and by scheduling relations, right. This is in fact sufficient information to reconstruct the entire control and data flow graphs of the program. All the abstracted information essentially consists of computations which can be used to decorate these graphs and regenerate the original program.

For instance, the instruction $\mathsf{T0} = (\mathsf{T1} = 0)$ is abstracted by the type $x_{\mathsf{L2}} \Rightarrow \hat{\mathsf{T1}} = \hat{\mathsf{T0}}$. It means: "when the block L2 is executed, T0 is present iff T1 is present". The scheduling constraint $x_{\mathsf{L2}} \Rightarrow \mathsf{T0} \rightarrow x_{\mathsf{L3}}$ additionally says that "$x'_{\mathsf{L3}}$ cannot happen before T0 at L2". Indeed, one first needs to examine the status of T0 before branching to L3.

| (dynamic type) | (static type) | (scheduling) |
|---|---|---|
| $x_{\mathsf{L2}} \Rightarrow \mathsf{T1} = \mathsf{idata}$ | $x_{\mathsf{L2}} \Rightarrow \hat{\mathsf{T1}} = \hat{\mathsf{idata}}$ | $\mathsf{T1} \leftarrow \mathsf{idata}$ |
| $\mathsf{T0} = (\mathsf{T1} = 0)$ | $\hat{\mathsf{T0}} = \hat{\mathsf{T1}}$ | $\mathsf{T1} \rightarrow \mathsf{T0}$ |
| $\mathsf{T0} \Rightarrow x'_{\mathsf{L3}}$ | $\mathsf{T0} \Rightarrow x'_{\mathsf{L3}}$ | $\mathsf{T0} \rightarrow x'_{\mathsf{L3}}$ |
| $\neg\mathsf{T0} \Rightarrow \mathsf{T2} = \mathsf{ocount}$ | $\neg\mathsf{T0} \Rightarrow \hat{\mathsf{T2}} = \hat{\mathsf{ocount}}$ | $\mathsf{T2} \leftarrow \mathsf{ocount}$ |
| $\mathsf{T3} = \mathsf{T1}\&1$ | $\hat{\mathsf{T3}} = \hat{\mathsf{T1}}$ | $\mathsf{T1} \rightarrow \mathsf{T3}$ |
| $\mathsf{ocount}' = \mathsf{T2} + \mathsf{T3}$ | $\hat{\mathsf{ocount}} = \hat{\mathsf{T2}} \wedge \hat{\mathsf{T3}}$ | $\mathsf{T2} \rightarrow \mathsf{ocount}' \leftarrow \mathsf{T3}$ |
| $\mathsf{idata}' = \mathsf{T1} >> 1$ | $\hat{\mathsf{idata}} = \mathsf{T1}$ | $\mathsf{T1} \rightarrow \mathsf{idata}'$ |
| $x'_{\mathsf{L2}}$ | $x'_{\mathsf{L2}}$ | $\mathsf{T0} \rightarrow x'_{\mathsf{L2}}$ |

*Figure 1.3.* Static abstraction of the behavioral type

The type associated with the block L2 uses the clocks denoted by the booleans $x_{\mathsf{L2}}$, T0 and $\neg$T0. Each clock denotes a branch in the control-flow graph of the block L2. The other clocks, e.g. $\hat{\mathsf{T1}}$, denote the presence of data. They are partially related to the "*label*" clocks $x_{\mathsf{L2}}$, T0 and $\neg$T0.

**Typed modules.** Section 3.4 develops the use of the dynamic or static information provided by the behavioral type inference system to perform design correctness checks. The most salient feature of the behavioral type system is yet the capability to reduce compositional design correctness verification to the validation of synthesized proof obligations. It is presented in the context of the inference system proposed for the SystemC module system, Section 3.3.

As an example, consider a class whose virtual fields are two clocks $x$ and $y$, and a procedure $f$. It defines an interface, named $m_0$, and will be used to type another class. Next, assume an explicit behavioral type declaration #TYPE$(f, Q)$ which associates the procedure $f$ with a description of its behavior: the proposition $Q$. Its aim is to associate the virtual class field $f$, a method, to the denotation of all possible implementations satisfying an expected functionality.

```
class m0 {
   virtual sc_clock x;
   virtual sc_clock y;
   virtual void f() {} #TYPE(f, Q)
};
```

Next, we associate the interface $m_0$ with the class parameter $m_1$ of a template class $m_2$. The interface $m_0$ now gives a behavioral type to the method $f$ in the class parameter $m_1$ expected by the module $m_2$. Indeed, the template class $m_2$ uses the class parameter $m_1$, that implements $m_0$, to launch a thread $m_1.f$ sensitive to $x$. The behavioral type $Q$, which gives an assumption on the behavior of $m_1.f$, is required to provide a guarantee on the behavior of the module $m_2$, produced by the template class.

```
template ⟨class m1⟩ #TYPE(m1, m0)
  SC_MODULE(m2) {SC_CTOR(m2) {
    SC_THREAD(m1.f) sensitive ≪ x
  }
};
```

**Proof obligations.** Let $m_3$ be a candidate parameter for the template class $m_2$. It structurally implements the interface $m_0$, because it provides the clocks $x$ and $y$ and defines the method $f$ by the program *pgm*. Using the type inference technique previously outlined, the program *pgm* is associated with a proposition $P$ that describes its behavioral type, and the class $m_3$ be decorated with the corresponding type declaration #TYPE$(f, P)$.

```
class m3 {
  sc_clock x; sc_clock y;
  void f() { pgm} #TYPE(f, P)
};
```

Finally, let $m_4$ be the class defined by the instantiation of the template $m_2$ with the actual parameter $m_3$. To check the compatibility of the actual parameter $m_3$ with the formal parameter $m_0$, we need to establish the containment of the behaviors denoted by the proposition $P$ (the behavioral type of the actual parameter) in the denotation of the proposition $Q$, the type abstraction declared in $m_0$.

$$m_2\langle m_3\rangle\, m_4 \text{ is type-safe iff } \models P \Rightarrow Q$$

This amounts to checking that $P$ implies $Q$. This proof obligation can either be implemented using model checking (if $P$ and $Q$ are *dynamic* interfaces) or using SAT checking, if $Q$ is a *static* interface, by calculating the static abstraction $\hat{P}$ of $P$ and by verifying that $\hat{P}$ implies $Q$.

## 2. A polychronous type system

Let us recall the behavioral type theory presented in [16] by giving an informal outline of the type system that will support and materialize the polychronous model of computation and type and sign the behavior of SystemC programs and classes. The key principles put to work in this notation are essentially borrowed from Pnueli's STS [12] and Dijkstra's guarded command language [3]. We call this notation the iSTS (for implicit synchronous transition systems)

In the iSTS, a process consists of simultaneous propositions that manipulate signals. A signal is an infinite flow of values that is sampled by a discrete series of instants or reactions. An event corresponds to the value carried by a signal during a particular reaction or instant.

Figure 1.4 puts together the main features of the behavioral type system. It describes the behavior of a counter modulo 2, noted $P$ ($\stackrel{\text{def}}{=}$ means "is defined by"), through a set of simultaneous *propositions*, labeled from (1) to (3).

- Proposition (1) is an invariant. It says that the initial value of the signal $s$, denoted by $s^0$, is false. This is specified by the proposition $\neg s^0$ (equivalent to $s^0 = 0$).
- Proposition (2) is a guarded command. It says that if the signal $x$ is present during a given reaction then the value of $s$ is toggled.
  - The left-most part of the sentence, the proposition $\hat{x}$, is a condition or a guard. It denotes the *clock* of $x$. It is true if the signal $x$ is present during the current reaction.
  - The right-most part of the sentence, the proposition $s' = \neg s$, is a transition. The term $s'$ refers to the next value of the signal $s$. The proposition $s' = \neg s$ says that the next value of $s$ is the negation of the present value of $s$.
- Proposition (3) is another guarded command. It says that if $s$ is true then $y$ is present.

$$
P \stackrel{\text{def}}{=} \begin{array}{lccll} & & & \neg s^0 & (1) \\ | & \hat{x} & \Rightarrow & s' = \neg s & (2) \\ | & s & \Rightarrow & \hat{y} & (3) \end{array}
$$

*Figure 1.4.* Specification of a counter modulo 2

Notice that, in proposition (3), the guard expects the signal $s$ to hold the value 1 but that its action does not actually specify the value of the signal $y$: it simply requires $y$ to be present. Proposition (3) is hence an *abstraction*: a proposition that partially describes the properties of the system under consideration without implying a particular implementation.

To implement a *function* or a *system*, this proposition needs to be compositionally *refined* by another one, saying what value $y$ should hold whenever it is present. To this end, one could for instance compose the counter $P$, Figure 1.4, with the proposition $Q \stackrel{\text{def}}{=} (y' = y + 1 \,|\, y^0 = 0)$.

Notice that the iteration specified by the proposition $Q$ is not guarded. This means that it is an invariant that describes the successive values of the signal $y$ in time, but not the particular time samples at which the signal $y$ should occur. The composition of $Q$ with $P$ has the effect of synchronizing the signal $y$ in $Q$ to the *clock* $s$ in $P$: to the time samples during which the signal $s$ holds the value true. This composition is called a refinement: the system obeys the specification denoted by the initial proposition $P$ and is constrained to further satisfy the additional conditions implied by $Q$.

$$P \text{ is an abstraction of } P \,|\, Q \qquad P \,|\, Q \text{ is a refinement of } P$$

## 2.1 Formal syntax of the type system

The formal syntax of propositions in the behavioral type system is defined by the inductive grammar $P$ in Table 1.1. A proposition or process $P$ manipulates boolean values noted $v \in \{0, 1\}$ and signals noted $x, y, z$. A location $l$ refers to the initial value $x^0$, the present value $x$ and the next value $x'$ of a signal. A reference $r$ stands for either a value $v$ or a signal $x$.

A clock expression $e$ is a proposition on boolean values. When true, a clock $e$ defines a particular period in time. The clocks 0 and 1 denote events that never/always happen. The clock $x = r$ denotes the proposition: "$x$ is present and holds the value $r$". Particular instances are:
- the clock noted $\hat{x}$ and defined by $x = x$ means that "$x$ is present"
- the clock noted $x$ and defined by $x = 1$ means that "$x$ is true"
- the clock noted $\neg x$ and defined by $x = 0$ means that "$x$ is false"

Clocks are propositions combined using the logical combinators of conjunction $e \wedge f$, to mean that both $e$ and $f$ hold, disjunction $e \vee f$, to mean that either $e$ or $f$ holds, and symmetric difference $e \setminus f$, to mean that $e$ holds and not $f$.

*Table 1.1.* Formal syntax of the type system

| | |
|---|---|
| (reference) | $r ::= x \,\|\, v$ |
| (location) | $l ::= x^0 \,\|\, x \,\|\, x'$ |
| (clock) | $e, f ::= 0 \,\|\, x = r \,\|\, e \wedge f \,\|\, e \vee f \,\|\, e \setminus f \,\|\, 1$ |
| (process) | $P, Q ::= 1 \,\|\, l = r \,\|\, x \to l \,\|\, e \Rightarrow P \,\|\, (P \,\|\, Q) \,\|\, P/x$ |

A process $P$ consists of the simultaneous composition of elementary propositions. 1 is the process that does nothing. The proposition $l = r$ means that "$l$ holds the value $r$". The proposition $x \to l$ means that "$l$ cannot happen before $x$". The process $e \Rightarrow P$ is a guarded command. It means: "if $e$ is present then $P$ holds". Processes are combined using synchronous composition $P \,|\, Q$ to denote the simultaneity of the propositions $P$ and $Q$. Restricting a signal name $x$ to the lexical scope of a process $P$ is written $P/x$. We refer to the *free variables* $\text{vars}(P)$ of a process $P$ as the set of signal names that occur free in the lexical scope of $P$.

**Notational conventions** In the formal presentation of the iSTS, we restrict ourselves to a subset of the elementary propositions in the grammar of Table 1.1, which we call atoms $a$:

$$\text{(atoms)} \quad a, b ::= x^0 = v \,|\, l = y \,|\, x \to l \; s.t. \, l ::= x \,|\, x'$$

Other propositions as well as additional syntactic shortcuts, used in the examples, can be defined by using this restricted subset as follows.

$$l = v \stackrel{\text{def}}{=} (l = x \mid x^0 = v \mid x' = x)/x \text{ iff } x \neq l \neq x' \qquad \hat{x} \stackrel{\text{def}}{=} (x = x)$$
$$l := x \stackrel{\text{def}}{=} (l = x | x \to l) \qquad l \stackrel{\text{def}}{=} (l = 1)$$
$$\hat{x} = \hat{y} \stackrel{\text{def}}{=} (\hat{x} \Rightarrow \hat{y} | \hat{y} \Rightarrow \hat{x}) \qquad \neg l \stackrel{\text{def}}{=} (l = 0)$$

## 3. Behavioral types for SystemC

We are now equipped with the required mathematical framework and formal methodology to address the modeling of GALS architectures described using SystemC. This model is described in terms of a type inference system and extended to the structuring elements of SystemC in terms of a module system. This framework allows to give a behavioral signature of systems components to compositionally check their correct composition to form architecture, to optimize the described software elements from the imposed hardware elements by, first, extracting the formal model from the functional architecture description and, second, using the model to regenerate an optimized software matching the requirements of the execution architecture. As a by-product, associating types with SystemC programs provides a formal denotational semantics implied by the interpretation of types in the polychronous model of computation.

### 3.1 Formal syntax of the SystemC core

We start with the definition of the core of the SystemC syntax relevant to the present study. A system consists of the composition of classes and modules *sys*, Table 1.2. A class declaration $\mathsf{class}\, m\, \{dec\}$ associates a class name $m$ with a sequence of fields *dec*. It is optionally parameterized by a class with $\mathsf{template}\, \langle \mathsf{class}\, m_1 \rangle$. To enforce a strong typing policy, we annotate the class parameter $m_1$ with $\#\text{TYPE}(m_1, m_2)$ to denote the type of $m_1$ with the virtual class $m_2$. A module $\text{SC\_MODULE}(m)$ is a class that defines an architecture component. Its constructor $\text{SC\_CTOR}(m)\ \{new; pgm\}$ allocates threads (e.g. $\text{SC\_THREAD}(f)$) and executes an initialization program *pgm*. While modules are sequentially declared in the program text, they define threads whose execution is concurrent.

Declarations *dec* associate locations $x$ with native classes or template class instances $m\langle m^*\rangle$, and procedures with a name $f$ and a definition *pgm*. For instance, int $x$ defines an integer variable $x$ while sc_signal⟨bool⟩ $x$ defines a boolean signal $x$. We assume $x$ to denote the name of a variable or signal and to be possibly prefixed as $m :: x$ by the name of the class it belongs to. We assume the relation $\leq$ to denote SystemC sub-typing, e.g., bool $\leq$ num or int $\leq$ num.

*Table 1.2.* Abstract syntax for SystemC

$$
\begin{array}{lll}
sys & ::= [\mathsf{template}\ \langle \mathsf{class}\ m_1\rangle\ \#\mathsf{TYPE}(m_1, m_2)]\ \mathsf{class}\ m\ \{dec\} & \text{(class)} \\
& \mid \mathsf{SC\_MODULE}(m)\ \{dec; \mathsf{SC\_CTOR}(m)\ \{new\}\} & \text{(module)} \\
& \mid sys; sys & \text{(sequence)} \\
dec & ::= m\langle m^*\rangle\ x & \text{(field)} \\
& \mid \mathsf{void}\ f()\ \{pgm\} & \text{(thread)} \\
& \mid dec; dec & \text{(sequence)} \\
new & ::= \mathsf{SC\_THREAD}(f)\ \mathsf{sensitive} \ll x^* \mid new; pgm & \text{(constructor)}
\end{array}
$$

The formal grammar of SystemC programs, listed in Table 1.3, is represented in static single-assignment intermediate form akin to that of the Tree-SSA package of the GCC project [9].

*Table 1.3.* Abstract syntax for SystemC programs in SSA form

$$
\begin{array}{llll}
\text{(program)} & pgm & ::= L{:}blk \mid pgm; pgm & \\
\text{(block)} & blk & ::= stm; blk \mid rtn & \\
\text{(instruction)} & stm & ::= x = f(x^*) & \text{(invoke)} \\
& & \quad \mid \mathsf{wait}\ x & \text{(lock)} \\
& & \quad \mid \mathsf{notify}\ x & \text{(unlock)} \\
& & \quad \mid \mathsf{if}\ x\ \mathsf{then}\ L & \text{(test)} \\
\text{(return)} & rtn & ::= \mathsf{goto}\ L & \text{(goto)} \\
& & \quad \mid \mathsf{return} & \text{(return)} \\
& & \quad \mid \mathsf{throw}\ x; & \text{(throw)} \\
& & \quad \ \ \mathsf{catch}\ x\ \mathsf{from}\ L & \text{(catch)} \\
& & \quad \ \ \mathsf{to}\ L\ \mathsf{using}\ L &
\end{array}
$$

SSA provides a language-independent, locally optimized intermediate representation (Tree-SSA currently accepts C, C++, Fortran 95, and Java inputs) in which language-specific syntactic sugar is absent. SSA transforms a given programming unit (a function, a method or a thread) into a structure in which all variables are read and written once and all native operations are represented by 3-address instructions $x = f(y, z)$.

A program *pgm* consists of a sequence of labeled blocks $L{:}blk$. Each block consists of a label $L$ and of a sequence of statements *stm* terminated by a return statement *rtn*. In the remainder, a block always starts with a label and finishes with a return statement: $stm_1; L{:}stm_2$ is rewritten as $stm_1$; goto $L$; $L{:}stm_2$. A wait is always placed at the beginning of a block: $stm_1$; wait $v$; $stm_2$ is rewritten as $stm_1$; goto $L$; $L$:wait $v$; $stm_2$.

Block instructions consist of native method invocations $x = f(x^*)$, lock monitoring and branches if $x$ then $L$. Blocks are returned from by either a goto $L$, a return or an exception throw $x$. The declaration catch $x$ from $L_1$ to $L_2$ using $L_3$ that matches an exception $x$ raised at block $L_1$ activates the exception handler $L_3$ and continues at block $L_2$.

EXAMPLE 1 *To outline the construction of the intermediate representation of a SystemC program, let us reconsider the example of Section 1.2 and detail the method that counts the number of bits set to* 1 *in a bit-array* epc.data.

```
void epc::ones () { sc_int<16> idata = 0, ocount = 0;
  while true { wait (epc.lock);
               idata = epc.data;
               ocount = 0;
               while (idata != 0) { ocount = ocount + (idata & 1);
                                    idata = idata >> 1; }
               epc.count = ocount;
               notify (epc.lock); }}}
```

*The method consists of three blocks. The block labeled* L1 *waits for the lock* epc.lock *before initializing the local state variable* idata *to the value of the input signal* epc.data *and* ocount *to* 0. *Label* L2 *corresponds to a loop that shifts* idata *right and adds its right-most bit to* ocount *until termination (condition* T0*). Finally, in the block* L3, ocount *is sent to the signal* epc.count *and* epc.lock *is unlocked before going back to* L1.

```
L1:wait (epc.lock);      L2:T1 = idata;              L3:epc.count = ocount;
   idata = epc.data;        T0 = T1 == 0;               notify (epc.lock);
   ocount = 0;              if T0 then goto L3;         goto L1;
   goto L2;                 T2 = ocount;
                            T3 = T1 & 1;
                            ocount = T2 + T3;
                            idata = T1 >> 1;
                            goto L2;
```

## 3.2 Behavioral type inference

The type inference function $\mathcal{I}[\![pgm]\!]$, Table 1.4, is defined by induction on the formal syntax of *pgm*. To define it, we assume that the finite set $\mathcal{L}_f$ of program labels $L$ defined in a given method $f$ respects the order of appearance in the text: $L_1 < L_2$ means that $L_1$ occurs before $L_2$.

To each block of label $L \in \mathcal{L}_f$, the function $\mathcal{I}[\![pgm]\!]$ associates an *input clock* $x_L$, an *immediate clock* $x_L^{imm}$ and an *output clock* $x_L^{exit}$. The clock $x_L$ is true iff $L$ has been activated in the previous transition (by emitting the event $x'_L$). The clock $x_L^{imm}$ is set to true to activate the block $L$ immediately. The clock $x_L^{exit}$ is set to true when the execution of the block labeled $L$ terminates. The default activation condition of this block is the clock $x_L \vee x_L^{imm}$ (equation (1) of Table 1.4). The block $L$ is executed iff the proposition $x_L \vee x_L^{imm}$ holds, meaning that the program counter is at $L$.

For a return instruction or for a block, the type inference function returns a type $P$. For a block instruction *stm*, the type inference function $\mathcal{I}[\![stm]\!]_L^{e_1} = \langle P \rangle^{e_2}$ takes three arguments: an instruction *stm*, the label $L$ of the block it belongs to, and an input clock $e_1$. It returns the type $P$ of the instruction and its output clock $e_2$. The output clock of *stm* corresponds to the input clock of the instruction that immediately follows it in the execution sequence of the block.

EXAMPLE 2 *Let us focus on the block* L2 *of the ones counter, as expressed in the SSA block below, left. On the first line, we associate the instruction* $\mathsf{T1} = \mathsf{idata}$ *of block label* L2 *to the proposition* $x_{\mathsf{L2}} \Rightarrow \mathsf{T1} = \mathsf{idata}$ *(in the behavioral type on the right). In this proposition, the new variable* $x_{\mathsf{L2}}$ *is a boolean that is true iff the label* L2 *is being executed. This means that, if the label* L2 *is being executed,* T1 *is always equal to* idata*. Otherwise, another proposition may hold. In our case, all subsequent propositions are conditioned by* $x_{\mathsf{L2}}$ *meaning that they hold when* L2 *is being executed.*

| (SSA code) | (behavioral type) |
|---|---|
| L2:T1 = idata; | $x_{\mathsf{L2}} \Rightarrow \mathsf{T1} := \mathsf{idata}$ |
| T0 = T1 == 0; | $\mathsf{T0} := (\mathsf{T1} = 0)$ |
| if T0 then goto L3; | $\mathsf{T0} \Rightarrow x'_{\mathsf{L3}}$ |
| T2 = ocount; | $\neg\mathsf{T0} \Rightarrow \mathsf{T2} := \mathsf{ocount}$ |
| T3 = T1 & 1; | $\mathsf{T3} := \mathsf{T1}\&1$ |
| ocount = T2 + T3; | $\mathsf{ocount}' := \mathsf{T2} + \mathsf{T3}$ |
| idata = T1 >> 1; | $\mathsf{idata}' := \mathsf{T1} >> 1$ |
| goto L2; | $x'_{\mathsf{L2}}$ |

*Next, consider the instruction* if T0 then L3. *It is likely that label* L3 *will, just as* L2*, perform some operation on the input* idata. *Therefore, we delay its execution until after the current reaction, and we refer to* $x'_{\mathsf{L3}}$ *as the next value of the state variable* $x_{\mathsf{L3}}$*, to indicate that it will be active during the next reaction. Hence, the proposition* $x_{\mathsf{L2}} \Rightarrow \mathsf{T0} \Rightarrow x'_{\mathsf{L3}}$ *says that the control will be passed to* L3 *at the next reaction* L2 *currently hold the control and* T0 *is true. The instructions that follow this test are conditioned by the negative* $\neg$T0*, this means: "in the block* L2 *and not in its branch going to* L3".

Table 1.4 defines the behavioral type inference system. Rules $(1-2)$ are concerned with the iterative decomposition of a program $pgm$ into blocks *blk* and with the decomposition of a block into $stm$ and $rtn$ instructions. In rule (2), the input clock $e$ of the block $stm$; *blk* is passed to $stm$. The output clock $e_1$ of $stm$ becomes the input clock of *blk*.

*Table 1.4.* Type inference

| | |
|---|---|
| (1) | $\mathcal{I}[\![L\!:\!blk;pgm]\!] = \mathcal{I}[\![blk]\!]_L^{x_L \vee x_L^{imm}} \,\vert\, \mathcal{I}[\![pgm]\!]$ |
| (2) | $\mathcal{I}[\![stm;blk]\!]_L^e = \text{let } \langle P\rangle^{e_1} = \mathcal{I}[\![stm]\!]_L^e \text{ in } P \,\vert\, \mathcal{I}[\![blk]\!]_L^{e_1}$ |
| (3) | $\mathcal{I}[\![\mathsf{if}\ x\ \mathsf{then}\ L_1]\!]_L^e = \langle \mathcal{G}_L(L_1, e\wedge x)\rangle^{e\wedge\neg x}$ |
| (4) | $\mathcal{I}[\![x = f(y^*)]\!]_L^e = \langle \mathcal{E}(f)(xy^*e)\rangle^e$ |
| (5) | $\mathcal{I}[\![\mathsf{notify}\ x]\!]_L^e = \langle e \Rightarrow (x' = \neg x)\rangle^e$ |
| (6) | $\mathcal{I}[\![\mathsf{wait}\ x]\!]_L^e = \langle e\wedge (x\neq x') \Rightarrow \hat{y} \,\vert\, e\setminus \hat{y} \Rightarrow x'_L\rangle^{\hat{y}}$ |
| (7) | $\mathcal{I}[\![\mathsf{goto}\ L_1]\!]_L^e = (e \Rightarrow x_L^{exit} \,\vert\, \mathcal{G}_L(L_1, e))$ |
| (8) | $\mathcal{I}[\![\mathsf{return}]\!]_L^e = (e \Rightarrow (x_L^{exit} \,\vert\, x_f^{exit}))$ |
| (9) | $\mathcal{I}[\![\mathsf{throw}\ x]\!]_L^e = (e \Rightarrow (x_L^{exit} \,\vert\, \hat{x}))$ |

where

$$\mathcal{G}_L(L_1, e) = \text{if } \mathcal{S}_L(L_1) \text{ then } e \Rightarrow x_{L_1}^{imm} \text{ else } \langle e \Rightarrow x'_{L_1}\rangle$$

$$\mathcal{E}(f)(xyze) = e \Rightarrow (\hat{y}\wedge\hat{z} \Rightarrow (\hat{x} \,\vert\, y\to x \,\vert\, z \to x)),\ \forall fxyze$$

$$\mathcal{I}[\![\mathsf{catch}\ x\ \mathsf{from}\ L\ \mathsf{to}\ L_1\ \mathsf{using}\ L_2]\!]_L^e = \mathcal{G}_L(L_2, \hat{x}\wedge x_L^{exit}) \,\vert\, G_{L_2}(L_1, x_{L_2}^{exit})$$

The input and output clocks of an instruction may differ. This is the case, rule (3), for an $\mathsf{if}\ x\ \mathsf{then}\ L_1$ instruction in a block $L$. Let $e$ be the input clock of the instruction. When $x$ is false, then control is passed to the rest of the block, at the output clock $e\wedge\neg x$. Otherwise, the control is passed to the block $L_1$, at the clock $e\wedge x$.

There are two ways of passing the control from $L$ to $L_1$ at a given clock $e$. They are defined by the function $\mathcal{G}_L(L_1, e)$: either immediately, by activating the immediate clock $x_{L_1}^{imm}$, i.e., $e \Rightarrow x_{L_1}^{imm}$; or by a delayed transition to $L_1$ at $e$, i.e., $e \Rightarrow x'_{L_1}$. This choice is decided by the auxiliary function $\mathcal{S}_L(L_1)$. It checks whether the block $L_1$ can be executed immediately after the block $L$. By definition, $\mathcal{S}_L(L_1)$ holds iff $L_1 > L$ ($L_1$ is after $L$ in the control flow) and $\mathcal{D}(L_1)\cap\mathcal{D}(L) = \emptyset$ (the set of variables defined in $L$ and $L_1$ are disjoint).

EXAMPLE 3 *In Example 1,* $\mathcal{D}(\mathsf{L1}) = \mathcal{D}(\mathsf{L2}) = \{\mathsf{ocount}, \mathsf{idata}\}$ *and* $\mathcal{D}(\mathsf{L3}) = \{\mathsf{count}, \mathsf{lock}\}$. *Hence, going from* L1 *to* L2 *requires a delayed transition because both* L1 *and* L2 *define* ocount *and* idata. *Conversely, going from* L2 *to* L3 *can be done immediately since* L3 *does not define neither* ocount *nor* idata.

Rule (4) is concerned with the typing of native and external method invocations $x = f(y^*)$. The generic type of $f$ is taken from an environment $\mathcal{E}(f)$. It is given the name of the result $x$, of the actual parameters $y^*$ and of the input clock $e$ to obtain the type of $x = f(y^*)$. On the right, the generic type of 3-address instructions $x = f(y, z)$ at clock $e$ is given by $\mathcal{E}(f)(xyze)$.

The wait-notify protocol (rules $(5 - 6)$) is modeled using a boolean flip-flop variable $x$. The notify method, rule (5), defines the next value of the lock $x$ by the negation of its current value at the input clock $e$.

The wait method, rule (6), activates its output clock $\hat{y}$ iff the value of the lock $x$ has changed at the input clock $e$: $e \wedge (x \neq x') \Rightarrow \hat{y}$. Otherwise, at the clock $e \setminus \hat{y}$, the control is passed to $L$ by a delayed transition $e \setminus \hat{y} \Rightarrow x'_L$.

EXAMPLE 4 *Consider the wait-notify protocol at the blocks labeled* L1 *and* L3 *in the* ones *counter. The type of the wait instruction defines the output clock* $\hat{y}$ *if* L1 *receives control at the clock* $x_{\mathsf{L1}}$*, and if the value of lock has changed (proposition* lock $\neq$ lock$'$*). If so, at the clock* $\hat{y}$*,* ocount *and* idata *are initialized and the control is passed to the block* L2 *by* $\mathcal{G}_{\mathsf{L1}}(\mathsf{L2}, \hat{y})$*. Otherwise, at the clock* $x_{\mathsf{L1}} \setminus \hat{y}$*, a delayed transition to* L1 *is scheduled:* $x_{\mathsf{L1}} \setminus \hat{y} \Rightarrow x'_{\mathsf{L1}}$.

| code | type |
|---|---|
| L1:wait (epc.lock); | $x_{\mathsf{L1}} \wedge (\mathsf{lock} \neq \mathsf{lock}') \Rightarrow \hat{y}$ |
| | $x_{\mathsf{L1}} \setminus \hat{y} \Rightarrow x'_{\mathsf{L1}}$ |
| ⋮ | ⋮ |
| goto L2; | $\hat{y} \Rightarrow x'_{\mathsf{L2}}$ |
| ⋮ | ⋮ |
| L3:epc.count = ocount; | $\widehat{\mathsf{ocount}} \wedge x_{\mathsf{L3}} \Rightarrow \widehat{\mathsf{count}}$ |
| notify (epc.lock); | $x_{\mathsf{L3}} \Rightarrow \mathsf{lock}' = \neg \mathsf{lock}$ |
| goto L1; | $x_{\mathsf{L3}} \Rightarrow x'_{\mathsf{L1}}$ |

All return instructions, rules $(7 - 9)$, define the output clock $x_L^{exit}$ of the current block $L$ by their input clock $e$. This is the right place to do that: $e$ defines the very condition upon which the block actually reaches its return statement.

A goto $L_1$ instruction, rule (7), passes control to block $L_1$ unconditionally at the input clock $e$ by $\mathcal{G}_L(L_1, e)$. A return instruction, rule (8), sets the exit clock $x_f$ to true at clock $e$ to inform the caller that $f$ is terminated.

A throw $x$ statement in block $L$, rule (9), triggers the exception signal $x$ at the input clock $e$ by $e \Rightarrow \hat{x}$. The matching catch statement, of the form catch $x$ from $L$ to $L_1$ using $L_2$ passes the control to the handler $L_2$ and then to the block $L_1$ upon termination of the handler.

This requires, first, to activate $L_2$ from $L$ when $x$ is present, i.e., $\mathcal{G}_L(L_2, \hat{x} \wedge x_L^{exit})$, and then to pass control to $L_1$ upon termination of the handler.

**Completion of the state logic** The encoding of Table 1.4 requires all entry clocks $x_L$, $x_L^{imm}$ and $x_f$ to be simultaneously present when the $f$ is being executed. Each signal $x_L$ holds the value 1 iff the block $L$ is active during a transition currently being executed. Otherwise, $x_L$ is set to 0. This default setting of the entry clocks requires a completion of the next-state logic by considering, for all $L \in \mathcal{L}_f$, the proposition $e_L \Rightarrow x'_L$ implied by the inferred type $P = \mathcal{I}[\![pgm]\!]$ and defines the default rule by $x_f \setminus e_L \Rightarrow \neg x'_L$. Completion is identical for the immediate and exit clocks $x_L^{imm}$ and $x_L^{exit}$ of the block $L$.

$$x_f \setminus e_L \Rightarrow \neg x'_L \text{ where } e_L \stackrel{\text{def}}{=} \bigvee \left(e \,|P \models e \Rightarrow x'_L\right)$$

**Modular extension to external method calls** The type inference scheme defined for wait, notify, and operations, rules $(4-6)$ can be extended to handle externally defined method calls in a modular and compositional way, depicted in Table 1.5.

*Table 1.5.* Modular extension of the inference function to separately defined methods

$$\begin{array}{ll}
(a) & \mathcal{I}[\![m\, f(x_{1..m})\, \mathsf{raises}\, y\{\, pgm\,\}]\!] = \lambda x_{1..m} x_f\, x_f^{exit} y.\, \left(\mathcal{I}[\![pgm]\!]\,|\, x_f \Rightarrow x_{\min \mathcal{L}_f}\right) / \mathcal{L}_f \\
(b) & \mathcal{I}[\![L : x_0 = f(x_{1..m})]\!]_L^e = e \Rightarrow (\mathcal{E}(f)(x_{1..m} exy)\,|\, e \setminus (\hat{y} \vee \hat{x}) \Rightarrow x'_L)^{e \wedge \hat{x}} \\
(c) & \mathcal{I}[\![\mathsf{return}\, x]\!]_L^e = (e \Rightarrow (x_L^{exit}\,|\, x_f^{exit} := x))
\end{array}$$

Consider a method $f$ with formal parameters $x_{1..m}$ (whose data-types are not displayed) and a result of type $m$, rule $(a)$. Let $y$ be an exception raised by the definition *pgm* of $f$ and escaping from it. The type of $f$ consists of a lambda abstraction whose arguments are the inputs $x_{1..m}$, the entry clock $x_f$, the exit clock $x_f^{exit}$, the return value $y_f$ and the exception $y$. It is used to parameterize the proposition $P$, which corresponds to *pgm*, with respect to these arguments.

The lambda abstraction is instantiated in place of a method invocation $L$ : $x_0 = f(x_{1..m})$, rule $(b)$, which needs to be placed at the beginning of a block (assuming that this block can take several transitions before termination). To model the method call, one just needs to activate the entry clock $x_f$ of the method at the input clock $e$.

The output signal $x$ is used to carry the value of the result. Its clock determines when the method has reached the corresponding return statement (rule $(c)$). When the method terminates, the exit clock of the method call is defined by $e \wedge \hat{x}$. Otherwise, if the exception $y$ is raised, a corresponding catch statement handles it. If $f$ has not finished at the end of the transition (at the clock $e \setminus (\hat{y} \vee \hat{x})$), a delayed transition to $L$ is performed $e \setminus (\hat{y} \vee \hat{x}) \Rightarrow x'_L$ in order to resume its execution at the next transition.

**Static interface of SystemC modules** The construction of a static abstraction from the behavioral type $P$ of a program is automatic, thanks to its clock inference system.

EXAMPLE 5 *For instance, the instruction* $\mathsf{T0} = (\mathsf{T1}! = 0)$ *is abstracted by the type* $x_{\mathsf{L2}} \Rightarrow \hat{\mathsf{T1}} = \hat{\mathsf{T0}}$ . *It means: "when the block* L2 *is executed,* T0 *is present iff* T1 *is present". The scheduling constraint* $x_{\mathsf{L2}} \Rightarrow \mathsf{T0} \rightarrow x_{\mathsf{L3}}$ *additionally says that "*$x'_{\mathsf{L3}}$ *cannot happen before* T0 *at* L2*". Indeed, one first needs to examine the status of* T0 *before branching to* L3.

*Table 1.6.* Abstraction of the behavioral type of block L2 by a static interface

| (dynamic type) | (static type) | (scheduling) |
|---|---|---|
| $x_{\mathsf{L2}} \Rightarrow \mathsf{T1} = \mathsf{idata}$ | $x_{\mathsf{L2}} \Rightarrow \hat{\mathsf{T1}} = \hat{\mathsf{idata}}$ | $\mathsf{T1} \leftarrow \mathsf{idata}$ |
| $\mathsf{T0} = (\mathsf{T1} = 0)$ | $\hat{\mathsf{T0}} = \hat{\mathsf{T1}}$ | $\mathsf{T1} \rightarrow \mathsf{T0}$ |
| $\mathsf{T0} \Rightarrow x'_{\mathsf{L3}}$ | $\mathsf{T0} \Rightarrow x'_{\mathsf{L3}}$ | $\mathsf{T0} \rightarrow x'_{\mathsf{L3}}$ |
| $\neg\mathsf{T0} \Rightarrow \mathsf{T2} = \mathsf{ocount}$ | $\neg\mathsf{T0} \Rightarrow \hat{\mathsf{T2}} = \hat{\mathsf{ocount}}$ | $\mathsf{T2} \leftarrow \mathsf{ocount}$ |
| $\mathsf{T3} = \mathsf{T1} \& 1$ | $\hat{\mathsf{T3}} = \hat{\mathsf{T1}}$ | $\mathsf{T1} \rightarrow \mathsf{T3}$ |
| $\mathsf{ocount}' = \mathsf{T2} + \mathsf{T3}$ | $\hat{\mathsf{ocount}} = \hat{\mathsf{T2}} \wedge \hat{\mathsf{T3}}$ | $\mathsf{T2} \rightarrow \mathsf{ocount}' \leftarrow \mathsf{T3}$ |
| $\mathsf{idata}' = \mathsf{T1} >> 1$ | $\hat{\mathsf{idata}} = \mathsf{T1}$ | $\mathsf{T1} \rightarrow \mathsf{idata}'$ |
| $x'_{\mathsf{L2}}$ | $x'_{\mathsf{L2}}$ | $\mathsf{T0} \rightarrow x'_{\mathsf{L2}}$ |

*The type associated with the block* L2 *uses the clocks denoted by the booleans* $x_{\mathsf{L2}}$, T0 *and* ¬T0. *Each clock denotes a branch in the control-flow graph of the block* L2. *The other clocks, e.g.* $\hat{\mathsf{T1}}$, *denote the presence of data. They are partially related to the "label" clocks* $x_{\mathsf{L2}}$, T0 *and* ¬T0.

## 3.3 A behavioral module system

We define a module system starting from the behavioral type inference function of Section 3.2.

*Table 1.7.* Behavioral types for modules

$$
\begin{array}{llll}
\text{(type)} & \mathcal{T} & ::= & \langle \mathcal{E}, P, \mathcal{C} \rangle \mid \mathcal{T} \rightarrow \mathcal{T} \\
\text{(context)} & \mathcal{E} & ::= & [] \mid \mathcal{E}[x : m] \mid \mathcal{E}[f : P] \mid \mathcal{E}[m : \mathcal{T}] \\
\text{(obligation)} & \mathcal{C} & ::= & 1 \mid P \Rightarrow Q \mid \mathcal{C} \wedge \mathcal{C}
\end{array}
$$

The type $\mathcal{T}$ of a module $m$, Table 1.7, consists of an environment $\mathcal{E}$ that associates its methods $f$ and fields $x$ with types, of a type $P$ that denotes the behavior of its constructor, and of a proof obligation $\mathcal{C}$.

The type $\mathcal{T}_1 \rightarrow \mathcal{T}_2$ denotes a template class that produces a module of type $\mathcal{T}_2$ given a parameter of type $\mathcal{T}_1$. A proof obligation is a conjunction of propositions of the form $P \Rightarrow Q$. A proof obligation $P \Rightarrow Q$ is incurred by the instantiation of a template class, whose formal parameter has type $P$, by an actual class parameter, of type $Q$.

The type inference function for modules, $\mathcal{I}[\![sys]\!]_{\mathcal{E}}$, Tables 1.8 and 1.9, assumes a type environment $\mathcal{E}$ that associates names with types. We write $\mathcal{E}(x)$ for the type of the location $x$ and $\mathcal{E}(m.x) = \mathcal{F}(m)(x)$ for the path $m$ to $x$ iff $\mathcal{E}(m) = \langle \mathcal{F}, P, \mathcal{C} \rangle$.

**Type inference for declarations** Rule $(a)$ sequentially processes the declarations $dec$ in a module. Class field declarations contribute to building the type $\mathcal{T}$ of a module: rule $(b)$ associates the location $x$ with the type name $m$ in the class-field $[x : m]$, rule $(c)$ associates the procedure $f$ with the class-field $[f : P]$. The type $\tau$ denotes a program that does nothing. It is neutral by composition.

In rule $(d)$, the initialization of a thread $\textsc{sc\_thread}(f)$ sensitive $\ll x$ in the constructor is associated with the behavior $\mathcal{E}(f)$ of the method $f$ it forks and with the type $\hat{x}_f \Leftarrow \hat{x}$, meaning that $x$ triggers $f$.

*Table 1.8.* Type inference for declarations

$$
\begin{array}{ll}
(a) & \mathcal{I}[\![dec_1; dec_2\}]\!]_{\mathcal{E}} = \text{let } \langle \mathcal{E}_1, P_1, \mathcal{C}_1 \rangle = \mathcal{I}[\![dec_1]\!]_{\mathcal{E}} \\
 & \quad \text{in } \langle \mathcal{E}_1, P_1, \mathcal{C}_1 \rangle \uplus \mathcal{I}[\![dec_2]\!]_{\mathcal{E}\mathcal{E}_1} \\
(b) & \mathcal{I}[\![m\, x]\!]_{\mathcal{E}} = \langle [x : m], \tau, 1 \rangle \\
(c) & \mathcal{I}[\![\mathsf{void}\ f()\ \{pgm\}]\!]_{\mathcal{E}} = \langle [f : \mathcal{I}[\![pgm]\!]_{\mathcal{E}}], \tau, 1 \rangle \\
(d) & \mathcal{I}[\![\textsc{sc\_thread}(f)\ \mathsf{sensitive} \ll x]\!]_{\mathcal{E}} = \langle [], \mathcal{E}(f) \,|\, (\hat{x}_f \Leftarrow \hat{x}), 1 \rangle
\end{array}
$$

**Type inference for modules** Rule $(e)$ processes a sequence of module declarations $sys_1; sys_2$. We write $\langle \mathcal{E}_1, P_1, \mathcal{C}_1 \rangle \uplus \langle \mathcal{E}_2, P_2, \mathcal{C}_2 \rangle = \langle \mathcal{E}_1\mathcal{E}_2, P_1 \,|\, P_2, \mathcal{C}_1 \wedge \mathcal{C}_2 \rangle$ to merge the types of $sys_1$ and $sys_2$. While processing is sequential, the composition of the behavioral type $P_1 \,|\, P_2$ is synchronous.

Rule $(f)$ first obtains the type $\mathcal{T}_1 = \langle \mathcal{E}_1, P_1, \mathcal{C}_1 \rangle$ of its class fields. Then, in the environment $\mathcal{E}$ extended with that of the class fields $\mathcal{E}_1$, the body $new; pgm$ of the constructor is processed to obtain its type $\mathcal{T}_2$. The type of the module becomes $m \cdot (\mathcal{T}_1 \uplus \mathcal{T}_2)$. The notation $m \cdot \langle \mathcal{E}, P, \mathcal{C} \rangle = \langle [m : \mathcal{E}], P, \mathcal{C} \rangle$ (resp. $m \cdot (\mathcal{T}_1 \to \mathcal{T}_2) = \mathcal{T}_1 \to (m \cdot \mathcal{T}_2)$) defines the type of the class $m$ from the type $\langle \mathcal{E}, P, \mathcal{C} \rangle$ of its class fields.

Rule $(g)$ determines the type of a template class $m_2$ whose formal parameter is a class $m_1$ that implements the virtual class $m$. The virtual class $m$ provides the type, and hence the expected behavior, of the formal parameter name $m_1$. It is obtained from the environment $\mathcal{E}$ by $m \cdot \mathcal{T} = \mathcal{E}(m)$. The body of the template (i.e. the field declarations $dec$ of the class $m_2$) is type-checked with the environment $\mathcal{E}$ extended with the association of $m_1$ to the type of the class fields $\mathcal{E}_1$ declared in $m$. This yields the type $m_2 \cdot \mathcal{T}_2$ of the class. The type of the template is defined by associating $m_2$ with the type $(m_1 \cdot \mathcal{T}_1) \to \mathcal{T}_2$ (and hence $m_1$ with the type $\mathcal{T}_1$).

Rule $(h)$ performs the instantiation $m_2\langle m\rangle\, x$ of a template class $m_2$ with an actual parameter $m$ to define the class name $x$. The type $(m_1 \cdot \mathcal{T}_1) \to \mathcal{T}_2$ of the template class $m_2$ and the type $m \cdot \mathcal{T}$ of the actual parameter $m$ are obtained from the supplied environment $\mathcal{E}$. Type matching between $\mathcal{T}$ and $\mathcal{T}_1$ requires the resolution of a sub-typing between $\mathcal{T}_1[m_2.m/m_1]$ and $\mathcal{T}_2[m_2.m/m]$. The term $\mathcal{T}_1[m_2.m/m_1]$ stands for the substitution of the name $m_1$ by the concatenation $m_2.m$ in $\mathcal{T}_1$. The resolution of the type matching constraints reduces to the synthesis of the proof obligation $\mathcal{C}$ by the algorithm $\mathcal{R}$. If $\mathcal{C}$ is satisfied, then the type of the location $x$ is $\mathcal{T}_2[m_2.m/m_1]$.

*Table 1.9.* Type inference for modules

$$(a)\quad \mathcal{I}[\![sys_1; sys_2]\!]_{\mathcal{E}} = \begin{array}[t]{l}\text{let } \langle \mathcal{E}_1, P_1, \mathcal{C}_1\rangle = \mathcal{I}[\![sys_1]\!]_{\mathcal{E}}\\ \text{in } \langle \mathcal{E}_1, P_1, \mathcal{C}_1\rangle \uplus \mathcal{I}[\![sys_2]\!]_{\mathcal{E}\mathcal{E}_1}\end{array}$$

$$(b)\quad \mathcal{I}\left[\!\!\left[\begin{array}{l}\textsf{SC\_MODULE}(m)\,\{\,dec;\\ \quad\textsf{SC\_CTOR}(m)\,\{new; pgm\}\}\end{array}\right]\!\!\right]_{\mathcal{E}} = \begin{array}[t]{l}\text{let } \langle \mathcal{E}_1, P_1, \mathcal{C}_1\rangle = \mathcal{I}[\![dec]\!]_{\mathcal{E}}\\ \text{and } \mathcal{T}_2 = \mathcal{I}[\![new; pgm]\!]_{\mathcal{E}\mathcal{E}_1}\\ \text{in } m \cdot ((\langle \mathcal{E}_1, P_1, \mathcal{C}_1\rangle \uplus \mathcal{T}_2)\end{array}$$

$$(c)\quad \mathcal{I}\left[\!\!\left[\begin{array}{l}\textsf{template}\,\langle\textsf{class}\, m_1\rangle\\ \quad\textsf{\#TYPE}(m_1, m)\\ \quad\textsf{class}\, m_2\,\{dec\}\end{array}\right]\!\!\right]_{\mathcal{E}} = \begin{array}[t]{l}\text{let } m \cdot \mathcal{T} = \mathcal{E}(m)\\ \text{and } \langle \mathcal{E}_1, \_, \_\rangle = m_1 \cdot \mathcal{T}\\ \text{and } \mathcal{T}_2 = \mathcal{I}[\![dec]\!]_{\mathcal{E}\mathcal{E}_1}\\ \text{in } \langle [m_2 : (m_1 \cdot \mathcal{T}_1) \to \mathcal{T}_2], \tau, 1\rangle\end{array}$$

$$(d)\quad \mathcal{I}[\![m_2\langle m\rangle\, x]\!]_{\mathcal{E}} = \begin{array}[t]{l}\text{let } (m_1 \cdot \mathcal{T}_1) \to \mathcal{T}_2 = \mathcal{E}(m_2)\\ \text{and } m \cdot \mathcal{T} = \mathcal{E}(m)\\ \text{and } \mathcal{C} = \mathcal{R}(\mathcal{T}_1[m_2.m/m_1], \mathcal{T}[m_2.m/m])\\ \text{in } x \cdot (\mathcal{T}_2[m_2.m/m_1]) \uplus \langle [], \tau, \mathcal{C}\rangle\end{array}$$

**Proof obligation synthesis** The resolution $\mathcal{R}(\mathcal{T}_1, \mathcal{T}_2)$ of sub-typing constraints is defined by induction on the structure of the pair $(\mathcal{T}_1, \mathcal{T}_2)$. It reduces to the proof of a conjunction of propositions of the form $P_1 \Rightarrow P_2$ (of denotation

$[\![P_1]\!] \subseteq [\![P_2]\!]$, Appendix 3.9).

$$
\begin{aligned}
\mathcal{R}([], []) &\Leftrightarrow 1 \\
\mathcal{R}(\mathcal{E}_1 \to \mathcal{T}_1, \mathcal{E}_2 \to \mathcal{T}_2) &\Leftrightarrow \mathcal{R}(\mathcal{E}_2, \mathcal{E}_1) \wedge \mathcal{R}(\mathcal{T}_1, \mathcal{T}_2) \\
\mathcal{R}(\mathcal{E}_1[x : t_1], \mathcal{E}_2[x : t_2]) &\Leftrightarrow \mathcal{R}(\mathcal{E}_1, \mathcal{E}_2) \wedge (t_2 \leq t_1) \\
\mathcal{R}(\mathcal{E}_1[f : P_1], \mathcal{E}_2[f : P_2]) &\Leftrightarrow \mathcal{R}(\mathcal{E}_1, \mathcal{E}_2) \wedge (P_2 \Rightarrow P_1) \\
\mathcal{R}(\mathcal{E}_1[m : \mathcal{T}_1], \mathcal{E}_2[m : \mathcal{T}_2]) &\Leftrightarrow \mathcal{R}(\mathcal{E}_1, \mathcal{E}_2) \wedge \mathcal{R}(\mathcal{T}_1, \mathcal{T}_2) \\
\mathcal{R}(\langle \mathcal{E}_1, P_1, \mathcal{C}_1 \rangle, \langle \mathcal{E}_2, P_2, \mathcal{C}_2 \rangle) &\Leftrightarrow \mathcal{R}(\mathcal{E}_1, \mathcal{E}_2) \wedge \mathcal{R}(P_1, P_2) \wedge \mathcal{R}(\mathcal{C}_1, \mathcal{C}_2)
\end{aligned}
$$

If $P_2$ is *static* (i.e. $P_2 \Leftrightarrow \hat{P}_2$) then the problem reduces to checking for satisfaction of the boolean proposition $\hat{P}_1 \Rightarrow \hat{P}_2$). If $P_2$ is *dynamic* then the problem reduces to verifying that $P_1 \Rightarrow P_2$ is an invariant of $P$, the type of the program, by using model-checking techniques. Both problems can be expressed and decided using the Polychrony workbench [17], as demonstrated in Appendix 3.9, where SystemC programs and their behavioral types are embedded in Signal.

## 3.4 Applications

We have introduced a type system allowing to model the control and data flow graphs of a given imperative program in SSA intermediate form and demonstrated that the expressive capability of the type system's semantics matched that of *de-facto* standard design languages (e.g. SystemC) and as well as that of related multi-clock synchronous formalisms (e.g. Signal). As such, applications of the proposed type system encompass optimization and verification issues encountered in related system design methodologies, yet with the following features that merit a highlight.

## 3.5 Scalability

Just as the theory of interfaces automata [4], types allow to scale the level of abstraction to be automatically obtained starting from the type inferred from a SystemC module within the simple formalism of Table 1.1. Behavioral types share with interface automata the capability to define *static* interfaces (boolean relations) and *dynamic* interfaces (a transition system). Behavioral types allows to relate a given proposition $P$ to a more abstract one, $Q$, in several ways:

- a transition $e \Rightarrow x' = y$ can be abstracted by a clock relation between $e$, $\hat{x}$ and $\hat{y}$;
- a bound signal $x$ in $P/x$ can be abstracted by any proposition $Q$ which contains $P$ and does not reference $x$;
- a free signal $x$ whose clock is not frequent (because it appears at a lower level in the clock hierarchy) can be abstracted by this clock in any $Q$ containing $P$.

All these examples are instances of a more generic abstraction pattern. In general, checking a user-specified abstraction $Q$ consistent with the type $P$ inferred from a given program amounts to the satisfaction of the containment relation $[\![\hat{P}]\!] \subseteq [\![Q]\!]$ (i.e. the denotation of $P$ is contained in the denotation of $Q$). If $P$ is a *static interface*, this amounts to the satisfaction of a boolean equation. Similarly, checking that a *dynamic interface* $Q$ is an abstraction of a process $P$ amounts to verifying that $Q$ simulates $P$ by model-checking.

## 3.6 Modularity

The main advantage of formulating a behavioral type system for SystemC is the formal foundation it offers to investigate modular and compositional design methodologies using separate compilation techniques. For instance, suppose that the declared type $P$ of a SystemC class template provides sufficient information about its formal parameter for its body to be checked controllable and compiled.

One may then provide it with an actual class parameter, of type $Q$, satisfying $Q \Rightarrow P$, without having the burden of fully instantiating the template code and recompile its code for that given instance.

EXAMPLE 6 *To exemplify the benefits of behavior type inference for SystemC modules, let us first consider the dynamic type $P$ of a counter modulo 2. It generates an output event $y$ upon two occurrences of the signal $x$. Its state $s$ is initially false. The signal $x$ triggers the calculation of the next value $s'$ of $s$ defined by the negation $\neg s$ of its current value. When $s$ is true then $y$ is triggered.*

$$P \stackrel{\text{def}}{=} (\neg s^0 \,|\, \hat{x} \Rightarrow (s' = \neg s) \,|\, s \Rightarrow \hat{y})$$

*Let $Q = (\hat{x} > \hat{y})$ be the static interface type of $P$ declared for the virtual class of the introductory example, Section 1.2. It just requires the clock $\hat{y}$ to be a sampling of $\hat{x}$. Let us associate the type $P$ and $Q$ with the procedure $f$ using the #TYPE annotation.*

```
class m0 {
    virtual sc_clock x;
    virtual sc_clock y;
    virtual void f() {} #TYPE(f, Q) };

template ⟨class m1⟩ #TYPE(m1, m0)
  SC_MODULE(m2) {SC_CTOR(m2) {
     SC_THREAD(m1.f) sensitive ≪ x
  }};
```

*Let us now briefly recall the example of Section 1.2. First, an interface $m_0$, of virtual class fields $x, y$, and $f$ is declared with the annotation #TYPE$(f, Q)$. The* *template class $m_2$ uses a class parameter $m_1$ that implements $m_0$ to launch a thread $m_1.f$ sensitive to $x$. The class $m_3$ implements the interface $m_0$ and defines the method $f$ by the program $pgm$ of type $P$. Finally, class $m_4$ is defined by the instantiation of the template $m_2$ with the parameter $m_3$.*

$$\begin{array}{l}\textsf{class}\, m_3 \,\{\\ \quad \textsf{sc\_clock}\, x; \textsf{sc\_clock}\, y;\\ \quad \textsf{void}\ f()\ \{\ pgm\}\ \textsf{\#TYPE}(f, P)\ \};\\ m_2\langle m_3\rangle\, m_4\end{array}$$

*Checking the guarantees of the actual parameter $m_3$ satisfy the assumptions of the formal parameter $m_0$ of the template $m_2$ amounts to verifying that $P$ implies $Q$. This is done by calculating the static abstraction $\hat{P} = \exists s. (\hat{s} = \hat{x} \,|\, \hat{y} = s))$ of $P$ and by checking that it implies $\hat{x} > \hat{y}$.*

## 3.7 Design checking

The proposed type system allows to easily formulate properties pertaining on common design errors the analysis of which has been the subject of numerous related works. Most of these approaches consist of proposing an ad-hoc type system for analyzing a specific pattern of design errors: race conditions, deadlocks, threads termination; and in a given programming language: Java, C, SystemC. By contrast, our behavioral type system provides a unified framework to perform both static verification via satisfaction checking or dynamic verification via model checking of behavioral properties of embedded systems described using imperative programming languages. The inference technique itself is language independent and the semantical peculiarities of language-specific runtime features and libraries can be modeled in the polychronous model of computation and its supportive type system.

**Termination.** One common design error found in embedded system design is the unexpected termination of a thread due to, e.g., an uncaught exception.

In our behavioral model of SystemC, the termination of the infinite loop of a thread $f$ can be represented by the property $x_f^{exit} = 1$. Unexpected termination can hence be avoided by model-checking the property that $x_f^{exit} = 0$ is an invariant of $f$:

$$P \models x_f^{exit} = 0$$

**Deadlocks.** Another common design error is a wait statement that does not match a notification and yields the thread to block. Let $x_{L_{1\dots n}}$ be the clocks of the blocks $L_{1\dots n}$ in which the lock $x$ is notified. Waiting for $x$ at a given label $L$ eventually terminates if

$$P \models x_L \wedge \neg(\wedge_{i=1}^{n} x_{L_i}) = 0$$

**Race conditions.** Similarly, concurrent write accesses to a variable $x$ shared by parallel threads can be checked exclusive by considering the input clocks $e_{1,..n}$ of all write statements $x = f(y, z)$ by verifying that

$$P \models (e_i \wedge (\vee_{j \neq i} e_j)) = 0, \; \forall i = 1, ..n$$

EXAMPLE 7 *For instance, consider checking exclusion between the transitions of the* ones *counter. The type* $P$ *of the counter implies the equations* $x'_{\mathsf{L2}} = (\hat{y}_1 \vee \hat{y}_2)$ *and* $x'_{\mathsf{L1}} = x_{\mathsf{L3}}$. *Verifying exclusion between them amounts to proving that* $(\hat{y}_1 \vee \hat{y}_2) \wedge x_{\mathsf{L3}} = 0$ *is an invariant of* $P$. *By construction of* $\overline{P}$, *we have:* $\hat{y}_1 = x_{\mathsf{L1}} \wedge (\mathsf{lock} \neq \mathsf{lock}')$, $\hat{y}_2 = x_{\mathsf{L2}} \setminus \mathsf{T0}$ *and* $x_{\mathsf{L3}} = \mathsf{T0}$. *The property follows by observing that* $\hat{\mathsf{T0}} = x_{\mathsf{L2}}$.

## 3.8 Design exploration

Just as the multi-clocked synchronous formalism Signal it is based upon, our type system allows for the refinement-based design methodologies considered in [15] to be easily implemented.

Checking the correctness of the refinement of an initial SystemC module, of type $P$, by its upgraded version, of type $Q$, amounts to verifying that the final type $Q$ satisfies the assumptions made by the initial specification. In the spirit of the refinement-checking methodology proposed in [15], this can be implemented by checking the upgrade $Q$ to be finitely flow preserving the initial design $P$.

In general, $Q$ may differ from $P$ by the insertion of a protocol between two components of $P$, by the adaptation of the services provided by $P$ with a new functionality implemented in $Q$. Along the way, one may abstract from the type $Q$ the signals and state variables introduced during the refinement in order to accelerate verification. In most cases, such upgrades may be checked incrementally, by checking the static containment relation between the static abstractions of $P$ and $Q$: $\hat{Q} \Rightarrow \hat{P}$.

## 3.9 Implementation in the Polychrony workbench

The Polychrony workbench [17] supports the synchronous multi-clocked data-flow programming language Signal in which the translation of the behavioral type system into Signal, Table 1.12, and the encoding of Signal by behavioral types, Table 1.11, are easily defined.

Signal belongs to the family of synchronous languages such as Esterel and Lustre. A Signal process $P$ consists of simultaneous equations over *signals*. A signal $x \in \mathcal{X}$ describes an infinite flow of values $v \in \mathcal{V}$. An equation $x := f(y, r)$ denotes a relation between the input signals $(y, r)$ and the output signal $x$ by a function or combinator $f$.

Signal has three primitive operators: the equation $x := y\$1\,\mathsf{init}\,v$ initially defines $x$ by $v$ and then by the previous value of $y$ in time, the equation $x := y\,\mathsf{when}\,z$ defines $x$ by $y$ when $z$ is true and the equation $x := y\,\mathsf{default}\,z$ defines $x$ by $y$ when $y$ is present and by $z$ otherwise. The synchronous composition $P\,|\,Q$ of two processes $P$ and $Q$ consists of the simultaneous solution of the system of equations $P$ and $Q$.

*Table 1.10.* Signal syntax core

| | | | |
|---|---|---|---|
| (process) | $P$ | $::=$ | $x := f(y, r) \mid (P\,\vert\,Q) \mid P/x$ |
| (combinator) | $f$ | $\in$ | $\{\$1\,\mathsf{init}\,v \mid v \in \mathcal{V}\} \cup \{\mathsf{when}, \mathsf{default}, \ldots\}$ |

Table 1.11 associates Signal equations with behavioral types, showing their close relationship within the polychronous model of computation.

*Table 1.11.* From Signal to behavioral types $\lceil\!\lceil P \rceil\!\rceil$

$$\lceil\!\lceil x := y\,\mathsf{when}\,z \rceil\!\rceil = (z \Rightarrow x := y) \qquad \lceil\!\lceil x := y\$1\,\mathsf{init}\,v \rceil\!\rceil = (x^0 = v \mid x' := y)$$

$$\lceil\!\lceil x := y\,\mathsf{default}\,z \rceil\!\rceil = (\hat{y} \Rightarrow x := y) \mid (\hat{z} \setminus \hat{y} \Rightarrow x := z) \qquad \lceil\!\lceil P\,|\,Q \rceil\!\rceil = \lceil\!\lceil P \rceil\!\rceil \mid \lceil\!\lceil Q \rceil\!\rceil$$

$$\lceil\!\lceil P/x \rceil\!\rceil = \lceil\!\lceil P \rceil\!\rceil / x$$

Table 1.12 completes the proof on the semantics equivalence between Signal and the behavioral type system. It defines an encoding of propositions by using partial equations $x ::= y\,\mathsf{when}\,z$, an additional feature of the Polychrony workbench, whose meaning match that of propositions $z \Rightarrow x := y$.

*Table 1.12.* From behavioral types to Signal $\lfloor\!\lfloor P \rfloor\!\rfloor$

$$\lfloor\!\lfloor x^0 = v \rfloor\!\rfloor = x := x'\$1\,\mathsf{init}\,v \qquad \lfloor\!\lfloor v \rfloor\!\rfloor = v$$

$$\lfloor\!\lfloor l := r \rfloor\!\rfloor^e = l ::= r\,\mathsf{when}\,\lfloor\!\lfloor e \rfloor\!\rfloor \qquad \lfloor\!\lfloor \hat{x} \rfloor\!\rfloor = \hat{}x$$

$$\lfloor\!\lfloor x \to l \rfloor\!\rfloor^e = x \to l\,\mathsf{when}\,\lfloor\!\lfloor e \rfloor\!\rfloor \qquad \lfloor\!\lfloor l = r \rfloor\!\rfloor = l = r$$

$$\lfloor\!\lfloor P\,|\,Q \rfloor\!\rfloor^e = \lfloor\!\lfloor P \rfloor\!\rfloor^e \mid \lfloor\!\lfloor Q \rfloor\!\rfloor^e \qquad \lfloor\!\lfloor e \wedge f \rfloor\!\rfloor = \lfloor\!\lfloor e \rfloor\!\rfloor\ \mathtt{when}\ \lfloor\!\lfloor f \rfloor\!\rfloor$$

$$\lfloor\!\lfloor e \Rightarrow P \rfloor\!\rfloor^f = \lfloor\!\lfloor P \rfloor\!\rfloor^{e \wedge f} \qquad \lfloor\!\lfloor e \vee f \rfloor\!\rfloor = \lfloor\!\lfloor e \rfloor\!\rfloor\ \mathtt{default}\ \lfloor\!\lfloor f \rfloor\!\rfloor$$

$$\lfloor\!\lfloor P/x \rfloor\!\rfloor^e = \lfloor\!\lfloor P \rfloor\!\rfloor^e / x \qquad \lfloor\!\lfloor e \setminus f \rfloor\!\rfloor = \mathtt{not}\ \lfloor\!\lfloor f \rfloor\!\rfloor\ \mathtt{default}\ \lfloor\!\lfloor e \rfloor\!\rfloor$$

The behavioral type system supports a direct translation into the data-flow notation Signal of the Polychrony workbench [17]. This translation allows for the complete embedding of SystemC modules into Polychrony, Table 1.13, to perform global design transformations, such as hierarchization or distribution and to perform a correct-by-construction design exploration towards the mapping of system functionalities onto a target execution architecture.

*Table 1.13.* Embedding the intermediate representation in Signal

$$
\begin{array}{rl}
\mathcal{C}[\![L\!:\!blk;\, pgm]\!] = & \mathcal{C}[\![L\!:\!blk]\!]_L^{x_L} \,|\, \mathcal{C}[\![pgm]\!] \\
\mathcal{C}[\![stm;\, blk]\!]_L^e = & \text{let } \langle \mathrm{P} \rangle^{e_1} = \mathcal{C}[\![stm]\!]_L^e \text{ in } \mathrm{P} \,|\, \mathcal{C}[\![blk]\!]_L^{e_1} \\
\mathcal{C}[\![rtn]\!]_L^e = & \mathcal{T}[\![\mathcal{I}[\![rtn]\!]_L^e]\!] \\
\mathcal{C}[\![stm]\!]_L^e = & \text{if } (stm \neq \text{"}x = f(x_{1...n})\text{"})) \\
 & \text{then let } \langle P \rangle^{e_1} = \mathcal{I}[\![stm]\!]_L^e \text{ in } \langle \mathcal{T}[\![P]\!] \rangle^{e_1} \\
 & \text{else let } \mathrm{e} = \mathcal{T}[\![e]\!] \text{ and } \mathrm{P} = \mathcal{T}[\![\mathcal{E}(f)(xx_{1...n}e)]\!] \\
 & \text{in} \left\langle \begin{array}{l} \mathtt{spec\,P} \\ \mathtt{pragmas\;CPP\_CODE\;"if\;e\;\{\;x = f(x_{1...n})\;\}"} \\ \mathtt{end\,pragmas} \end{array} \right\rangle^{e}
\end{array}
$$

The translation consists of representing guarded commands $e \Rightarrow (x' = v)$ and $e \Rightarrow \hat{x}$ by partial equations $\mathtt{x} ::= \mathtt{v\;when\;e}$ and $\hat{}\mathtt{x} ::= \mathtt{when\;e}$, of identical meaning, and of embedding native method invocations $x = f(x_{1...n})$ in wrappers (the CPP_CODE part) visible from Signal via a behavioral type (the spec part). Since the previous value of a signal $x$ is noted $x\$1$ in Signal, we assume the default equation $x := x'\$1$ to define the current value $x$ of a variable (of type bool or int in SystemC).

EXAMPLE 8 *As an example, embedding the* ones *counter into Signal consists of emulating control by partial equations and of wrapping computations using typed* pragmas *statements.*

| (SSA code) | (Signal) |
|---|---|
| L1:wait (epc.lock); | $x_1$:= (lock $\neq$ lock$'$) when $x_{L1}$ |
| idata = epc.data; | $x'_{L1}$::= when not $x_1$ |
| ocount = 0; | idata::= data when $x_1$ |
| goto L2; | ocount::= 0 when $x_1$ |
| | $x'_{L2}$::= when $x_1$ |
| ⋮ | ⋮ |

*This embedding allows to operate global architectural transformations on the initial program, such as hierarchization or distributed protocol synthesis, using the Polychrony platform [17] or perform both static (SAT-checking) or dynamic (model-checking) verification of its design properties, whose spectrum is outlined next.*

| (SSA code) | (Signal) |
|---|---|
| ⋮ | ⋮ |
| L2:T1 = idata; | T1:= idata$1 when $x_{L2}$ |
| T0 = T1 == 0; | T0:= (T1 = 0) when $x_{L2}$ |
| if T0 then goto L3; | $x_{L3}$::= when T0 |
| T2 = ocount; | $x_2$:= when not T0 |
| T3 = T1 & 1; | T2:= ocount$1 when $x_2$ |
| ocount = T2 + T3; | T3:= T1 when $x_2$ |
| idata = T1 >> 1; | ocount::= T2 + T3 when $x_2$ |
| goto L2; | idata::= $f$(T1 when $x_2$) |
| | $x'_{L2}$::= when $x_2$ |
| L3:epc.count = ocount; | count::= ocount when $x_{L3}$ |
| notify (epc.lock); | lock::= not lock when $x_{L3}$ |
| goto L1; | $x'_{L1}$::= when $x_{L3}$ |

*The completion of the state-logic is implemented by the aggregation of partial state equations. Notice that* L1 *and* L2 *are always activated by a delayed transition, whereas* L3 *is always immediate when* T0 *holds. Therefore:*

$$
\begin{aligned}
&x_{L1} := x'_{L1}\$1 \text{ init true}\\
&x_{L2} := x'_{L2}\$1 \text{ init false}\\
&x'_{L1} := \text{true when } (\text{not } x_1 \texttt{ default } x_{L3}) \texttt{ default } \text{false}\\
&x'_{L2} := \text{true when } (x_1 \texttt{ default } x_2) \texttt{ default } \text{false}\\
&x_{L3} := \text{true when T0} \texttt{ default } \text{false}\\
&x_f\hat{\ } = x_{L1}\hat{\ } = x_{L2}\hat{\ } = x_{L3}
\end{aligned}
$$

*We inlined native operations on integer-sized bit-arrays in the Signal code except the call to the native method >> that corresponds to an external function $f$ defined by:*

```
function f = ( ? i1 ! i2 )  spec (| ^i1 = ^i2 | i1→i2 |)
                            pragmas C_CODE "i2 = i1 >> 1" end pragmas;
```

Thanks to the polychronous model of computation, the SystemC scheduler, used to interleave the execution of threads, does not have to be specified. Indeed, SystemC scheduling amounts to determining a fixed-point to inconsistently propagated signal values using the notion of $\delta$-cycle, until a fixed-point is reached (Example 9). Fixed-point of $\delta$-cycles relate to instants in the synchronous semantics using the notion of Kantor metrics of Lee et al. [6]. By contrast, the polychronous model of computation relies on the notion of synchrony to denote this fixed-point directly. Its implementation makes use of type-based scheduling information to determine a static scheduling of elementary instructions.

EXAMPLE 9 *Whereas the simulation semantics of SystemC operates by performing δ-delays to determine a scheduling of instructions into micro-steps, synchrony allows to directly capture the fixed-point of this iterative process. For instance, consider two methods simulating* and-or *gates connected via the signals $a, b, c, d$, and $e$.*

```
void andgate () { while true {d=a&&b;wait()}}
void orgate  () { while true {e=d||c;wait()}}
```

*The synchronous semantics of the circuit has all signals available instantaneously, whereas its SystemC simulation requires knowledge on the* and *gate to be sensitive to $a, b$ and then on the* or *gate to be sensitive to $c, d$ in order to schedule the* and *operation first and the* or *operation second. In the real circuit, as in the polychronous model of computation, no δ-cycles are needed: scheduling is resolved at compile-time.*

## 4. Case study of a Finite Impulse Response filter

This section exemplifies the presented approach with the design of a finite impulse response filter (FIR). It details the decomposition of a full featured SystemC specification into an SSA representation. The different analysis steps are demonstrated down to the final typed Signal representation.

As a starting point, we use the SystemC model of an FIR (an example from the SystemC 2.0.1 distribution [18]) and translate it into SSA code. We show how this SSA code is analyzed and how clock and scheduling information can be extracted. In Section 4.5 the corresponding Signal type is presented and it is shown how to obtain it with the preceding information.

### 4.1 The SystemC model

The SystemC model of the FIR filter is taken from an example of the SystemC 2.0.1 distribution [18]. The Filter itself is one functional block, it is surrounded by a testbench consisting of a Stimulus that generates input values and a Display that receives the output and displays it on the screen (Figure 1.5).

The FIR unit is implemented as an SC_THREAD that is triggered on the positive clock edge. The other blocks are SC_METHODS. The left hand side of Table 1.14 displays the SystemC code of the entry function for the FIR block. The first 10 lines just handle the initialization of variables. Then there is an infinite *while* loop that contains the actual filter functionality.

Roughly speaking, it waits until there is a valid input available, reads this input, processes it, writes it to an output, and then notifies its environment that the result is available. At the end of each *while* loop it suspends itself until the next positive clock edge. The FIR result is the sum of the last 15 input values weighted with 15 coefficients. This is done in two *for* loops. The first one does the weighting and the second one is shifting the buffer array containing the last inputs.

*Figure 1.5.* Structure of the FIR filter with testbench

Communication with the environment is done via *enable* signals. The Stimulus indicates with the signal *input_valid* that a new value is available. In the same way, the Display is sensitive to the variable *output_data_ready* that is set when a new output value is available.

## 4.2 Obtaining an SSA representation

The Single Static Assignment form (SSA) is an intermediate representation in which every variable is assigned exactly once. It is particularly used for high level compiler optimization and will be part of the upcoming version 3.5 of the Gnu Compiler Collection (GCC), scheduled te be released this summer. For our approach, the usage of SSA as an intermediate step proves to be very useful: its simplicity makes the extraction of behavior straightforward.

The right hand side of Table 1.14 shows the SSA code that corresponds to the SystemC FIR. We use the tree-ssa [20] branch of the Gnu Compiler Collection [19] (as long as the 3.5 version of GCC is not released yet).

For the generation of a clean SSA representation we follow three steps. First, preprocessing of the SystemC code, second, translation to SSA with GCC, and third, postprocessing of the generated SSA code. The direct generation of SSA from SystemC can be done, but it results in very large and hardly readable code.

A closer look reveals that most of this bloating is due to the SystemC types and statements, which are implemented as macros and get translated as well. If we replace the SystemC types by corresponding C++ types, e.g. *sc_int* is changed to *int* or *unsigned*, in a simple preprocessing step, the size of the generated code shrinks drastically.

More complex statements such as *wait(signal)*, however, still cause a considerable increase of the code size compared to the original SystemC code. We decide to simply comment these out in the SystemC source so they are ignored by the compiler and can later be taken care of separately in a postprocessing step.

*Table 1.14.* SystemC code [18] and corresponding SSA code for the FIR

```
void fir::entry() {
  sc_int<8> sample_tmp;
  sc_int<17>pro;
  sc_int<19>acc;
  sc_int<8>shift[16];
  result.write(0);
  output_data_ready.write(false);

  for (int i=0; i<=15; i++)
    shift[i] = 0;
  wait();

  while(1) {
    output_data_ready.write(false);
    wait_until(input_valid.delayed() == true);
    sample_tmp = sample.read();
    acc = sample_tmp*coefs[0];
    for(int i=14; i>=0; i–) {
      pro = shift[i]*coefs[i+1];
      acc += pro;
    };
    for(int i=14; i>=0; i–)
      shift[i+1] = shift[i];
    shift[0] = sample_tmp;
    // write output values
    result.write((int)acc);
    output_data_ready.write(true);
    wait();
  };
}
```

```
void fir::entry() {
     int shift[16], i, acc, sample_tmp;
     i=0; goto L1;
     this → result = 0;
     this→output_data_ready = 0;
L0:  shift[i]= 0
     i = i + 1;
L1:  t_i = (i<=15)
     if (t_i) goto L0;
     else goto L1a;
L1a:wait (clk_pos);
L3:  this→output_data_ready = 0;
L3a:wait_until(input_valid == true);
L3b:sample_tmp = this→sample;
     acc = this→coefs[0] * sample_tmp;
     i = 14;
     goto L5;
L4:  acc = acc + shift[i]* this→coefs[i+1];
     i = i - 1;
L5:  if (i >= 0) goto L4;
     else goto L13;
L13:i = 14;
     goto L8;
L7:  shift[i + 1] = shift[i];
     i = i - 1;
L8:  if (i>=0) goto L7;
     else goto L9;
L9:  shift[0] = sample_tmp;
     this→result = acc;
     this→output_data_ready = 1;
L9a:wait (clk_pos);
     goto L3; }
```

## 4.3 SSA code size

Table 1.15 illustrates the different code sizes for different compiler options and preprocessing measures. Plain gimplifying of 39 lines of SystemC code results in over 2000 lines of gimple code. This can be reduced by using different compiler options for additional optimization steps. The option *-fdump-tree-optimized* results in almost 5 times less code, but still over 10 times the original SystemC code. Using no SystemC types and leaving the handling of the *wait* statements to the postprocessing results in an SSA code that is only slightly bigger than the original SystemC code.

During post processing we replace the *wait(signal)* statements by corresponding SSA statements. Logically a *wait* statement is similar to an *if* branch. Depending on a condition something is executed, else something else. The condition is the signal that we are waiting for (e.g. *input_valid == true*. If there is no signal given, the process waits for the signal it is sensitive to (this is the positive slope of the clock in this example).

*Table 1.15.* Different SSA code sizes for FIR entry function

| | |
|---|---|
| original SystemC | 39 lines |
| gcc -fdump-tree-gimple | 2217 lines |
| gcc -fdump-tree-ssa | 1310 lines |
| gcc -fdump-tree-optimized | 446 lines |
| + without SystemC types | 142 lines |
| + without wait() | 55 lines |

In order to be able to execute the *wait* statement separately, we have to introduce a separate label for it. As we can see on the right hand side of Table 1.14, for *L1*, *L1a* is introduced since the wait statement is not at the beginning of the block, and for *L3*, there are two additional labels, *L3a* and *L3b* because this wait statement is in the middle of a block.

## 4.4 Extracting clock and scheduling information

Though slightly bigger in size, the SSA representation has several advantages in respect to automated analysis and conversion. On the one hand it consists of very simple, highly repetitive statements, and on the other hand it is separated into blocks of sequential execution that do not contain branches and where no variable gets assigned twice. The extracted behavioral type information can be separated into two parts, control and data flow.

Table 1.6 displays this information separately for the FIR. On the left hand side figures the control flow in the form of clock relations and on the right hand side the data flow information in the form of scheduling dependencies. In order to understand how these clock dependencies are obtained, we have to take a look at the SSA form in Table 1.14. $x_{\mathsf{L0}} \Rightarrow \hat{shift}$ means that whenever block L0 is entered, the signal *shift* has to be present. Transitions from one block to another are represented like this: $x_{\mathsf{L4}} \Rightarrow x_{\mathsf{L5}}$ . However, if in the following block a signal is assigned that has already been assigned in the current block, it cannot be executed in the same cycle. The time has to be advanced, this is expressed in $x_{\mathsf{fir}} \Rightarrow x'_{\mathsf{L1}}$, where the $'$ indicates the next value for this signal. For *if* statements - such as in block L1 - the value of a boolean signal decides which of the two targets is taken. The unio n ('$\wedge$') and the difference ('/') operator suffice to express all required action.

$x_{\mathsf{fir}} \Rightarrow \hat{i}$
$x_{\mathsf{fir}} \Rightarrow x'_{\mathsf{L1}}$
$x_{\mathsf{fir}} \Rightarrow \widehat{result}$
$x_{\mathsf{fir}} \Rightarrow \widehat{output\_data\_ready};\ x_{\mathsf{L0}} \Rightarrow \widehat{shift}$
$x_{\mathsf{L0}} \Rightarrow \hat{i};\ x_{\mathsf{L0}} \Rightarrow x_{\mathsf{L1}}$
$x_{\mathsf{L1}} \Rightarrow \widehat{t\_i};\ x_{\mathsf{L1}} \Rightarrow \hat{i};\ x_{\mathsf{L1}} \wedge t\_i \Rightarrow x'_{\mathsf{L0}}$
$x_{\mathsf{L1}}/x'_{\mathsf{L0}} \Rightarrow x_{\mathsf{L1a}}$
$x_{\mathsf{L1a}} \wedge (clk \neq clk') \Rightarrow \hat{y}$
$x_{\mathsf{L1a}} \wedge \hat{y} \Rightarrow x_{\mathsf{L3}};\ x_{\mathsf{L1a}}/\hat{y} \Rightarrow x'_{\mathsf{L1a}}$
$x_{\mathsf{L3}} \Rightarrow \widehat{output\_data\_ready};\ x_{\mathsf{L3}} \Rightarrow x_{\mathsf{L3a}}$
$x_{\mathsf{L3a}} \wedge input\_valid \Rightarrow \hat{y}$
$x_{\mathsf{L3a}} \wedge \hat{y} \Rightarrow x_{\mathsf{L3b}};\ x_{\mathsf{L3a}}/\hat{y} \Rightarrow x'_{\mathsf{L3a}}$
$x_{\mathsf{L3b}} \Rightarrow \widehat{sample\_tmp};\ x_{\mathsf{L3b}} \Rightarrow \widehat{sample};$
$x_{\mathsf{L3b}} \Rightarrow \widehat{acc};\ x_{\mathsf{L3b}} \Rightarrow \widehat{coefs}$
$x_{\mathsf{L3b}} \wedge \hat{i};\ x_{\mathsf{L3b}} \Rightarrow x_{\mathsf{L5}}$
$x_{\mathsf{L4}} \Rightarrow \widehat{acc};\ x_{\mathsf{L4}} \Rightarrow \widehat{shift};\ x_{\mathsf{L4}} \Rightarrow \widehat{coefs}$
$x_{\mathsf{L4}} \Rightarrow \hat{i}$
$x_{\mathsf{L4}} \Rightarrow x_{\mathsf{L5}}$
$x_{\mathsf{L5}} \wedge (i \mathrel{\hat{\geq}} 0) \Rightarrow x'_{\mathsf{L4}}$
$x_{\mathsf{L5}} \wedge (i \mathrel{\hat{<}} 0) \Rightarrow x'_{\mathsf{L13}}$
$x_{\mathsf{L13}} \Rightarrow \hat{i}$
$x_{\mathsf{L13}} \Rightarrow x_{\mathsf{L8}}$
$x_{\mathsf{L7}} \Rightarrow \widehat{shift};\ x_{\mathsf{L7}} \Rightarrow \hat{i};$
$x_{\mathsf{L7}} \Rightarrow x_{\mathsf{L8}}$
$x_{\mathsf{L8}} \Rightarrow \hat{i}$
$x_{\mathsf{L8}} \wedge (i \geq 0) \Rightarrow x'_{\mathsf{L7}}$
$x_{\mathsf{L8}}/x'_{\mathsf{L7}} \Rightarrow x_{\mathsf{L9}}$
$x_{\mathsf{L9}} \Rightarrow \widehat{shift};\ x_{\mathsf{L9}} \Rightarrow \widehat{sample\_tmp}$
$x_{\mathsf{L9}} \Rightarrow \widehat{result};\ x_{\mathsf{L9}} \Rightarrow \widehat{acc}$
$x_{\mathsf{L9}} \Rightarrow \widehat{output\_data\_ready};\ x_{\mathsf{L9}} \Rightarrow x_{\mathsf{L9a}}$
$x_{\mathsf{L9a}} \wedge (clk \neq clk') \Rightarrow \hat{y};$
$x_{\mathsf{L9a}} \wedge \hat{y} \Rightarrow x'_{\mathsf{L3}};\ x_{\mathsf{L9a}}/\hat{y} \Rightarrow x'_{\mathsf{L9a}}$

$x_{\mathsf{L0}} \Rightarrow i \rightarrow shift$
$x_{\mathsf{L1}} \Rightarrow i \rightarrow t\_i \rightarrow x'_{\mathsf{L0}}$
$x_{\mathsf{L3b}} \Rightarrow sample \rightarrow sample\_tmp$
$x_{\mathsf{L3b}} \Rightarrow coefs \rightarrow acc \leftarrow sample\_tmp$
$x_{\mathsf{L4}} \Rightarrow i \rightarrow shift \rightarrow acc$
$x_{\mathsf{L4}} \Rightarrow i \rightarrow coefs \rightarrow acc$
$x_{\mathsf{L7}} \Rightarrow i \rightarrow shift \rightarrow x_{\mathsf{L8}}$
$x_{\mathsf{L8}} \Rightarrow i \rightarrow x'_{\mathsf{L7}}$
$x_{\mathsf{L9}} \Rightarrow sample\_tmp \rightarrow shift$
$x_{\mathsf{L9}} \Rightarrow acc \rightarrow result \rightarrow output\_data\_ready$

*Figure 1.6.* Clock and scheduling relations for the FIR

Figure 1.7 graphically details this control flow. There are several the small loops, such as the one between L1 and L2, representing the manipulation of an array of values.

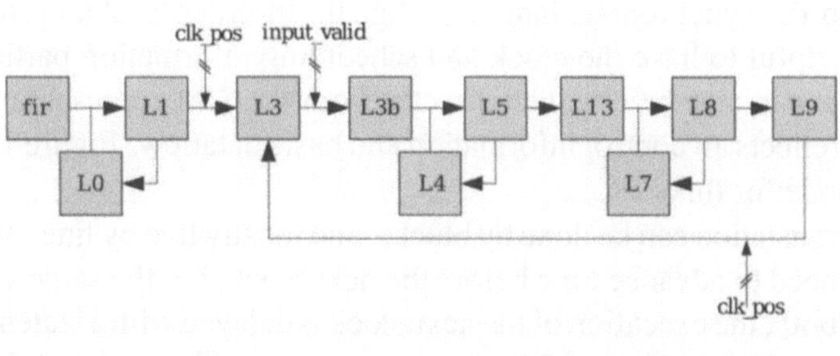


*Figure 1.7.* Control flow of the FIR filter

The big loop between L3 and L9 represents the actual program execution loop. Everything before that deals with initialization. After initialization the program waits for the next positive clock slope. At the beginning of the execution it is waiting for a valid input value. Then the calculations are executed and it subsequently waits for the next positive clock slope before resuming execution at block L3.

Data flow dependencies for the FIR are displayed on the right hand side of Table 1.6. The structure of these dependencies is very simple, the arrow (a→b) designing that *a* has to be present before *b* can be evaluated. Overall we see that for the FIR example, the control part largely outweighs the data flow part. Figure 1.8 illustrates the data flow of the example. We see that for the execution loop, the major data activity takes place in blocks L3b, L4, L7, and L9. L3b reads the external inputs *coefs* and *sample*, L9 eventually produces the outputs *output_data_ready* and *result*.

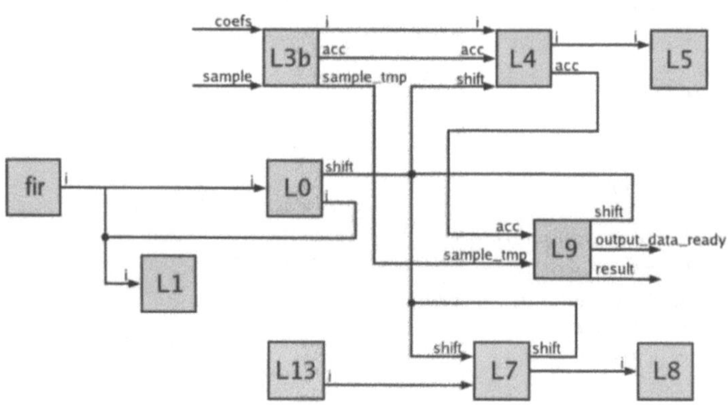


*Figure 1.8.* Data Flow of the FIR

## 4.5 The equivalent Signal program

As described earlier, the combination of the control and dataflow can be expressed in the synchronous language Signal. In order to obtain this Signal code, it is helpful to have the clock and scheduling information particularized earlier, but it can also be obtained directly from the SSA representation. The signal type reflects all control information and basic dataflow. Figure 1.9 details the signal code for this FIR.

At first, translation can be done by blocks, and mostly line by line. Whenever there is the need to advance time before the next block, i.e. the same variable is assigned in both, the execution of the next block is delayed with a statement such as *xL1 := xnL1$1 init false*. *xnL1* represents the next value and *xL1* the current value. The Signal language strictly prohibits multiple assignments of a variable within one block and in the same instant. So if the time is not advanced between two assignments, the Signal compiler indicates that there is an error. Obviously this is only needed, where we have delayed transitions, such as fir→L1, or loops such as in L0→L1→L0.

```
process fir =
  (? boolean input_valid, clk1; integer sample
   ! integer result, boolean out_ready)
(|coefs :=[16,15,14,13,12,11,10,9,8,7,6,5,4,3,2,1]
 | xfir := (xfir$ init true) AND false
 | i := 0 when xfir
     default (i$ init 0) +1 when xL0
     default 14 when xL3b
     default (i$ init 0) -1 when xL4
     default 14 when xL13
     default (i$ init 0) -1 when xL7
     default (i$ init 0)
 | result := 0 when xfir
     default acc when xL9
     default result$ init 0
 | out_ready := false when xfir
     default false when xL3
     default true when xL9
     default out_ready$ init false
 | xnL1 := xfir
 | xL1 := when xL0
     default xnL1$1 init false
 | shift := [k to 15:((0 when xL0)
     default (shift[k+1]$ when xL7 when k=i)
     default (sample_tmp when k=0 when xL9)
     default shift[k]$)]
 | t_i := (i <= 15) when xL1 default false
 | xnL0:= t_i
 | xL0 := xnL0$1
 | xL3 := true when xL1a when not(clk1 = clk1$)
     default when xL9a when not(clk1 = clk1$)
     default false
 | xnL1a := when not t_i
     default when xL1a when (clk1 = clk1$)
     default false
 | xL1a := xnL1a$
 | xL3a := when xL3
     default xnL3a$1 init false
 :
 :
 | xL3b := when xL3a
   when input_valid==true
     default false
 | xnL3a := when xL3a when input_valid==false
     default false
 | sample_tmp := sample when xL3b
     default sample_tmp$ init 0
 | acc := coefs[0] * sample_tmp when xL3b
     default acc + shift[i] * coefs[i]
     when xL4
     default acc$ init 0
 | xL5 := xL3b
 | xnL4:= when i>=0 when xL5 default false
 | xL4 := xnL4$
 | xnL13:= when i<0 when xL5 default false
 | xL13 := xnL13$
 | xL8 := xL13
 | xnL7 := when i<=0 when xL8 default false
 | xL7 := xnL7$
 | xL9 := when i>0 when xL8 default false
 | xnL9a := when xL9
     default when xL9a when (clk1 = clk1$ init 0)
     default false
 | xL9a := xnL9a$
 | t_i ≙ sample ≙ input_valid ≙ sample_tmp
   ≙ acc ≙ shift ≙ coefs ≙ result ≙ xfir
   ≙xL1 ≙xL3 ≙xL3b ≙xL9
   ≙xnL1a ≙xnL3a ≙xnL4 ≙xnL13 ≙xnL7
   ≙xnL9a ≙ clk1 ≙ out_ready|)
where
  boolean t_i;
  integer i, acc, sample_tmp;
  [16] integer shift; [16] integer coefs;
  event xfir, xL0, xL1, xL1a, xL3, xL3a,
  xL3b, xL4, xL5, xL7, xL8, xL9, xL9a, xL13,
  xnL0, xnL1, xnL1a, xnL3a, xnL4, xnL13,
  xnL7, xnL9a;
end;
```

*Figure 1.9.* Signal type for the FIR filter

Variables that are assigned in several blocks can, in a first step, be defined with partial equations (e.g output_data_ready ::= false when xL3 | output_data_ready ::= true when xL9). Partial equations are a source of errors since it is difficult to make sure that there are no conflicts between them. This is why we subsequently combine these partial equations to complete equations (e.g output_data_ready := false when xL3 default true when xL9). In the code in Figure 1.9 six partial equations for variable *i* have been combined and it is easy to see that they are not in conflict.

In this example we do not have any complex data manipulations. Where these occur, it can be unreasonably complicated to describe them in Signal, so for these cases they are wrapped into external functions, leaving them in the original code. This permits the handling of data flow intensive applications without much additional cost.

### 4.5.1 Abstracting the signal type.

The type in Figure 1.9 implements the FIR (it is an exact Signal mirror of the original SystemC implementation). For many purposes, however, all functionality is not needed in order to evaluate the validity of a condition.

```
process fir =
  (? boolean input_valid; integer sample
   ! integer result, boolean out_ready)
(| result := 0 when xfir
    default 1 when xL9
 | output_data_ready := false when xfir
    default false when xL3
    default true when xL9
 | xL1a := xfir default when xnL1a$
    default false
 | xL3 := when xL1a when not(clk = clk$)
    default when xL9a when not(clk = clk$)
 | xnL1a := when not t_i
    default when xL1a when (clk = clk$)
 | xL3a := xL3
    default xnL3a$1 init false
 | xL3b := when xL3a when input_valid==true
    default false
 | xnL3a := when xL3a when input_valid==false
    default false
 | xL9 := xL3b
 | xnL9a := when xL9
     default when xL9a when not xL3
    default false
 | xL9a := xnL9a$
 | xfir ≙ ≙xL3 ≙xL3b
    ≙xnL1a ≙xnL3a ≙xnL9a|)
where
  event xfir, xL1a, xL3,
  xL3b, xL9, xL9a,  xnL1a, xnL3a,xnL9a;
end;
```

*Figure 1.10.* Abstracted Signal type

An abstracted type that does not contain data manipulations is much lighter and still can serve to check conditions such as deadlocks, termination, and race-conditions. Figure 1.10 depicts the code for a possible abstraction for the fir type. The light weight for this type allows for much faster verification of properties, and, therefore makes it possible to check for these properties on a higher level, possibly comprising the whole system. For detailed checks including correctness of data manipulations or range-checks, still the complete type can be used.

## 4.6 Status

To get this process smoothly to work, there are some obstacles. As we have seen in Section 4.2, it is not obvious to have a clean SSA code. Therefore substantial effort has to be put into the generation of clean and reasonably short SSA code. This can be done with compiler optimization on the one hand and pre- / postprocessing on the other hand.

When generating clock dependencies and the Signal type respectively, the crucial point is the advancement of time. It has to be made sure that if blocks assign the same variable, there has to be an advancement of time in between. As we have different branches, this problem breaks down to graph coloring and is not trivial.

The presented approach illustrates its applicability for C++ and the same applies for JAVA . However, for SystemC, there are some additional issues to consider. As for now, we treated only one entry function of a SystemC program. In order to type a whole SystemC application consisting of several modules in the same way, multiple entry functions would have to be treated as well as the architecture and connectivity between them. This is ineffective to do in SSA because the change to the lower level will obstruct the higher level hierarchy structures. Consequently the preprocessing step has to be designed to be more sophisticated in order to handle the structure correctly. Also adequate Signal equivalence for certain SystemC constructs such as *sc_fifo* or *sc_semaphore* have to be defined.

The approach we presented shows how to obtain a behavioral Signal type from SystemC components. The passage through the SSA form allows for a relatively straightforward translation to the formal synchronous description. When handling plain C++ or Java code the pre-and-post-processing steps can be omitted, significantly simplifying the transformation process.

# 5. Related works

Our approach is based on a high-level specification methodology that ensures compositional correctness through a type theory capturing behavioral aspects of component interfaces. The proposed system builds upon previous work on scalable design exploration using refinement and abstraction-based design methodologies [15], implemented in the Polychrony workbench [17], and on a layered Component Composition Environment, Balboa [5], which allows the specification of SystemC [18] components with mixed levels of structural and behavioral details, and a high degree of concurrency and strict timing requirements. The capture of the behavior of a system through a type theoretical framework relates our technique to the work of Rajamani et al. [13], and many others, on abstracting high-level and concurrent specifications, e.g.

the $\pi$-calculus, by using a formalism, e.g. Milner's CCS, in which, primarily, checking type equivalence, e.g. bisimulation, is decidable.

Our contribution contrasts from related studies by the capability to capture scalable abstractions of the type-checked system. In our type system, scalability ranges from the capability to express the exact meaning of the program, in order to make structural transformations and optimizations on it, down to properties expressed by boolean equations between clocks, allowing for a rapid static-checking of design correctness properties (as demonstrated in the examples of Section 3.4, especially Example 7). Our system further allows for a wide spectrum of correct-by-construction design abstractions and refinement patterns to be applied on a model, e.g. abstraction of states by clocks, abstraction of existentially quantified clocks, hierarchic abstraction (as demonstrated Example 5), in the aim of choosing an optimal degree of abstraction for a faster verification.

We share the aim of a scalable and correct-by-construction exploration of abstraction/refinement of system behaviors with the work of Henzinger et al. on interface automata [4]. Our approach primarily differs from interface automata in the data-structure used in the Polychrony workbench: clock equations, boolean propositions and state variable transitions express the multi-clocked synchronous behavior of a system. Compared to an automata-based approach, our declarative approach allows to hierarchically explore abstraction capabilities and to cover design exploration with the methodological notion of refinement along the whole design cycle of the system, ranging from the early requirements specification to the latest sequential and distributed code-generation [15, 7].

## 6. Conclusions and future works

The main novelty in our approach is the use of a multi-clocked synchronous formalism to support the construction of a scalable behavioral type inference system for the *de facto* standard system-design language SystemC, and the materialization of a companion refinement-based design methodology imposed through the strong typing policy of a module system, that reduces compositional design correctness verification to the validation of synthesized proof obligations. The proposed type system allows to capture the behavior of an entire system-level design and to re-factor it, allowing to generate an optimized scheduling using hierarchization, allowing to modularly express a wide spectrum of static and dynamic behavioral properties, and to automatically or manually scale the desired degree of abstraction of these properties for efficient verification. The type system is presented using a generic and language-independent intermediate representation. It operates transformations implemented in the platform Polychrony, to perform refinement-based design exploration and directly yields to verification tools using SAT checking and model checking allowing for an efficient verification of expected design properties and an early discovery of design errors.

## References

[1] AMAGBEGNON, T. P., BESNARD, L., LE GUERNIC, P. "Implementation of the data-flow synchronous language SIGNAL". In *Conference on Programming Language Design and Implementation.* ACM Press, 1995.

[2] J.T. BUCK, S. HA, E.A. LEE AND D.G. MESSERSCHMITT. Ptolemy: A Framework for Simulating and Prototyping Heterogeneous Systems. In *International Journal of Computer Simulation, special issue on "Simulation Software Development"*v. 4, pp. 155-182. Ablex, April 1994.

[3] E. DIJKSTRA "A Discipline of Programming". Prentice Hall, 1976.

[4] DE ALFARO, L., HENZINGER, T. A. "Interface theories for component-based design". *International Workshop on Embedded Software.* Lecture Notes in Computer Science v. 2211. Springer-Verlag, 2001.

[5] F. DOUCET, S. SHUKLA, AND R. GUPTA. "Typing abstractions and management in a component framework". *Asia and South Pacific Design Automation Conference*, January 2003.

[6] LEE, E. A., SANGIOVANNI-VINCENTELLI, A. "A framework for comparing models of computation". In *IEEE transactions on computer-aided design*, v. 17, n. 12. IEEE Press, December 1998.

[7] LE GUERNIC, P., TALPIN, J.-P., LE LANN, J.-L. Polychrony for system design. In *Journal of Circuits, Systems and Computers. Special Issue on Application-Specific Hardware Design.* World Scientific, 2002.

[8] MOUSAVI, M., R., LE GUERNIC, P., TALPIN, J.-P., SHUKLA, S., BASTEN, T. Modeling and validation of asynchronous systems in synchronous frameworks. In *Digital Automation and Test Europe.* IEEE Press, February 2004.

[9] NOVILLO, D. "Tree SSA, a new optimization infrastructure for GCC". GCC developers summit, 2003.

[10] NOWAK, D., BEAUVAIS, J.-R., TALPIN, J.-P. "Co-inductive axiomatization of a synchronous language". In *International Conference on Theorem Proving in Higher-Order Logics.* Lecture Notes in Computer Science, Springer Verlag, October 1998.

[11] NOWAK, D., TALPIN, J.-P., LE GUERNIC, P. "Synchronous structures". In *International Conference on Concurrency Theory.* Lecture Notes in Computer Science, Springer Verlag, August 1999.

[12] PNUELI, A., SHANKAR, N., SINGERMAN, E. Fair synchronous transition systems and their liveness proofs. *International School and Symposium on Formal Techniques in Real-time and Fault-tolerant Systems.* Lecture Notes in Computer Science v. 1468. Springer Verlag, 1998.

[13] S. K. RAJAMANI AND J. REHOF, "A BEHAVIORAL MODULE SYSTEM FOR THE $\pi$-CALCULUS". *Static Analysis Symposium.* Lecture Notes in Computer Science. Springer Verlag, July 2001.

[14] TALPIN, J.-P., GAMATIÉ, A., LE DEZ, B., BERNER, D., LE GUERNIC, P. "Hard real-time implementation of embedded systems in JAVA". *International Workshop on Scientific Engineering of Distributed JAVA Applications.* Lecture Notes in Computer Science. Springer Verlag, November 2003.

[15] J.-P. TALPIN, P. LE GUERNIC, S. K. SHUKLA, R. GUPTA, AND F. DOUCET. "Polychrony for formal refinement-checking in a system-level design methodology". *Application of Concurrency to System Design.* IEEE Press, June 2003.

[16] J.-P. TALPIN, P. LE GUERNIC "Algebraic theory for behavioral type inference". In *Formal Methods and Models for System Design (this volume).* Kluwer Academic Publishers, June 2004.

[17] The Polychrony website. http://www.irisa.fr/espresso/Polychrony, 2004.

[18] The OSCI SystemC website. http://www.SystemC.org, 2004.

[19] The Gnu Compiler Collection (GCC). http://http://gcc.gnu.org, 2004.

[20] The GCC Tree-SSA Branch. http://gcc.gnu.org/projects/tree-ssa, 2004.

IV

# OPTIMIZING SYSTEM MODELS

Chapter 11

# OPTIMIZATIONS FOR FASTER EXECUTION OF ESTEREL PROGRAMS

Dumitru Potop-Butucaru*

*IRISA, Campus de Beaulieu, 35042 Rennes, France*

Dumitru.Potop@irisa.fr

Robert de Simone

*INRIA, 2004 rte. des Lucioles, 06902 Sophia Antipolis Cedex, France*

Robert.De_Simone@inria.fr

**Abstract** The fine-grained parallelism and the need for determinism are traditional issues in the design of real-time embedded software. In addition, the increasing complexity of the specifications requires an increasing use of higher level formalisms. The Esterel language offers natural solutions to all these problems, but its compilation proved challenging, so that efficient compilation techniques have only recently been proposed. Consisting essentially in direct simulation of the reactive primitives of the language, these techniques now need to be accommodated with traditional issues of Esterel: the definition of formal semantics, the constructive causality, and the design of analysis and optimization methods that are both efficient and correct.

We address these problems by defining a new intermediate model, called GRC, for the representation of Esterel programs. The GRC representation preserves much of the initial program structure while making the control flow pattern and the hierarchical state structure explicit. It fully complies with the semantics of Esterel, and, most important, it is a good support for efficient analysis, optimization, and code generation algorithms based on static analysis.

The use of a GRC-based approach results in significant efficiency and generality gains. The approach could be extended to the hardware compilation of Esterel and to the compilation of other synchronous languages with fine-grained parallelism.

*Partly supported by the ARTIST European project



*R. Gupta et al. (eds.), Formal Methods and Models for System Design,* 285-315.

**Keywords:** Causality, code generation, compiler, embedded system, Esterel, intermediate representation, optimization, static analysis, synchronous.

## 1. Introduction

Sequential languages like C and assembly are still widely used in the development of reactive real-time embedded systems, even for applications that are most naturally described as concurrent systems. Real-time operating systems (RTOS) are used in these cases to schedule the resulting sequential processes, thus providing the needed concurrency. Two problems arise from here. The first is that a RTOS can be unpredictable, making functional and timing verification difficult. The second is that RTOS-level scheduling can be very inefficient for applications involving fine-grained parallelism, due to context switching overhead.

The *synchronous approach* [BB91, Hal93, BCE$^+$03] uses classical hardware concepts in order to provide deterministic concurrency and better verification capabilities. The execution of a synchronous system is cyclic, driven by a *global clock*. At each clock cycle the system reads the inputs, computes its *reaction*, and outputs the result. The code executed during a clock cycle is *loop-free* and the clock is operated by the system itself, allowing precise timing control. The underlying synchronous model also allows the use of the well-studied functional verification techniques developed for synchronous circuits, so that the approach is a good basis for the development of safety-critical embedded software.

On the other hand, the high semantic standards behind the *synchronous languages* (*e.g.* Esterel [BG92], Lustre [HCRP91], Signal [LGLL91]) made their correct and efficient implementation a challenging task. A particular difficulty is here that all the informations which might affect a given variable at a given clock cycle must be compiled *before* the variable is used. A typical example is the *signal absence*, which can only be decided after negative information about all potential emissions has been propagated. The fine-grained parallelism and the intricate instruction ordering make run-time scheduling inefficient, due to the context switching overhead. In consequence, implementation code is often generated by compiling away all internal concurrency[1], while proceeding in a single-loop periodic evaluation of all the instructions of the program. The synchronous specification is transformed into a single sequential (C) *reaction function* where all the operations have been *statically scheduled* using auxiliary variables and tests. A small run-time executive interfaces with the asynchronous environment to read the inputs and decide when to call the reaction function.

[1] In distributed, multi-processor implementations like the one proposed by Caspi *et al.* [CGP99], some concurrency is preserved between specification parts operating on different processors

A program written in a concurrent synchronous language like Esterel and its desired sequential implementation are very different objects. For this reason, all Esterel compilers use intermediate representations, and each intermediate representation determines in turn the capabilities of the associated compiling technique. The first implementations of Esterel [BG92, Est00] used mathematical models (Mealy machines and digital circuits) as intermediate representations. Consequently, the resulting C code was either needlessly large, or slow. More recent compilers [Edw02, WBC$^{+}$00] use control-flow graphs that are easily translated into well-structured C code, but are too far from Esterel to represent its exact semantics or to support a number of efficient analysis and optimizations.

This paper presents a new approach to the compilation of the Esterel language which achieves significant performance and generality improvements by using the GRC (*GRaph Code*) intermediate representation.

The main contribution of the paper is the definition of the GRC code, which uses two graph-based structures – a *hierarchical state representation* (HSR) and a *concurrent control-flow graph* (CCFG) – to preserve most of the structural information of the Esterel program while making the control flow explicit with few graph-building primitive nodes. The hierarchical state can be seen as a structured data memory that preserves state information across reactions. In each instant, a set of activation conditions (triggers) is computed from this memory state, to drive the execution towards active instructions (for historical reasons such activation triggers are generally called *clocks* in reactive synchronous terminology). Because of the formal synchronous semantics these clocks enjoy nice properties (a computation is only triggered when its outcome will be consumed, and vice-versa). In the current GRC scheme, the computation of *activation clocks* from the current state has to be performed first thing in each reaction. But then the actual data instruction blocks will only be activated at proper pace.

On the new format, we introduce several transformations, mostly semantics-preserving optimization algorithms based on static analysis methods. Due to the mathematical nature of the GRC model these optimization can be formally proved sound, although we shall not insist on this aspect in the current paper. Benchmarks indicate the strong interest of the approach in terms of potential optimizing gains. The overall compilation scheme can be seen either as an attempt to produce efficient software code, or efficient simulation code for hardware representations (a goal equally followed by the simulation semantics of VHDL and Verilog hardware description languages, but without our strong synchronous semantics).

We focus here on the generation of software simulation code by defining a GRC-to-C translation scheme that handles all *GRC-acyclic* and *circuit-acyclic* Esterel programs. We shall disregard here the topics of simulating cyclic GRC specifications, or of determining program correctness in the sense of constructive causality at GRC level. Advances in that direction can be found in [PB02]. The efficiency of our approach is supported by benchmarks comparing our GRC-based compiler `grc2c` to existing compilers.

The paper is organized as follows. Section 2 is devoted to an informal presentation of the Esterel language, including the description of the small example which will be used throughout the paper. A review of the existing Esterel compilers is given in section 3. The next three sections describe our new approach. In section 4 we define the GRC intermediate representation: its primitives, their semantics, and the Esterel translation to GRC. We also insist here on some causal correctness (acyclicity) issues relating GRC and the digital circuits. Section 5 presents the GRC-level optimization algorithms that are currently used in our compiler. The software code generation technique, based on efficient state encoding and static scheduling, is described in section 6. Section 7 presents experimental results comparing `grc2c` with existing compilers, and we conclude with some suggestions about how to extend this work.

## 2. The Esterel language

Esterel [BG92, Est00, BdS91] is a design language for the representation of reactive real-time embedded systems. It has a full-fledged formal operational semantics [Ber99, PB02] which solves tricky modeling issues such as the proper definition of global synchronous reactive behaviors and the constructive causality in control flow propagation. Esterel specifications can be translated into sequential software code, digital synchronous circuits, or a combination of both, but our work only concerns its software compilation.

Through its intuitive, highly-readable imperative style, Esterel tries to combine strong modeling features together with efficient compilation purposes. As illustrated by the small example of fig. 11.1, Esterel features the control flow operators of a language like C (sequence, loop), but also provides concurrency, preemption, and a synchronous, clock-driven execution model. In each clock cycle the inputs are read, and the program computes its *reaction* by resuming the control threads and running them until suspension. The communication is done through *broadcast signals* which are either *present* or *absent* in each clock cycle. The signal `S` is present if a statement "`emit S`" is executed during the cycle, and absent otherwise.

Our small example has four input signals and one output signal. Meant to model a cyclic computation like a communication protocol, the core of our example is the loop which awaits the input `I`, emits `O`, and then awaits `J` before instantly restarting. The local signal `END` signals the completion of loop cycles.

```
module Example: input I,J,KILL,SUSP; output O;
suspend
  trap T in %exception handler, performs the preemption
    signal END in
      loop %basic computation loop
        await I;emit O;await J;emit END
      end
    ||
      %preemption protocol, triggered by KILL
      await KILL;await END;exit T
    end
  end;
when SUSP %suspend signal
end module
```

*Figure 11.1.* A simple Esterel program modeling a cyclic computation (like a communication protocol) which can be interrupted between cycles and which can be suspended

When started, the `await` statement waits for the *next* clock cycle where its signal is present. The computation of all the other statements present in our example is performed during a single clock cycle, so that the `await` statements are the only places where control can be suspended between reactions (they preserve the *state* of the program between cycles). A direct consequence is that the signals `I` and `J` must come in different clock cycles in order not to be discarded.

| clock | inputs | outputs | comments |
|---|---|---|---|
| 0 | any | | all inputs discarded |
| 1 | I | O | |
| 2 | KILL | | preemption protocol triggered |
| 3 | | | nothing happens |
| 4 | J,SUSP | | suspend, J discarded |
| 5 | J | | END emitted, T raised, program terminates |

*Figure 11.2.* A possible execution trace for our example

The loop is preempted by the exception handling statement `trap` when "`exit T`" is executed. In this case, `trap` instantly terminates, control is given in sequence, and the program terminates. The preemption protocol is triggered by the input signal `KILL`, but the exception `T` is raised only when `END` is emitted. The program is suspended – no computation is performed and the state is kept unchanged – in clock cycles where the `SUSP` signal is received. A possible execution trace for our program is given in fig. 11.2.

The operations that compose a reaction are *causally ordered* by the control flow and by the signal communication. A signal `S` can be tested in a clock cycle only after its status has been determined. The signal becomes present as soon as it has been emitted. It becomes absent when we can determine that no "`emit S`" statement can be executed during the current clock cycle. The signal causality may lead to deadlocks in the computation of a reaction despite the fact that the code of a reaction is loop-free. Programs that can deadlock (*e.g.* `present T then emit T end`) are incorrect and must be rejected at compilation.

Determining whether a program can deadlock or not is not easy. The semantics of Esterel is *constructive*, in the sense that the absence of signals must be determined without any kind of guess. To do so, one has to notice that according to decisions already made in the current reaction all the emissions of the signal are invalidated. The recursive fact propagation is exemplified with the following (correct) statement:

```
emit S;
present T then
  present S else emit T end
end
```

When the statement is started, the signal `S` is emitted and the control is blocked on the test on signal `T`, whose status is still unknown. The *present* status of `S` is then propagated into the not yet executed code. There, it invalidates the `else` branch of the test on `S`, so that `T` becomes *absent*, and execution can proceed. Note that our execution involved the evaluation of code (the test on `S`) that will never be executed.

Due to its constructive semantics, Esterel has a natural semantic model as Synchronous Digital Circuits, at schematic gate level, endowed with intuitionistic constructive semantics [Ber99]. In turn, such circuits have a natural operational interpretation in terms of Finite State Mealy Machines. Thus, all Esterel modeling primitives are provided meaning as hierarchical constructs allowing to build compound objects in each of these domains. Actually, this goes for reactive control flow structuring primitives, while the language also offers data handling of a classical imperative nature (assignments).

## 3. Related work

Four approaches have been proposed for the compilation of Esterel into sequential code. Historically the first, **the automata-based approach** [BG92, Bre02] basically follows the Structural Operational Semantics of the language to produce a *flat, global automaton* that is then encoded in C. Performed by means of exhaustive symbolic simulation, the automaton expansion also determines the causal correctness of the specification and solves all the interleaving problems that are due to constructivity. The automata-based C code is theoretically the fastest possible, as only semantically active code is executed in each state. It can also be exponentially larger than the source, due to state-wise code replication.

Several approaches have been proposed to reduce the size of the generated code. The automata compiler of the Polis group [BCG+99] uses binary decision diagrams to identify code that can be shared between states. The HIPPCO automaton optimizer of Castelluccia *et al.* [CDO97] uses code sharing, inlining, and re-ordering of tests to improve not only the code size, but also the execution time (on average and for privileged paths). The Esterel compiler of Bres [Bre02, Est00] does not use the unoptimized automaton representation. It directly generates code where common sub-trees are shared inside a state. While these techniques improve the quality of the generated code, the number of states still explodes for most meaningful examples to the point where the generation of the automaton is intractable.

Surprisingly, benchmarks show that `grc2c`-generated code is often equally fast or faster than the theoretically-optimal automata-based code. The reason is probably twofold. First, the limited size of the processor caches penalizes the usually larger automata-based executables. Second, the GRC code optimizations leave only few redundancies, and the generated code is well-structured. The code generated by `grc2c` can also be exponentially smaller because the translation does not involve state-wise code duplication.

**The circuit-based compilers** [Est00] start by structurally translating the Esterel program into a sequential digital circuit, at netlist level. The compiler then generates C code that is a *levelized compiled logic simulator*. At each reaction, the code emulates circuit activity by evaluating in an ordered manner all the gates of the circuit. The circuit-based code is compact, quasi-linear in the size of the initial Esterel program. It is also slow because all the gates are evaluated in each clock cycle.

The circuit-based code is always slower than its `grc2c`-generated counterpart, sometimes by factors of 50. The encoding of all program operators using Boolean gates also leads to a larger code size, except for the small examples where aggressive *circuit optimizations* can be applied.

**The Saxo compiler** of Closse *et al.* [WBC+00, CPP+02] uses a discrete-event interpretation of Esterel to generate a *compiled event-driven simulator.* The compiler flow is similar to that of VeriSUIF [FLLO95], but Esterel's synchronous semantics are used to highly simplify the approach. An *event graph* intermediate representation is used here to split the program into a list of *guarded procedures*. The guards intuitively correspond to events that trigger computation. At each clock cycle, the simulation engine traverses the list once, from the beginning to the end, and executes the procedures with an active guard. The execution of a procedure may modify the guards for the current cycle and for the next cycle.

The resulting code is much faster than the circuit-based one, as the guards prevent the evaluation of semantically-inactive code. At the same time, two reasons make it slower than its `grc2c`-generated counterpart: First, it does not exploit the hierarchy of exclusion relations determined by switching statements like the tests. Second, optimization is less effective because the program hierarchy is lost when the state is (very redundantly) encoded using guards.

**The EC compiler** of Edwards [Edw02] has been a major source of inspiration in our work. It treats Esterel as having control-flow semantics (in the spirit of [LPM99, Muc97]) in order to take advantage of the initial program hierarchy and produce efficient, well-structured C code. The Esterel program is first translated here into a *concurrent control-flow graph* (CCFG) representing the computation of a reaction. The translation makes the control flow explicit and encodes the state access operations using tests and assignments of integer variables. The back-end accepts only acyclic CCFGs. Its *static scheduling* algorithm takes advantage of the mutual exclusions between parts of the program and generates code that uses *program counter* variables instead of simple Boolean guards. The result is therefore faster than its Saxo-generated counterpart.

EC and `grc2c` share many common aspects. Both use control-flow graphs as intermediate representations and both use static scheduling to generate well-structured code. Moreover, the GRC-level CCFG and the CCFG of the EC compiler are very similar in form because they are obtained through similar Esterel-specific demultiplexing and code duplication. Essential differences remain, however: EC takes a syntax-driven approach in determining the internal concurrency of the Esterel program. Parallel threads are identified on the Esterel source and hard-coded in the CCFG-level state encoding with tests and assignments over integer variables. The approach of `grc2c` is semantics-driven, aimed at achieving better optimization. The GRC code preserves the state structure of the initial Esterel program and uses static analysis techniques to determine redundancies in the activation pattern. Thus, it is able to simplify both the final state representation and the CCFG. For this reason, `grc2c` generates faster code than EC.

## 3.1 Program correctness issues. Acyclicity

The constructive semantics of Esterel, introduced in section 2, is based on fact propagation schemes traversing at each instant not only the active parts of the program, but also inactive ones. This comes in contradiction with the philosophy of efficient code generation. For this reason, the constructive causality is usually subsumed in this context by a more restrictive requirement, namely *acyclicity*.

Acyclicity cannot be formally defined at the syntactic level of Esterel programs, as it largely depends on connections drawn between signal emissions and corresponding receptions. It was first defined at the level of circuits, and was simply rephrased on event graphs and CCFGs. Acyclicity is then dependent on both the intermediate representation and the translation scheme. Regardless of the intermediate representation, acyclicity represents the (strict) *restriction of the constructive model* where the evaluation process can be performed in the same order at each clock cycle. Thus, it supports the generation of efficient statically-scheduled C code and represents a cheap, syntactic correctness criterion.

Among the compiling techniques that are able to handle large Esterel programs, the circuit-based approach is the only one currently able to analyze and compile cyclic programs (by transforming cyclic parts into equivalent acyclic ones). However, the procedure is expensive, requiring symbolic state space exploration and circuit resynthesis [SBT96, Tom97, Edw03]. In practice, acyclicity is the most used correctness criterion for Esterel programs, and the only working on large specifications.

Most meaningful programs are acyclic, regardless of the intermediate compiler representation. However, differences subsist, the circuit-based compilers accepting more correct programs due to the finer grain of their intermediate representation. The GRC-based approach supports a refinement technique making the GRC-level acyclicity equivalent to the circuit-level one. Thus, we pave the way towards defining acyclicity as a representation-independent correctness criterion.

## 4. The GRC intermediate representation

The translation of the Esterel source into GRC (for *GRaph Code*) makes the control flow explicit while preserving an important part of the structural information. The resulting GRC model consists of two objects: a *concurrent control-flow graph* (CCFG) representing the behavior of the Esterel program and a *hierarchical state representation* (HSR), historically called *selection tree*. The CCFG represents in an operational fashion the computation of an instant (the transition function). During each reaction, the dynamic CCFG operates on the static HSR by marking/unmarking component subtrees with "active" tags as they are activated or deactivated by the semantics.

### 4.1 The Hierarchical State Representation

The hierarchical state representation of a GRC model is an abstraction of the syntax tree of the initial Esterel program. It preserves the modular structure of the program while forgetting about behavioral statements. The HSR of our example is presented in fig. 11.3, along with intuitive tags showing the correspondence between HSR nodes and Esterel statements.

The square-shaped HSR leaves correspond there to simple program statements, like `await`, where control can be suspended between clock cycles. The oval-shaped intermediate nodes correspond to composed statements. Specific HSR tags mark the combination mode in nodes having more than one child. *Exclusive nodes*, tagged with #, correspond to sequences and to if-then-else Esterel statements. *Parallel nodes*, tagged with ||, correspond to parallel Esterel constructs. Simple statements that cannot retain control between clock cycles (*e.g.* `emit`, `exit`) are discarded in the construction of the HSR.

The HSR nodes of indices 0 and 1 do not correspond to program statements. They are automatically generated at program level in order to represent the start state of the program, where no statement is active, but the program is.

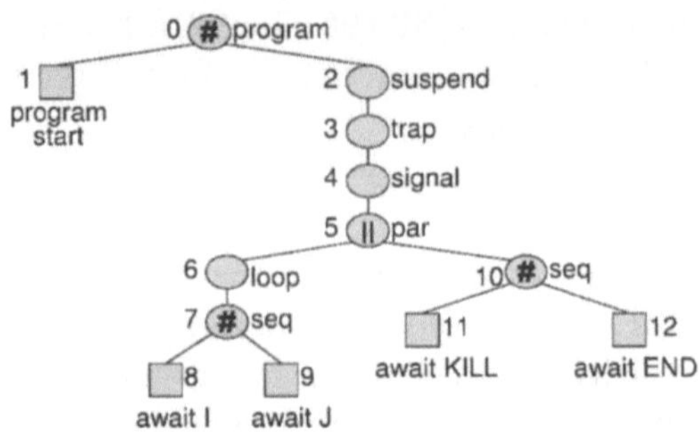


*Figure 11.3.* The hierarchical state representation (HSR) of our example

Each node of the state representation is endowed with a *selection status* flag (a Boolean data variable) which records at run-time the current activity status of the corresponding subprogram. Selection statuses are checked and modified by *state access operations* performed from the CCFG (and described in the next section). We say that a state representation node is *active* if its selection status is *true*. The definition naturally extends to the associated Esterel statement.

We shall see later that optimization methods will introduce more HSR tags, computed by static analysis of both the HSR and the CCFG. In essence these tags will record redundancy in the selection statuses. For instance it is often the case that a branch of parallel construct is always active, in which case the associated selection status and the CCFG code managing it can safely be removed. Thus, the HSR acts both as a state representation and as a high-level information repository.

## 4.2 The Concurrent Control-Flow Graph

We pictured in fig. 11.4 the CCFG associated with our small example. Intuitively, control enters the flowgraph at each execution instant through the topmost node. Then, it follows the edges until complete traversal of the flow-graph.

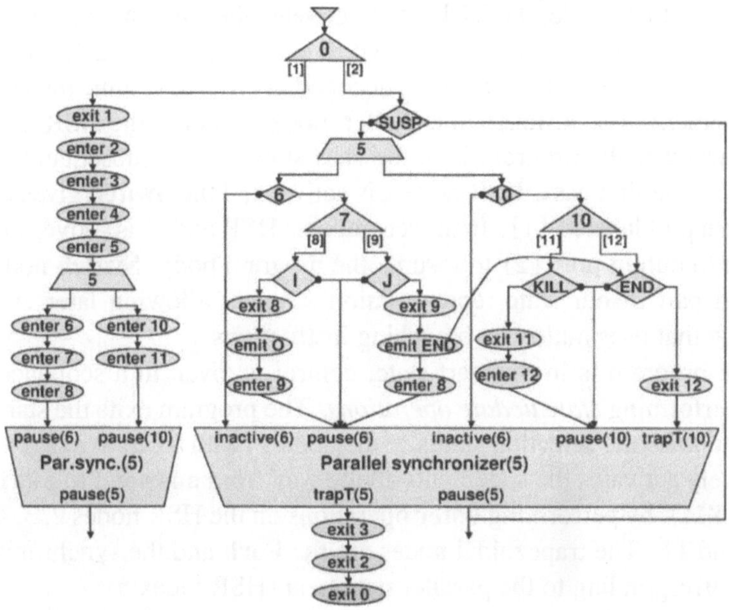


*Figure 11.4.* The concurrent control-flow graph (CCFG) of our example

All the edges of our example are oriented *control edges* representing control flow and signal `emit/present` dependencies. Control edges propagate the control and the signal statuses between *typed computation nodes*. Esterel programs involving data manipulations may also require the use of *data dependency edges*, representing scheduling constraints not implied by the control ones.

A computation node has a number of *named input and output ports* to which edges will be connected according to the translation rules. Each control edge connects exactly one output port to one input port. Several edges (or none) can be connected to every given port. Most nodes also contain a *data access operation*, allowing them to test/update user-defined variables or selection statuses. During a reaction, a node is executed when it receives enough control and signal information through its input ports. It performs then its computation, which may involve the activation of its data operation, and propagates the results through its output ports. We now describe the nodes and the specific data access operations by exemplifying on fig. 11.4. Note that we dropped (for space reasons) most input and output port names from our graphical representation. Explanations will make up for this inconvenience.

A unique **Tick** node (topmost in our example) serves in every CCFG as unique control entry point for the behavioral reactions. It bears a single output port, called *cont*.

At each clock cycle, the **Tick** node activates the **Switch** node decoding the status of the HSR node of index 0. **Switch** nodes represent the locations where the status of exclusive HSR nodes is decoded in order to resume the appropriate sub-statement. For instance, the **Switch** node decoding the HSR node 0 will decide whether the program is in the start state or in a subsequent execution instant. In the first case, HSR node 1 is active, and the **Switch** gives control to the output port labeled `[1]`. In the second case HSR node 2 is active, and control is given to output port `[2]` to resume the program body. **Switch** nodes are an essential part of our state representation scheme, allowing later an efficient encoding that uses multiway branching instructions.

If the program is in the start state, control is given to a sequence of **Call** nodes performing *state update operations*. The program exits the start state by setting to *false* the selection status of the node 1 (with a call to **exit 1**). Then, it recursively activates the statements all the way from `suspend` to `await I` and `await KILL` by performing **enter** operations on the HSR nodes 2, 3, 4, 5, 6, 7, 8, 10, and 11. The trapezoidal nodes are the **Fork** and the synchronizer **Sync** nodes corresponding to the parallel statement (HSR index 5).

If the program is not in the start state, then we have to perform the suspension test on signal `SUSP`. The associated **Test** node has one input port, named *go*, which represents the test trigger. It has no signal input port because the signal `SUSP`, which is not emitted within the program, is treated like a data variable. The node has two output ports, named *then* and *else* (the latter is identified with a dot). If the test succeeds, the hierarchical state decoding is interrupted and control (given to the *then* port) leaves the flowgraph thereby suspending the execution of the program until the next clock cycle. We also say in this case that the program *pauses*.

If the test fails, the sub-statements of `suspend` are recursively resumed. In practice, this amounts to resuming the branches of the parallel statement *i.e.* checking that they are active (with the **Test** nodes reading the selection statuses of the HSR nodes 6 and 10), decoding their internal status (with the **Switch** nodes on the HSR nodes 7 and 10), and giving control to the appropriate `await` signal tests. Note how the internal signal communication on `END` is encoded using a control edge that connects its emission with its test.

When a parallel branch completes its execution for the current clock cycle, it reports to the parallel synchronizer **Sync** node its completion code (inactive, paused, terminated, or exited through some exception). When all the branches complete the execution of the cycle, the **Sync** node can compute the completion code of the parallel statement[2] and activate the appropriate output port. In our case, the parallel can pause (suspend until the next clock cycle) or raise the exception `T`. In the first case, the control directly leaves the flowgraph. In the second case the program terminates, as **exit 0** sets the program status to inactive.

**CCFG semantics** We have presented here the intuitive, control-flow semantics of the GRC flowgraph. Its formal *constructive simulation semantics* [PB02] is based on the constructive circuit semantics of Berry [Ber99]. Under constructive semantics, each control edge of the CCFG has a status of $\perp$, `true`, or `false`. The value $\perp$ means "not yet resolved, not stabilized". In the beginning of each clock cycle, the status of all control edges (with the exception of the inputs) is set to $\perp$. The computation of a reaction consists in incrementally assigning a value of `true` (to represent execution or signal presence) or `false` (to represent inactivity or signal absence) to every control edge. There are technical intricacies such as the fact that a signal becomes *present* (`true` value) as soon as it has been emitted, while it can be turned to *absent* (`false` value) only when all emissions have been provably discarded by control-flow choices (the sequential implementation will avoid this problem by properly scheduling the corresponding assignments so that all emissions precede all signal tests).

The incremental computation of the statuses is performed by the CCFG nodes, which change their output as soon as their inputs change. The computation of a node may have side-effects that model state access and data access operations. Good properties of the computation nodes (monotony) guarantee that once the Boolean status of a control edge is established, it does not change for the current reaction. Thus, the computation converges in finite time. In so-called *constructively causal* programs, signal values are all defined (i.e., not $\perp$) in the end of each reaction from every reachable configuration of the system.

**Well-formedness properties** Not every GRC specification built with the previously defined nodes is meaningful. Otherwise said, while the CCFG of a GRC specification is not explicitly structured, it has to satisfy a number of well-formedness properties insuring its correct operation. Here are some of the most important:

- apart from the internal signal and data dependencies, the CCFG must be acyclic

[2] If at least a parallel branch has reported an exception code, then the parallel itself completes with an exception (the highest-level among those generated by branches). If no branch reported an exception code, but at least one paused, then the parallel pauses. Otherwise, the parallel terminates its execution and gives control in sequence.

- the state decoding part of the CCFG has to be hierarchic, following the HSR structure. Moreover, all state update nodes must occur *after* the state decoding nodes.

These properties are insured by the translation scheme defined in the following section and must be preserved by all GRC optimization algorithms.

## 4.3 Esterel to GRC translation

Most Esterel statements have distinct start and resumption behaviors. For instance, the signal test is performed by "`await S`" only when the statement is resumed. This multiplexing of behaviors produces intuitive, high-level constructs, very useful in the development of large applications. At the same time, it prohibits most control flow optimizations, as the start and resume behaviors are usually subject to different redundancies. The main purpose of the Esterel translation to GRC is to make the control flow explicit and allow its optimization.

In an approach borrowed from Berry's circuit translation [Ber92, Ber99], the translation of Esterel into GRC is based on a structural unfolding using well-defined patterns associated to the primitive statements. The translation involves two phases. The *structural phase* builds the HSR, the graph nodes, and the control dependencies not corresponding to signals. The *link phase* properly connects with control edges the ports corresponding to signal emission and reception, and connects with data dependency edges the data operations that access the same shared variable.

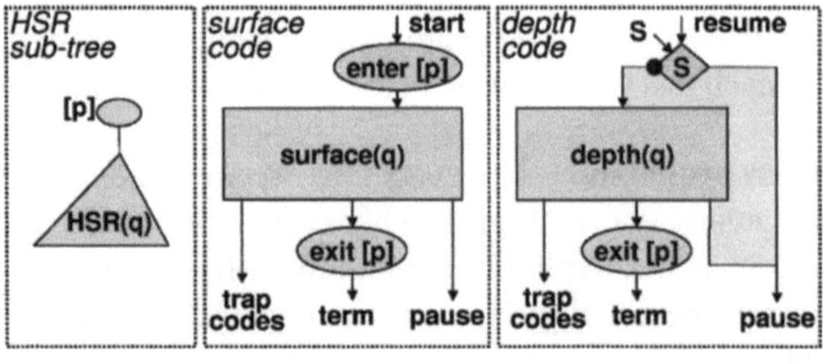


*Figure 11.5.* Translation of $p$ = `suspend` $q$ `when S`

For each statement three objects are generated: the *HSR sub-tree*, the *surface code*, representing the start behavior, and the *depth code*, representing the resumption behavior of the statement. For instance, the `suspend` statement is translated using the pattern pictured in fig. 11.5. To obtain the GRC representation, the three components are then put together at global level using the pattern of fig. 11.6.

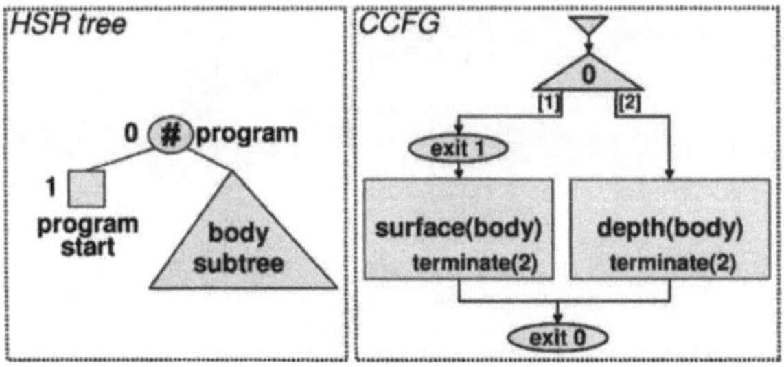


*Figure 11.6.* The global context

Well-studied semantic reasons [Ber99] often *require* a certain separation between the start and the resume code associated with a statement. Performing it, as we do, in a systematic and thorough way has the disadvantage of increasing the GRC code size. However, the duplication may also reduce the number and scope of signal and data dependencies, thus helping the optimization and software code generation algorithms. More parsimonious code duplication schemes like that of Edwards [Edw02] are supported by GRC, but we did not evaluate them in practice.

## 4.4 Acyclic GRC specifications

In section 3.1 we introduced the notion of cyclic program, showing that it depends on the chosen intermediate representation and compiling scheme. We also explained why it is important in practice to unify the circuit-level and GRC-level notions of acyclicity. We now derive a partial solution to this problem. More exactly, we use the similarity between the Esterel translations to GRC and digital circuits in order to determine the origin of the differences between the two notions of acyclicity. Based on this analysis, we define a minimal transformation of the GRC code that unifies the two notions, but still allows the generation of efficient C code.

The result is weak in formulation, as it relies specific, unoptimized translation schemes. It is nevertheless important, because it offers a road-map for developing flowgraph- and circuit-based compilers that accept the same classes of programs.

The translation of Esterel into digital circuits is usually performed by encoding the control flow with wires (and gates), while the state is represented using Boolean registers. Without losing generality, we shall use throughout this section the circuit translation of Potop-Butucaru [PB02]. This one, a close variant of the "reference" circuit translation of Berry, is exemplified in fig. 11.7.

The circuit translation is very similar in form to our GRC translation: It is based on patterns associated to the primitive statements of Esterel, and the

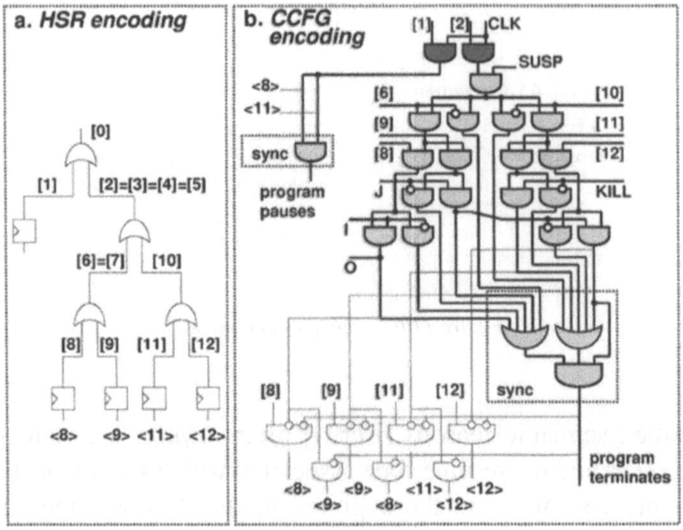


*Figure 11.7.* The circuit translation of our example. Unfilled gates perform the circuit-level state encoding. To emphasize the similarity with the GRC representation of fig. 11.3 and 11.4, we divided the circuit in two parts. To obtain the global circuit, connect the wires with identical $< num >$ or $[num]$ labels.

final circuit is obtained by structurally combining and linking these patterns. In fact, one can consider the GRC code as an intermediate step in the translation of Esterel into digital circuits, a step where the basic Esterel structures (state, tests, synchronizations) have not yet been encoded with logic gates. For instance, most of the gates and wires of fig. 11.7 can be obtained from the corresponding GRC representation (fig. 11.3 and 11.4) through simple expansion operations. We emphasized with darker gates the expansion of the topmost **Switch** node, and with dotted boxes the expansion of the parallel synchronizers.

In both circuits and GRC specifications generated from Esterel programs we can identify a distinct class of components whose role is exclusively to encode the state of the program for the next instant. In GRC, these components are the **enter** and **exit** state update operations. In the circuit of fig. 11.7 they are the unfilled gates and thin wires that drive the registers. These state encoding components never determine cycles, so that we can remove them in order to perform our analysis using simplified versions of the CCFG flowgraph and of the circuit. Fig. 11.8 gives the simplified CCFG and circuit associated with a small example that has the particularity of being GRC-cyclic, but circuit-acyclic.

After the simplification of the unneeded state encoding components, a very simple correspondence exists between the GRC nodes and edges and the circuit gates and wires. The simplified GRC code is in fact an abstracted version of

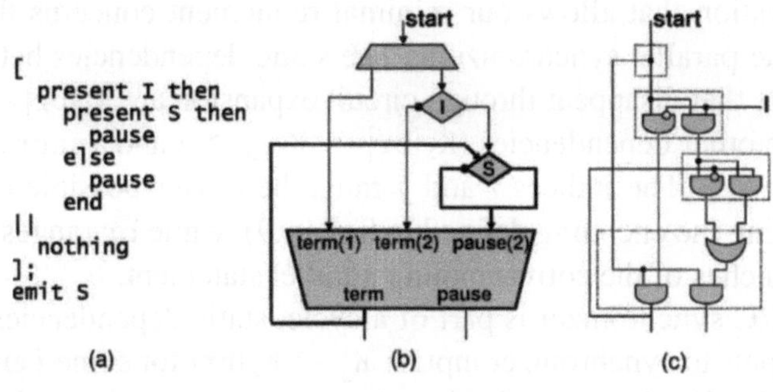


*Figure 11.8.* Esterel example (a) that is GRC-cyclic (b), but circuit-acyclic (c).

the simplified circuit code, which means that a GRC-acyclic Esterel program is always circuit-acyclic.

At this point, we are able to see that expanding into logic gates a node of type **Switch** or **Test** cannot lead to the removal of a cycle, as the resulting logic gates bear the same static dependencies[3] as the initial node. Thus, only the circuit expansion of a parallel synchronizer **Sync** node can remove causal dependencies.

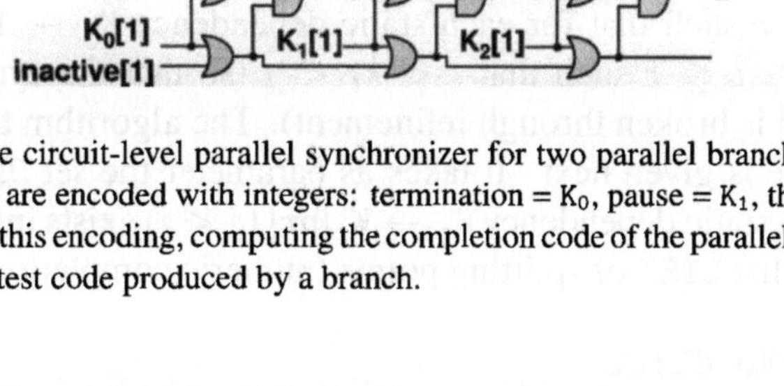


*Figure 11.9.* The circuit-level parallel synchronizer for two parallel branches (0 and 1). The completion codes are encoded with integers: termination = $K_0$, pause = $K_1$, the $n^{th}$ surrounding trap = $K_{n+1}$. With this encoding, computing the completion code of the parallel statement consists in taking the greatest code produced by a branch.

Meanwhile, the circuit expansion of a synchronizer, given in fig. 11.9, is quite complex and does not offer support for the generation of efficient sequential code. To allow a better encoding, we shall try to keep the gates grouped into larger blocks. We define here a technique that allows us, in cases where the circuit is acyclic, but the GRC code not, to split the synchronizer into a minimal number of components of a form that allows a simple software encoding.

---

[3]The static dependency relation between ports (input and output, without distinction) of nodes of a CCFG graph, denoted with $\rightarrow$, is the smallest transitive relation such that (1) if $i$ is an input port and $o$ an output port of the same node, then $i \rightarrow o$ and (2) if a control arc connects in the CCFG the output port $o$ with the input port $i$, then $o \rightarrow i$. The CCFG is cyclic if $p \rightarrow p$ for some port $p$.

The observation that allows our minimal refinement concerns the internal structure of the parallel synchronizers: The static dependencies between synchronizer ports that disappear through circuit expansion are $\mathtt{K_i[br]} \rightarrow \mathtt{K_j}$, with $j < i$. All the other dependencies ($\mathtt{K_i[br]} \rightarrow \mathtt{K_j}, j \geq i$ and $\mathtt{Inactive[br]} \rightarrow \mathtt{K_j}$) are preserved. The indices $i$ and $j$ range here over possible completion codes (following the encoding defined in fig. 11.9), while br ranges over HSR indices of branches of the corresponding parallel statement.

When a GRC synchronizer is part of a cycle, static dependencies link synchronizer outputs to synchronizer inputs: $\mathtt{K_j} \rightarrow \mathtt{K_i[br]}$ for some $i$ and $j$. If the circuit expansion breaks the cycle, then $j < i$. Otherwise, the cycle cannot be broken by circuit expansion. We are then looking for the minimal expansion of a synchronizer, under given constraints of the form $\mathtt{K_j} \rightarrow \mathtt{K_i[br]}, j < i$.

We shall be splitting the synchronizers into parts that correspond to sets of consecutive completion codes. A parallel synchronizer SYNC handling completion codes from 0 to $n$ can be split into $\mathtt{SYNC[0, k_1]}, \ldots, \mathtt{SYNC[k_r, n]}$ with $0 < k_1 < \ldots < k_r \leq n$. Here, $\mathtt{SYNC[k_i, k_{i+1}]}$ receives the completion code wires $\mathtt{K_i[br]}, k_i \leq i < k_{i+1}$, produces the outputs $\mathtt{K_i}, k_i \leq i < k_{i+1}$, and:

- if $k_i = 0$, then $\mathtt{SYNC[k_i, k_{i+1}]}$ receives the inputs Inactive[br].

- if $k_i \neq 0$, then $\mathtt{SYNC[k_i, k_{i+1}]}$ receives from $\mathtt{SYNC[k_{i-1}, k_i]}$ one carry wire per parallel branch.

Thus, our problem is to determine a sequence $0 < k_1 < \ldots < k_r \leq n$ of minimal length $r$ such that for each static dependency $\mathtt{K_j} \rightarrow \mathtt{K_i[br]}$ ($j < i$) there exists $1 \leq t \leq r$ such that $i < k_t < j$ (so that the cycle determined by $\mathtt{K_j} \rightarrow \mathtt{K_i[br]}$ is broken through refinement). The algorithm that determines such a sequence is given next. It takes as parameter the set DEPEND of pairs $(j, i)$ such that a static dependency $\mathtt{K_j} \rightarrow \mathtt{K_i[br]}$ ($j < i$) exists in the GRC code, and returns the list LIST of splitting points (numeric completion codes):

```
LIST:=NewVoidList ;
for code:=0 to n do
  if (i,code) in DEPEND then
    Append(LIST,code);
    RemoveDependenciesBrokenByNewSplitPoint(DEPEND,code)
  end
end
```

We do not formally prove here the minimality of the resulting split. Intuitively, it is determined by the fact that split points are only appended to the list when this is required. For instance, if only one dependency exists (like in fig. 11.8), the synchronizer shall be split in two parts.

# 5. Analysis and optimizations

The translation of Esterel into both circuits and flowgraph-based formats like GRC usually results in code that is far from optimal. There are good reasons for this: Translations have to be structurally defined, so that the code resulting from a subprogram includes elements allowing it to be handled in *any* surrounding reactive context. For instance, provision is made in the translated code to let *any* subcomponent be started, resumed, or preempted. In addition, the translated form maintains the activity information everywhere, consuming memory elements for the state representation. Very often, large parts of the program do not need such a heavy complete encapsulation (because they are always active, to mention a simple case), and the description can be drastically simplified.

For FSMs and digital circuits well-studied optimization techniques exist, based on mathematical properties of the underlying formal models: bisimulation minimization in the first case, combinational and sequential logic optimization in the second. These techniques come together with very efficient algorithms and software tools. However, they are mostly of a global nature, so that they are generally not able to handle industrial-size specifications. At GRC level one can hope to define optimizations based on static analysis techniques, less complete but more efficient in complexity of analysis (low-exponent polynomial). One main prospect in the introduction of the GRC format was indeed to be able to describe and formally justify such optimizations on a well-defined representation model. Some work has already been done in this direction in the existing flowgraph-based compilers. Still, results lack generality and formal support, due essentially to limitations of the intermediate compiler representations.

The optimization of a GRC specification is performed in two phases. First, potential redundancies between state encoding elements are detected through static analysis of the GRC code, and represented with tags on the HSR. The tags are then used in the second phase, where actual optimizations (by removal of edges and nodes) are performed on the CCFG. Actually, the tags are also used to improve software encoding, but this aspect shall be presented in section 6.

The CCFG simplification phase may unveil more redundancies between state elements, so that the optimization can be iterated to further savings. Current practice shows that three optimization cycles (tag computation followed by flowgraph simplification) are enough on our largest examples to reach full optimization.

## 5.1 HSR tag computation

The HSR serves two purposes in a GRC representation. It completes the semantics of the flowgraph by keeping track of the state hierarchy and it also is a repository for tags representing potential redundancies in the state representation. The choice of tags is given by their utility (experimentally determined) in the optimization and encoding process. In addition to # and ||, that were already defined, we currently use the tags: `nt:` and `void:`. Here is their meaning:

- `void:` indicates that the tagged HSR node never remains active at the end of a reaction. The associated statement is either never started, or instantly preempted upon start.
- `nt:` (for *non-terminating*) indicates that the selection status of the tagged HSR node is always equal to that of its parent. The tag is of specific interest on direct children of parallel nodes. In general, a parallel branch can terminate while other branches (and thus the parallel itself) remain active. This, however, is not the case when the branch never terminates, or when it only does so by raising an exception, thereby aborting the whole parallel construct.

Note that other tags can be easily defined for use with different analysis and optimization techniques. Promising analysis algorithms need, for instance, information about loop scopes, or the refinement of the exclusion relation into sequence and conditional exclusion. Our set of tags is only a first version, with already good practical results.

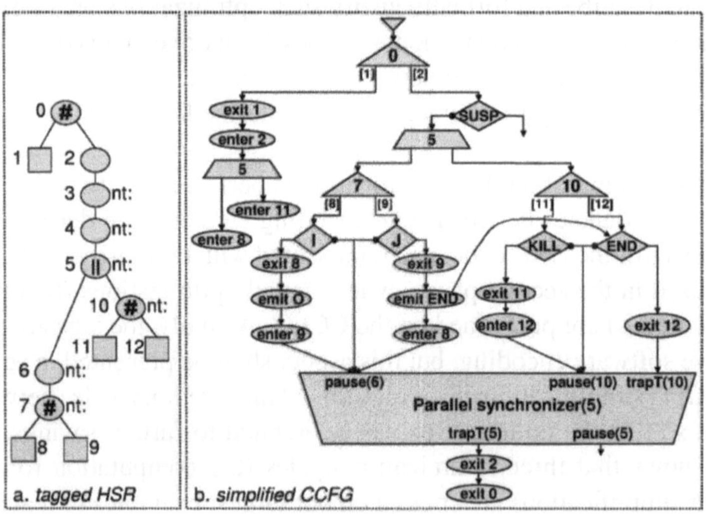


*Figure 11.10.* Tagged HSR and optimized CCFG for our example

The *computation of tags* currently relies on algorithms of linear complexity in the size of the GRC specification. We set the `void:` tag on the HSR node of index $n$ if it falls in one of the following cases: (1) The CCFG contains no **enter**($n$) operation. (2) All of $n$'s children are tagged with `void:`. (3) The parallel node $n$ has at least one child tagged with both `void:` and `nt:`. (4) The node $n$ is child of a parallel node itself tagged with `void:`.

The `nt:` flag is set on all the children of unary HSR nodes. In addition, it is set onto direct children of parallel nodes based on the following analysis of the CCFG parallel synchronizers: *Let p be a parallel HSR node, let c be one of its children, and let $s_1, \ldots, s_n$ be the parallel synchronizers associated with p. Then, we set the* `nt:` *tag on c if none of $s_1, \ldots, s_n$ receives a control edge representing the normal termination of the parallel branch c* (i.e. *if the translation did not generate such an edge or if it has been optimized out).* For our example, the result of the tagging phase is given in fig. 11.10(a).

## 5.2 CCFG optimizations

We apply on GRC flowgraphs two types of simplifications: (1) state access/update protocol simplifications that take into account the new HSR tags and (2) "internal" graph simplifications that do not use state representation information.

**5.2.1 Simplifications based on HSR tags.** These simplifications take into account the state representation redundancies that are represented with tags in order to simplify the state access/update protocol by removing unneeded operations. In many cases, maintaining the coherency of the flowgraph also requires the removal of the driver flowgraph nodes. The following simplifications are currently performed:

- the **Switch** nodes are simplified by erasing the outputs corresponding to `void:` branches. If only one output remains, the node is deleted, and the output is directly connected to the *go* input.
- **Test** nodes whose target is tagged with `void:` or `nt:` are deleted. If only one of the tags is set on the target, then the *go* input is directly connected to either *then*, if the tag is `nt:`, or *else*, for a `void:` tag.
- state update operations whose target is tagged with `void:` or `nt:` are deleted. The associated **Call** node is also deleted, and its output is directly connected to its input.

**5.2.2 Internal flowgraph simplifications.** Derived for the most part from classical control-flow optimization techniques, the "internal" graph simplifications delete dead code, reduce the number of dependencies, and simplify the nodes. We mention three techniques that proved very effective in practice:

- the *constructive sweep* intuitively corresponds to constant propagation and to dead code removal. Dead code is identified by performing a symbolic execution of the GRC code where all accessible control paths are explored in parallel (linear complexity). Then, we can delete control edges that have not been traversed. We also delete some of the adjacent nodes:

  - **Call**, **Test**, and **Switch** nodes whose *go* input port receives no more edges, as well as **Sync** nodes whose input edges have all been deleted.

  - **Test** nodes whose expression has been uniformly set to *true* or *false* due to the removal of signal edges.

- the *dependency simplification* consists of removing certain signal and data dependencies between exclusive control branches (*e.g.* branches of a test). We currently perform the simplification on tests (**Test** or **Switch** nodes) whose expression only involves input signals or state tests.

- the *useless code simplification* consists of deleting code that does not drive actions on data (state changes, output signal emissions, or user data update).

Figure 11.10(b) shows the result of the simplifications on the control-flow graph of our example.

## 6. Code generation

Once tagged and optimized, a GRC model is translated into sequential C code. There are two important steps in this translation, to further improve efficiency. First, a state encoding is choosed so that various **enter** and **exit** operations of the CCFG can be collapsed into fewer C-level data operations. Second, a proper static scheduling of the parallel threads in the CCFG is synthesized. We recall that our approach currently handles only GRC-acyclic and circuit-acyclic specifications, due essentially to the static scheduling algorithm which allows an efficient, low-overhead encoding of the context switches.

### 6.1 State encoding

The state encoding must allow the hierarchic (top-down) identification of the active program statements, through successive tests corresponding to the state decoding nodes of the CCFG. Our encoding technique basically tries to minimize the number and complexity of the state access operations performed in a clock cycle, but without performing code duplication (like the automaton expansion does).

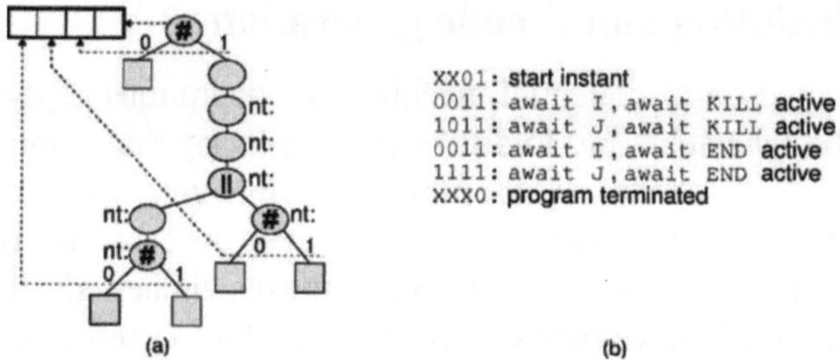


*Figure 11.11.* State encoding for our example: (a) the bit allocation, optimized according to the HSR tags and (b) the encoding of the reachable states of the program.

The encoding algorithm starts by associating an integer test variable to HSR nodes where a state decoding decision may be necessary:

- exclusive selection nodes having more than one child not tagged with `void:`
- children of a parallel node that are not tagged with `nt:`

In the first case, the $n$ children that are not tagged with `void:` are assigned indices from 0 to $n-1$, and the variable is set to $i$ to mark the activation of the branch of index $i$ (and the de-activation of all the other). In the second case, the variable is set to 1 when the parallel branch is active. Otherwise, it is 0. The variable assignment can in fact be further optimized, as the activation bits are in many cases redundant.

The second encoding phase uses exclusiveness properties to minimize the *size* of the representation. The basic remark is here that variables associated to nodes on different sub-trees of an exclusive selection node will never be decoded in the same reaction. Then, they can share the same implementation variable to minimize the executable size and the cache-miss probability. The multiplexing algorithm is actually performed at bit level, in a bottom up fashion. This phase defines the final mapping of the state operations onto the state variables of the implementation.

Figure 11.11 pictures the bit allocation pattern of our example and gives the encoding of the reachable states. For instance, 0011 represents an active program (bit0=1) that is not in the start reaction (bit1=1) and that awaits the signals `I` and `KILL`.

In order to match the HSR encoding, the state access primitives are transformed into assignments and tests of the integer state variable. This step is essential in the generation of efficient code, as each assignment operation usually encodes multiple state update primitives.

## 6.2 Scheduling and C code generation

To be executed on a sequential machine, the operations of the concurrent CCFG must be *scheduled*, *i.e.* totally ordered in a way that complies with the causality relation. A scheduling algorithm works by *interleaving* the concurrent branches of the CCFG using *context switches* where the execution of a branch is suspended and another branch is resumed. Context switches take time, and efficient schedules should have as few of them as possible. Determining a schedule with a minimal number of context switches is NP-complete, but our approach produces acceptable code using a simple depth-first scheduling algorithm.

The CCFG represents with edges all control and data causality, so that any topological order is a valid schedule. To avoid complex, data-dependent scheduling issues, only acyclic GRC specifications are accepted for code generation. In these cases, a topological order always exists and any topological order gives a *static schedule*, *i.e.* one that works at each clock cycle.

```
if(state&binary(0001)){   /*program is active*/
  if(state&binary(0010)){ /*not boot instant*/
    if(!SUSP) {
      int par = 1 ;          /*parallel completion code, init. with the minimal code*/
      bool cs = false ;      /*context switch variable*/
      bool END = 0 ;         /*the internal signal, initially absent*/
      /*parallel branch 2 starts the execution*/
      if(state&binary(0100)) { cs=true ; /*suspend execution on branch 2*/}
      else if(KILL) { state&=binary(1011) ; }
      /*parallel branch 1 starts the execution*/
      if(state&binary(1000)){ if(J){ END=1 ; state&=binary(0111) ; } }
      else{ if(I){ O=1 ; state|=binary(1000) ; } }
      /*parallel branch 2 is resumed*/
      if(cs){ if(END){ par=2 ; } }
      /*parallel synchronization code*/
      if(par==2){ emit_O() ; state=binary(0000) ; }
    }
  }else{ /*boot instant*/ state=binary(0010) ; }
}else{ error(''program already terminated'') ; }
```

*Figure 11.12.* Resulting C code. Note the Boolean *context switch variable* `cs` that is used to resume the second parallel branch after the first one is executed. The global variable `state` is initialized with 0001.

To determine an acceptable static schedule, the code generator uses a simple "greedy" approach. Throughout the scheduling process, the list of concurrent branches is cyclically traversed from the beginning to the end. The code generator then uses a depth-first traversal technique to schedule as much as possible of the current concurrent branch before, if necessary, suspending it and going to the next after performing a context switch. Fig. 11.12 shows the reaction function generated for our example. The variable `state` holds the state representation. The context switches are encoded using Boolean guards. The example needs only one (cs) in order to resume, if necessary, control on the second parallel branch. In general, the resulting code is well-structured, as guards corresponding to lower-level concurrent threads are subsumed to higher-level ones.

# 7. Results

To measure the efficiency of our approach we developed a prototype GRC-based optimizing compiler, called `grc2c`, that plugs into the Esterel compiler system [Est00] developed at INRIA/CMA. The prototype compiles all GRC-acyclic Esterel programs and implements the optimizations described in this paper. In-depth technical information concerning the implementation is provided in [PB02].

| example | description | Esterel(LC) | CCFG nodes | | |
|---|---|---|---|---|---|
| | | statements | initial | optimized | |
| 1-tcint | turbochannel bus | 516 | 798 | 375 | 46.9% |
| 2-wristwatch | berry's example[Est00] | 533 | 834 | 366 | 43.8% |
| 3-atds100 | video generator | 1122 | 2119 | 1059 | 49.9% |
| 4-mca200 | shock absorber[CEG+96] | 3769 | 4562 | 3596 | 78.8% |
| 5-Chorus | OS model[WBC+00] | 4751 | 6299 | 3539 | 56.1% |
| 6-Fuel | avionics fuel controller | 4986 | 8544 | 4516 | 52.8% |
| 7-cabine | avionics display[BBF+00] | 10991 | 18872 | 8037 | 42.5% |
| 8-global | avionics controller | 15831 | 18852 | 13253 | 70.3% |

*Table 11.1.* Testbench examples: description and GRC optimization results

We compared the output of `grc2c` with the output of four other compilers: The automata compiler of Bres[Bre02], the circuit-based compiler of Berry [Est00], the EC compiler of Edwards [Edw02], and the Saxo compiler of Closse *et al.* [WBC+00]. A state-of-the-art sequential circuit optimizer (SIS/blifopt [SSL+92]) has been used whenever possible on the intermediate netlist representation of the circuit-based compiler.

The examples of the testbench are, `wristwatch` excepted, industrial examples of controllers, either directly programmed in Esterel or obtained through automatic translation from SyncCharts [And96] specifications. Table 11.1 gives the initial and optimized specification sizes, and a short description for each of our 8 examples. The LC statement count is good estimation of the Esterel specification size. The unoptimized number of nodes in the GRC flowgraph is always larger than the LC count due to the demultiplexing and to the structural code duplication. However, despite variations in the optimization ratios, the optimized figures always fall under the LC count, meaning that our systematic code duplication pays off. Recall that the optimization algorithms are all linear, so that the optimization phase does not represent a bottleneck in the compilation process. On the test machine (a Linux/PIII/1GHz/128M) it takes less than 1

| example | automata | circuit | EC | Saxo | grc2c |
|---|---|---|---|---|---|
| 1-tcint | 0.30 | 0.66[5] | **0.25** | 0.36 | **0.25** |
| 2-wristwatch | **0.95** | 1.40[5] | 1.26 | 1.29 | **0.95** |
| 3-atds100 | **0.05** | 3.69[5] | 0.14 | 0.12 | 0.08 |
| 4-mca200 | –[4] | 48.19[5] | 2.61 | 3.68 | **1.88** |
| 5-chorus | –[4] | 54.29 | 2.76 | 5.30 | **1.10** |
| 6-fuel | –[4] | 37.66 | –[7] | 16.81 | **15.65** |
| 7-cabine | –[4] | –[6] | –[7] | 31.97 | **29.26** |
| 8-global | –[4] | –[6] | –[7] | 57.45 | **43.27** |

*Table 11.2.* Generated code speed (sec/1Mcycle)

minute for the largest example, depending quasi-linearly on the specification size.

| example | automata | circuit | EC | Saxo | grc2c |
|---|---|---|---|---|---|
| 1-tcint | 72.8 | **11.4**[5] | 11.5 | 15.3 | 17.5 |
| 2-wristwatch | 14.8 | **11.2**[5] | 13.3 | 16.1 | 14.7 |
| 3-atds100 | 28.8 | 31.3[5] | **21.7** | 33.9 | 33.7 |
| 4-mca200 | –[4] | 171.8[5] | 70.2 | 78.9 | **67.7** |
| 5-chorus | –[4] | 230.3 | 99.5 | 104.1 | **98.6** |
| 6-fuel | –[4] | 201.4 | –[7] | **147.5** | 168.7 |
| 7-cabine | –[4] | –[6] | –[7] | 256.1 | **196.8** |
| 8-global | –[4] | –[6] | –[7] | 309.1 | **273.9** |

*Table 11.3.* Object code size (Kbytes)

For each example, the C routines generated by the 5 compilers have been compiled using "`gcc2.96 -O`" and then run for 1 million cycles to determine the average reaction time. The input sequences were either provided by the programmer of the application or (when none provided) generated pseudorandom inputs satisfying the given environmental constraints. The results are presented in the tables 11.2 and 11.3.

The three flowgraph-based compilers (EC, Saxo, and `grc2c`) produce similar figures in both speed and size. Among them, `grc2c` generates faster code, while the differences in code size are not relevant. This seems to indicate that (1) removing redundant state tests (like we do) greatly improves the code speed

[4] Out of memory or timeout (1 hour) during automaton expansion

[5] The intermediate circuit representation has been aggressively optimized using SIS/blifopt prior to C code generation (only possible on small- to medium-sized specifications).

[6] `gcc -O` exhausted system memory during compilation.

[7] The code was not available for tests.

and (2) the C language encoding of `grc2c` could be improved to match on small examples the EC-generated code size.

As expected, the code generated by the circuit-based compiler is slow, even when the intermediate circuit representation has been aggressively optimized. The circuit-based code is also very large and difficult to compile.

Surprisingly, only one example results into `grc2c`-generated code that is slower than its automata-based counterpart. This is probably due to the fact that the (small) processor caches penalize larger programs, and to the fact that GRC-level optimizations leave only few redundancies in the state representation. The automata-based compilation is restricted to small programs, even more than circuit optimization. For all but 3 examples in our testbench, we interrupted the automaton expansion after 1 hour.

The global `grc2c` compilation time, including GRC code generation, optimization, and C encoding, is negligible. It increases from several seconds to approximately 2 minutes, following the size of the GRC specification. The `grc2c`-generated C code is easily compiled by `gcc`.

## 8. Conclusion and Future work

We introduced in this paper a new intermediate model, called GRC, for the representation of Esterel programs. The GRC representation makes the control flow of the Esterel program explicit, but also preserves information about its initial structure. Thus, its supports a variety of efficient analysis, optimization, and sequential code generation algorithms based on low-cost static analysis. Benchmarks show that the GRC-based `grc2c` compiler generates faster code than the other high-capacity compilers, for a similar code size.

The GRC representation has well-defined formal semantics, allowing us to (1) prove the correctness of the software code with respect to the constructive semantics of Esterel and (2) maintain a close relation between GRC and circuit translations of an Esterel program. We were therefore able to define a notion of acyclic GRC description and match it with the corresponding one on circuits. The result is of special practical interest, since acyclicity is the main correctness criterion used for Esterel programs.

The `grc2c` compiler has been implemented as a module of the INRIA/CMA Esterel version 5 compiler system. At the same time, some of the optimizations defined in this paper have been incorporated into Esterel Studio, an industrial Esterel-based environment for the design and verification of complex systems-on-a-chip (SoC).

Other applications are possible, in addition to generating software code for Esterel. The GRC-based approach could be easily adapted to compile other synchronous languages and formalisms with fine-grained parallelism, such as ECL [LS99] or the SyncCharts [And96]. However, the most obvious application of Esterel is the generation of optimized digital circuit code for Esterel: The analysis and optimizations presented in this paper are not software-specific and some experiments [PB01] suggest that applying them to Esterel-generated circuits would result in important simplifications. Such Esterel-specific simplifications should be performed *before* the main circuit optimization phase in order to improve its output and (usually long) execution time. However, applying the simplifications on flat netlists would be inefficient. Instead, the GRC code should be used as a *target-independent representation* for (Esterel-specific) optimizations based on high-level structural information. The optimized GRC code should then be translated into a netlist and passed to a state-of-the-art circuit optimizer. Such a GRC-based optimized circuit generator is under construction at Columbia University under the supervision of Stephen Edwards.

Many directions remain to be explored. First, it is essential to pursue the development of new GRC-level analysis and optimization techniques. The domain seems promising due to the large quantity of Esterel structural information not yet exploited. An interesting problem is here to determine the amount of code duplication that results in the best size/speed ratio, and to reduce the unneeded duplication by sharing identical branches.

A second direction that we would like to investigate is the generation of efficient simulation code for *correct cyclic* GRC specifications, thereby raising the applicability to the same class of programs as the automata- and circuit-based compilers. Once cycles have been precisely located and proved correct, an idea would be to run them locally under the 3-valued constructive semantics of circuits. Some advances in this direction can be found in Potop-Butucaru [PB02]. Other schemes would require substituting the cyclic GRC subgraph by another, equivalent acyclic one. Of particular interest is here the resynthesis technique of Edwards [Edw03], which performs a controlled code duplication that preserves many of the structures of the initial circuit. The technique could be extended to perform a similar GRC-level resynthesis while still using the current circuit-level analysis. Another question, related but nevertheless distinct, would be to find GRC-level static analysis approximation methods that give sufficient conditions for correctness in cyclic cases, and then try to derive a corresponding, more dynamic scheduling scheme which could cope with different ordering according to different state configurations.

Finally, we are investigating the possibility of a union, at intermediate representation level, between Esterel and the synchronous data-flow language Signal, for analysis and efficient simulation purposes. The main goal would be here to use Esterel for specifying behaviors and Signal for specifying architectural aspects. The high-level intermediate structures of the Signal compiler – the *clock tree* and the *conditional dependency graph* – are respectively similar in function and form to the HSR and the CCFG of the GRC code. The first represents the hierarchy of the activation conditions, while the second represents the actual control and data flow. However, important differences exits: The HSR of a GRC specification represents the internal structure of the Esterel program, closely followed by the state encoding/decoding part of the CCFG. In the Signal compiler, the activation conditions of the clock tree are synthesized from (clock) constraints on the input and output signals. Thus, the union would require the ability to assign/derive complex interface constraints to/from an Esterel specification.

## References


[And96] Charles André. Representation and analysis of reactive behaviors: A synchronous approach. In *Proceedings CESA'96*, Lille, France, July 1996. Also available as I3S technical report RR 95-52.

[BB91] Albert Benveniste and Gérard Berry. The synchronous approach to reactive and real-time systems. *Proceedings of the IEEE*, 79(9):1270–1282, 1991.

[BBF+00] Gérard Berry, Amar Bouali, Xavier Fornari, Emmanuel Ledinot, Eric Nassor, and Robert de Simone. Esterel: A formal method applied to avionic software development. *Science of Computer Programming*, 36:5–25, 2000.

[BCE+03] Albert Benveniste, Paul Caspi, Stephen Edwards, Nicolas Halbwachs, Paul LeGuernic, and Robert de Simone. The synchronous languages 12 years later. *Proceedings of the IEEE*, 91(1):64–83, 2003.

[BCG+99] F. Balarin, M. Chiodo, P. Giusto, H. Hsieh, A. Jurecska, L. Lavagno, A. Sangiovanni-Vincentelli, E. Sentovich, and K. Suzuki. Synthesis of software programs for embedded control applications. *IEEE Transactions on Computer-Aided Design*, 18(6):834–849, 1999.

[BdS91] Frédéric Boussinot and Robert de Simone. The Esterel language. *Proceedings of the IEEE*, 79(9):1293–1304, 1991.

[Ber92] Gérard Berry. Esterel on hardware. *Philosophical Transactions of the Royal Society of London, Series A*, 339:87–104, 1992.

[Ber99] Gérard Berry. The constructive semantics of Pure Esterel. Draft book available at http://www.esterel-technologies.com/, 1999.

[BG92] Gérard Berry and Georges Gonthier. The Esterel synchronous programming language: Design, semantics, implementation. *Science of Computer Programming*, 19(2):87–152, 1992.

[Bre02] Yannis Bres. *Exploration implicite et explicite de l'espace d'états atteignables de circuits logiques Esterel (Implicit and explicit ex-*

*ploration of the reachable state space of Esterel logic circuits).* PhD thesis, Ecole des Mines de Paris, december 2002.

[CDO97] Claude Castelluccia, Walid Dabbous, and Sean O'Malley. Generating efficient protocol code from an abstract specification. *IEEE/ ACM Transactions on Networking*, 5(4):514–524, 1997.

[CEG+96] M. Chiodo, D. Engels, P. Giusto, H. Hsieh, A. Jurecska, L. Lavagno, K. Suzuki, and A. Sangiovanni-Vincentelli. A case study in computer-aided co-design of embedded systems. *Design Automation for Embedded Systems*, 1:51–67, 1996.

[CGP99] Paul Caspi, Alain Girault, and Daniel Pilaud. Automatic distribution of reactive systems for asynchronous networks of processors. *IEEE Transactions on Software Engineering*, 25(3):416–427, 1999.

[CPP+02] Etienne Closse, Michel Poize, Jacques Pulou, Patrick Vernier, and Daniel Weil. Saxo-RT: Interpreting Esterel semantic on a sequential execution structure. In Florence Maraninchi, Alain Girault, and Eric Rutten, editors, *Electronic Notes in Theoretical Computer Science*, volume 65. Elsevier, 2002.

[Edw02] Stephen Edwards. An Esterel compiler for large control-dominated systems. *IEEE Transactions on Computer-Aided Design of Integrated Circuits and Systems*, 21(2), February 2002.

[Edw03] Stephen Edwards. Making cyclic circuits acyclic. In *Proceedings of the 40th conference on Design automation*, Anaheim, CA, USA, 2003.

[Est00] The Esterel manual, part of the Esterel version 5.92 distribution. Available for download at www.esterel-technologies.com, 2000.

[FLLO95] Robert French, Monica Lam, Jeremy Levitt, and Kunle Olukotun. A general method for compiling event-driven simulations. In *Proceedings of the 32nd Design Automation Conference*, San Francisco, CA, USA, 1995.

[Hal93] Nicolas Halbwachs. *Synchronous programming of reactive systems*. Kluwer Academic Publishers, 1993.

[HCRP91] Nicolas Halbwachs, Paul Caspi, Pascal Raymond, and Daniel Pilaud. The synchronous data flow programming language Lustre. *Proceedings of the IEEE*, 79(9):1305–1320, 1991.

[LGLL91] Paul LeGuernic, Thierry Gauthier, Michel LeBorgne, and Claude LeMaire. Programming real-time applications with Signal. *Proceedings of the IEEE*, 79(9):1321–1336, 1991.

[LPM99] Jaejin Lee, David Padua, and Samuel Midkiff. Basic compiler algorithms for parallel programs. In *Proceedings of the 7th ACM*

*SIGPLAN Symposium on Principles and Practice of Parallel Programming*, Atlanta, GA, USA, 1999.

[LS99] Luciano Lavagno and Ellen Sentovich. ECL: A specification environment for system-level design. In *Proceedings of the 36th Design Automation Conference*, New Orleans, LA, 1999.

[Muc97] Steven Muchnick. *Advanced Compiler Design and Implementation*. Morgan Kaufmann Publishers, 1997.

[PB01] Dumitru Potop-Butucaru. Fast redundancy elimination using high-level structural information from Esterel. RR 4330, INRIA, Sophia Antipolis, France, 2001.

[PB02] Dumitru Potop-Butucaru. *Optimizations for faster simulation of Esterel programs*. PhD thesis, Ecole des Mines de Paris, November 2002.

[SBT96] Tom Shiple, Gérard Berry, and Hervé Touati. Constructive analysis of cyclic circuits. In *Proceedings of the International Design and Testing Conference*, Paris, France, 1996.

[SSL+92] Ellen Sentovich, Kanwar Jit Singh, Luciano Lavagno, Cho Moon, Rajeev Murgai, Alexander Saldanha, Hamid Savoj, Paul Stephan, Robert Brayton, and Alberto Sangiovanni-Vincentelli. SIS: A system for sequential circuit synthesis. Memorandum M92/41, UCB, ERL, 1992.

[Tom97] Horia Toma. *Analyse constructive et optimisation séquentielle des circuits générés à partir du langage synchrone réactif Esterel (Constructive analysis and sequential optimizations of circuits generated from the synchronous reactive language Esterel)*. PhD thesis, Ecole des Mines de Paris, September 1997.

[WBC+00] Daniel Weil, Valérie Bertin, Etienne Closse, Michel Poize, Patrick Vernier, and Jacques Pulou. Efficient compilation of Esterel for real-time embedded systems. In *Proceedings CASES2000*, San Jose, CA, USA, 2000.

Chapter 12

# OPTIMIZING SYSTEM MODELS FOR SIMULATION EFFICIENCY

Shekhar A. Sharad, and Sandeep Kumar Shukla
*Department of Electrical and Computer Engineering*
*Virginia Tech*
*Blacksburg, Virginia, USA*
{ssa,shukla}@vt.edu

**Abstract** This chapter describes algorithmic ways to transform a multi-threaded SystemC model into a non-threaded model for simulation efficiency. Such a transformation must be correctness preserving, and the transformed model must simulate the original model. We show simulation results to establish the efficiency enhancement by the systematic transformation.

**Keywords:** SystemC Threads, Methods, Simulation Efficiency, Transformation, Optimization

## 1. Introduction

Embedded System Design has entered a new era with several key developments such as fabrication technology hitting the submicron levels, improved algorithms for validation and testing, and better tool support for designing Embedded Systems. Miniaturization has been the epitome of development with no compromise on speed or capacity making the design-productivity gap an area of concern for the Embedded Systems community. SystemC [15], SpecC [5], SystemVerilog [1] have been posed as effective ways to mitigate the challenge of the *design productivity gap*. However, for system level modelling to be acceptable to the design community, the use of such new modelling paradigm must show some concrete benefits. One of the benefits that translates into a faster time-to-market is simulation efficiency at the system level due to a raised level of abstraction [7]. Another significant propellant of any modelling paradigm in Embedded Systems is that the system must be synthesizable from the system level model.



*R. Gupta et al. (eds.), Formal Methods and Models for System Design,* 317-330.

Inspite of the observation that this may not be possible in all the cases, it is desirable that if it is not synthesizable by automated means, it should bear enough structural resemblance to the final implementation, so that the mapping of the system level model onto a preconceived structure should not be too difficult, or time consuming. These observations are a result of Industrial Design Flow concerns and hence, an area that calls for attention. The motivation from these factors led to the work on *concurrency re-assignment* discussed in [7, 11, 12]. The idea was that in order for the system level model to be close to the implementation structurally, one needs to express the hardware concurrency naturally in the system model. This implied that threads must be used, because in programming languages, threads are a convenient means to express concurrency and system level languages such as SystemC provides facilities for threaded modules by the use of wait and wait_until constructs. Another system level language that provides the same kind of support is the widely popular VHDL, which also uses wait statements. Similarly, other languages also have process structures which are akin to user-level threads in order to depict concurrency at the system level.Threads in SystemC for example, allow designers to express the design along the hardware module boundaries, which in real hardware, work truly concurrently. However, threads imply context-switch overheads, leading to inefficient simulation. It was recognized in [11], that the reason for inefficiency is more due to the fact that the amount of computation between context switches is too low because most hardware modules compute only one clock worth of computation per invocation. This problem was solved in [11, 12] by analyzing the thread interaction patterns followed by restructuring and creating longer computation sequences based on such analysis. These long computation sequences are then mapped onto kernel level threads such as POSIX threads. This showed improvements in the order of 100% on multiprocessor platforms for some examples [11]. However, such restructuring of the system level model does not bear much resemblance to the final implementation as it is merely scheduling of SystemC code onto a kernel level thread. In such a scenario, the task of proving the equivalence of the converted model to the original model, is cumbersome. It is further difficult to visualize a systematic approach to converting this scheduled SystemC code into actual implementation without loss of time or effort owing to the lack of resemblance between the converted and original model. These are the problems that this work addresses. This work addresses the area of simulating concurrent systems efficiently even on single-processor systems. The two algorithms that are presented achieve this goal through a series of systematic transformations that make sure that a transformed model can simulate identical to the original model. This work also addresses the concern of bearing resemblance to the final implementation, such that developing an automated means to synthesizing the converted systems is an achievable goal.

This work on concurrency elimination may seem comparable to *quasi-static scheduling* [13]. It is worth noting, however, that our work and quasi-static scheduling have some major differences in their objective and methods. First, quasi-static scheduling is done to schedule a concurrent specification onto a single thread of execution, with the possibility of data-dependent switch between multiple schedules. The techniques used for quasi-static scheduling are quite complex, and often based on formalisms of free-choice Petri-nets [13]. On the other hand, system-level languages such as SystemC have their own scheduler built into the simulation kernel, which schedules either threads or methods [15]. Concurrency re-assignment needs to somehow make the scheduler work in a way that makes the simulation faster.

Also note that our synthesis of control state machines is not the same as synthesis of explicit state machines [9, 6], where a behavior is synthesized into an explicit state machine. Such mapping leads to a hardware implementation of the behavior at the RTL level, but leads to slower simulation time. Our state machines keep the level of abstraction of the model intact, but are designed to make the discrete event simulation kernel work in a more efficient manner.

Also note the difference between this work and work presented in [3]. In [3], the idea is to maintain a reconstruction tree to compile into concurrent control flow graphs and make efficient concurrent code from Esterel. Our objective as well as method is very different. We compile away all threading to avoid user-level thread usage. One might think that if SystemC was implemented with POSIX or other kernel level threads, we could use the techniques of [3], however, our past experiences in replacing a systemC kernel with POSIX threads did not show any improvement in performance unless multi-processor system was used. So the work presented here shows how to improve simulation efficiency without multi-processors and changes in the SystemC simulation kernel.

As mentioned already, in comparison with [11, 12], our work shows that one can synthesize special user level state machines, that co-operate with the scheduler to achieve the objectives of sequencing the computations in a particular fashion so as to achieve simulation efficiency. This work does not require mapping onto kernel level threads, nor does it depend on multi-processor machines to realize the efficiency gains. Another important aspect of our transformation is that it is completely *localized.* In the previous work a global analysis of the original model leads to the re-threading, whereas our work only needs to look at each individual thread to carry out the transformation. This also leads to easier implementation of the transformation.

The rest of the chapter is organized as follows: In Section 2, we formally explain the problem of concurrency re-assignment for simulation efficiency and provide the basic definitions which we use in describing our localized transformations. In Section 3, we describe our algorithm for control state machine synthesis. In Section 4, we describe the experimental results to show the gains that our algorithm achieved over the original model.

## 2. Problem Statement and Definitions

At it has been noted earlier, system level languages use several structures to depict concurrency, such as VHDL *processes* (with no sensitivity list but a set of wait statements), or Verilog *always* blocks, or the *threaded structures* in SystemC. Without loss of generality, it can be stated that whatever be the mode of expressing concurrency. such concurrency assignment may be mapped onto software threads for simulation. For example, in SystemC, user-level threads, specified using the SC_THREAD and SC_CTHREAD constructs, are directly used to express them. However, an executing thread $T$ needs to be suspended to let other threads or processes to perform their computation. This condition holds even in the presence of one concurrent construct along with several other non-concurrent constructs. The objective behind such a behavior is attributed to the fact that each hardware block needs to perform computation in each cycle before the scheduler can move on to the next cycle in time. To put it in a more hardware-oriented sense, in actual hardware, at every cycle, all hardware blocks have an opportunity to perform an operation(s) and many of them do so. The key feature of any system level language is to capture this behavior accurately in order to allow for realistic simulation and synthesis. Another caveat to this clause is that many a times, a thread needs to be suspended because some condition needs to be satisfied by computation done in another module. While languages like VHDL allow this behavior to be depicted using a boolean combination of conditions on the wait statement, SystemC handles this kind of suspension using the wait_until(boolean_expression) construct. Motivated by the fact that such a threaded structure is subject to context-switching overheads which in turn lead to an increase in simulation time and consequently a decrease in simulation efficiency, it is clear that it is advantageous to develop an algorithm(s) that operates on these threaded structures, without sacrificing the basic structure and flow of the same, while at the same time, resulting in a non-threaded structure that is cycle-true and accurate in simulation which would be faster and more efficient than the original one. The increase in efficiency as we shall show, is dependant a lot on the kind of system being modelled. Nevertheless, any system undergoing such a transformation is likely to show an improvement in simulation efficiency because of the decrease in overheads. We shall now define certain key-terms we will be using in explaining the transformation algorithm that we have developed. It may be noted that though the underlying notion of these concepts is applicable to all system-level design languages, this work is dedicated towards implementing these concepts in SystemC, because, unlike other HDL languages such as VHDL and Verilog which compile into executable code, SystemC has a wrapper library to communicate with in order to exhibit the hardware-behavior. Implementing such a methodology into SystemC greatly improves simulation efficiency as it reduces the number of interactions of the Kernel with the module.

DEFINITION 12.1 ***Suspension Points*** *in a thread of computation are defined as the the places with either a* $wait\_until(boolean\_expr)$ *or* $wait()$ *statement. In case wait() statements with an argument, such as wait(n) are encountered, they are broken down into* ***n*** *wait() statements, and hence represent n consecutive suspension points.*

In essence, a suspension point represents at that point of execution, the states of all the variables, signals and channels that a particular process writes to. Though it is possible that no change in state occurs to any of the signals or channels, as would be the case, when a wait() statement is encountered, it is also intuitive that when the thread resumes from a suspension point, the only signals and channels that may have changed their state are the ones that this process is sensitive to. So at the resumption point, the state of all channels, signals and variables can be called the *resumption state* of that suspension point.

Note that, if the computation $f_{ij}$ following a resumption from a suspension point contains an if-then-else branching, then the direction that branching takes is determined by the resumption state of the suspension point.

It is important now to define a *distance value* $d(s)$ of a *wait*() type suspension point. Though the following definition is not complete owing to the incomplete coverage of branching constructs, it serves to provide the general idea of the what we mean by a *distance value*.

DEFINITION 12.2 $d(s) = 1$ *if s is the first suspension point after the start of the thread body. Now if s is the first suspension point after a branching, then* $d(s) = d(s') + 1$ *where* $s'$ *is the last suspension point in the program text order before this branching. When two branches join, the first wait statement s after the join has* $d(s) = min_{s'}\{d(s') + 1\}$ *where* $s'$ *ranges over all the last suspension points of all the joining branches.*

We illustrate the concept of distance values using a simple module with branching constructs with suspension points. Note that the *else branch has one suspension point lesser than the* if branch. Consequently, the suspension point following the if-else block takes on the (minimum + 1) of the two branches, which happens to be 4 in this case.

```
void ex_module::ex_thread()
{
   /*thread illustrating distance values*/
  while(1)
  {
   c = a.read();
   d = b.read();
   wait();            // d(s) = 1
   op_port.write(c+d);
   wait_until(ack.delayed() == true); //d(s) = 2
   if(c+d > 10)
   {
    op_port1.write(c+d-5);
    wait();          //d(s) = 3
    ack_signal.read();
    sig_ip = ip_port1.read();
    wait_until(data_accept.delayed() == true);
    //d(s) = 4
   }
   else
   {
    op_port1.write(c+d*20-2);
    wait();  //d(s) = 3
    c = c+5;
    d = d-2;
   }
   final_result_ready.write(true);
   wait();  //d(s) = min(3,4) + 1 = 4
   final_result_ready.write(false);
   wait(); //d(s) = 5
  }
}
```

By virtue of this definition, it is now possible for us to assign a number to each suspension point. It may be noted that there exists a possibility that a suspension point is never reached. This situation arises due to the presence of suspension points in branching constructs such as if-then-else and switch-case statements.

Simply put, a SystemC thread $t_i$ can be thought to have multiple computation blocks (which during simulation is carried out in delta-cycles or zero-time) $f_{ij}$'s, where any two consecutive computation blocks have a suspension point between them. In case we have a linear system, i.e., a system that does not have branching constructs, by visualizing this system as a series of computation blocks separated by suspension points, we can observe a finite state machine, $S$ with $k_i + 1$ states where $k_i$ is the number of suspension points in the system. State $s_0$ is the state at the start of the thread, and $s_j$ is the state at the suspension point $j$. The transition from state $s_j$ to $s_{j+1}$ for $j \in \{0..n-1\}$ triggers the action $f_{i(j+1)}$. According to the simulation semantics of SystemC[15], this action occurs in zero-time. From state $s_n$, there is a transition back to $s_1$ with the action $f_{i(k_i+1)}; f_{i1}$ where ; is sequential composition of the actions $f_{i(k_i+1)}$ and $f_{i1}$ which also happens in zero time in simulation.

If the system is not linear, i.e., branching constructs such as if-then-else and switch-case exist, then for any computation block $f_{ij}$ that contains a branching construct, one can think of that $f_{ij}$ as a block of multiple alternatives represented by ${f^{11}}_{ij}, {f^{12}}_{ij}, ..., {f^{1l}}_{ij}$ for the suspension points on one arm of the branch, and ${f^{21}}_{ij}, {f^{22}}_{ij}, ..., {f^{2l}}_{ij}$ for the suspension points on the second arm of the branch and so on until the all the branches are covered. It is apparent that once a branch $y$ where $y \in \{0..h\}$ is selected, then for that particular run of the module, the suspension points and hence the computation blocks in the other $h-1$ branches are not taken thus resulting in a unique sequence of computations for the branching construct. However, one can think of this as a state machine, with multiple outgoing transitions from a suspension point, based on the resumption state of that suspension point.

Note that the actual simulation kernel executes all the threads that are ready to run in the beginning of a simulation cycle, and each thread is then executed until the next suspension point from the resumption state. If the suspension point is a $wait()$ statement, then the thread $t_i$ will only get a chance in the next simulation clock cycle. If the thread is waiting for an event, and that event occurs during the delta cycles, the thread may get a chance to run to the next suspension point during the same simulation clock cycle. In any case, we can then mark the execution of the threaded model as the points where a thread gets suspended, irrespective of whether the suspension is resumed at a delta cycle within the same clock cycle or the next one.

Let us also note, by the semantics of the scheduler in the simulation kernel, two threads may interact only at the suspension points by affecting each other's resumption state. One can think of this updating of signals/channels to affect each other as a synchronization action forced by the kernel. So if we now consider the possible execution of each thread as described above as a state machine, we may observe, that the interaction between the state machines can be described by classical parallel composition operation in CCS or CSP [8, 10] style.

## 3. The Algorithms

Our algorithm involves restructuring each individual thread $t_i \in T$ to convert it into its equivalent method $m_i \in M$. We also describe a modification of this algorithm to optimize performance even further. The objective of the first algorithm is to add finite state controllers to each thread while eliminating suspension points. The modified version implements the same functionality of the method $m_i \in M$ in the form of a centralized controller. These two algorithms will be developed in this section.

DEFINITION 12.3 ***Functional Blocks:*** *The set of statements between two consecutive suspension points is called a functional block, F. A functional block can also be null, as in case of wait(n) constructs decomposed into n wait() statements.*

DEFINITION 12.4 ***Guard:*** *Each Functional block is preceded by a guard G, which is a unique value of a counter and sometimes, a boolean expression. The boolean expression statements are usually encountered when there is a branch construct.*

**Multiple-Controller Algorithm**

1 For all threads, Identify all variables declared in the thread and make them global. If they are initialized at the time of declaration, initialize them in the constructor.

2 For each thread $t_i$ in $T$ in the system do the following

  (a) Identify all the suspension points of $t_i$. Let the number of suspension points in this thread, $t_i$, be $k_i$

  (b) Associate a unique static counter, $C_i$ with $t_i$. Initialize to 0. Since methods execute at once (for now assume no use of $next\_trigger()$ for simplicity, but it can be handled as well.), the values of the counters will be used as a guard to prevent the whole method to execute at once. In each functional block between two $wait()$ type suspension points, add statement to increment the counter.

  (c) Replace each $wait()$ type suspension point $s$ with an appropriate guard $G$, such that $G = (C_i = d(s))$. For $wait\_until(boolean\_expr)$ type suspension point $s$, add a guard $G = (boolean\_expr \&\& (C_i = d(s)))$ if $d(s)$ is the distance of the suspension point as defined before.

  (d) Modify the definition of the module to include the SC_METHOD() object instead of the the SC_THREAD() object making it sensitive to the appropriate signals.

3 If all threads are converted, end else goto step 2

**Single Controller Algorithm:**This is a slight modification of the Multiple-Controller algorithm by bringing the guards out in to global scope of the system level model, and then creating function pointers associated with each functional block and attaching them to the guards. So at every simulation cycle guards are evaluated and the corresponding functions are called. For some cases this modification shows further simulation performance improvement. However, this is just a C++ trick to avoid engaging the simulation kernel entering a module when guards are not true. In the experimental results section, we show the simulation performance comparisons between original threaded model, model converted with Multiple-Controller algorithm, as well as the model converted by Single-Controller algorithm.

We illustrate the Multiple-Controller algorithm with the following example. The conversion of this code snippet by Single-Controller algorithm is done in a similar manner.

Original Code:

```
while(1) {
 c=a.read() + b.read();
 wait_until(flag.delayed()==true);
 input_freq=output_freq.read() + 50.0;
 if(input_freq > 100.0)
 {
  c_value=input_freq * 0.09 * c;
  wait();
  cap_value.write(c_value);
  wait();
 }

 else
 {
   c_value=input_freq * 0.18 * c;
   wait();
   cap_value.write(c_value);
   wait();
 }
}
```

Transformed Code:

```
if(counter==0) {
   c=a.read() + b.read();
   counter++;
} else if(counter==1 && flag.read()==true) {
   input_freq=output_freq.read() + 50.0;
   if(input_freq >100.0)
   {
     c_value=input_freq * 0.09 * C;
     counter++;
   }
   else
```

```
    {
      c_value=input_freq * 0.18 * C;
      counter++;
    }
  } else if(counter==2) {
    cap_value.write(c_value);
    counter=0;
  } else return; }
```

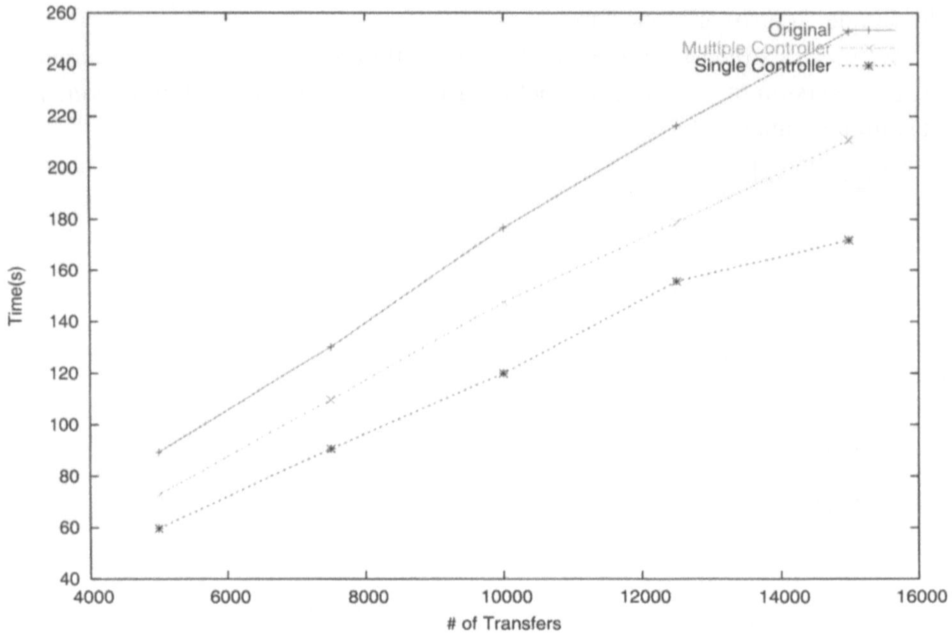


*Figure 12.1.* DMA Transfer Rate

# 4. Experimental Results

The two algorithms have been successfully used to convert four models with varying characteristics and complexity. These converted models are available at our website for download. The models chosen include a DMA system, a Sobel Operator model, a FFT model and a RISC processor model. The description of these systems are available in [2, 14, 4] The results of the increase in simulation efficiency are shown in figures12.1,12.2,12.3 and 12.4 respectively.

By carrying out the simulation on these four models, we found that the Single Controller algorithm worked better than the Multiple-Controller algorithm only for the DMA model and equalled the latter for the Sobel Operator model. In the other two models, the Multiple-Controller algorithm was found to perform better. We believe this phenomenon is due to the increased overhead of referring

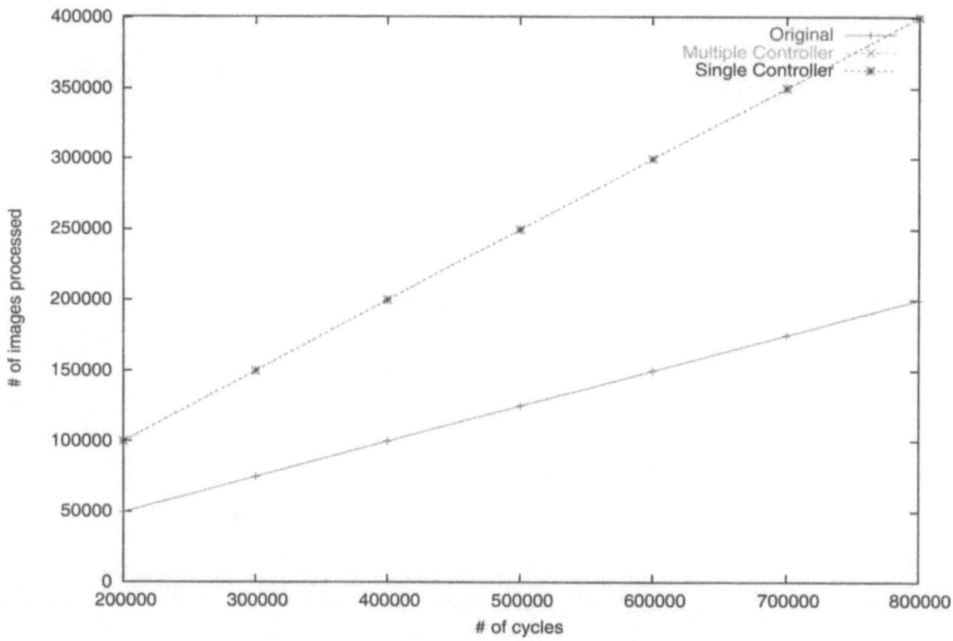


*Figure 12.2.* Images Processed by Sobel

to the structure for every variable in the Single-Controller algorithm which is not needed in the Multiple-Controller algorithm. The performance characteristics of the DMA model is shown in Figure 12.1. An improvement of 1.2x and 1.5x over the original model was observed after applying algorithm 1 and 2 respectively. The Sobel operator model, whose performance curves are shown in Figure 12.2, showed an improvement of 2x over the original model after conversion. This improvement was observed for both algorithm 1 and 2. The performance measurement here was the number of images processed. The Sobel algorithm did not have any branches involved. So this behavior was as expected. The FFT system showed an improvement of 3.6x and 1.8x after it was subject to algorithms 1 and 2 respectively. The performance measurements for the FFT model are shown in Figure 12.3. Finally the RISC model involving the characteristics of all the above models showed an improvement of 1.65x and 1.25x by algorithm 1 and 2 respectively as shown in Figure 12.4. All the simulations were carried out on a P-IV 2.0 Ghz system with 512 MB RAM running the Linux OS (Kernel ver. 2.4.19) with SystemC 2.0.1.

The implementation of this algorithm using hand-written means is tedious, time-consuming and may lead to errors. In order to make this process less-tedious and standardized so that every transformation is carried out in the same fashion, we are in the process of developing an automated transformer, EPT.

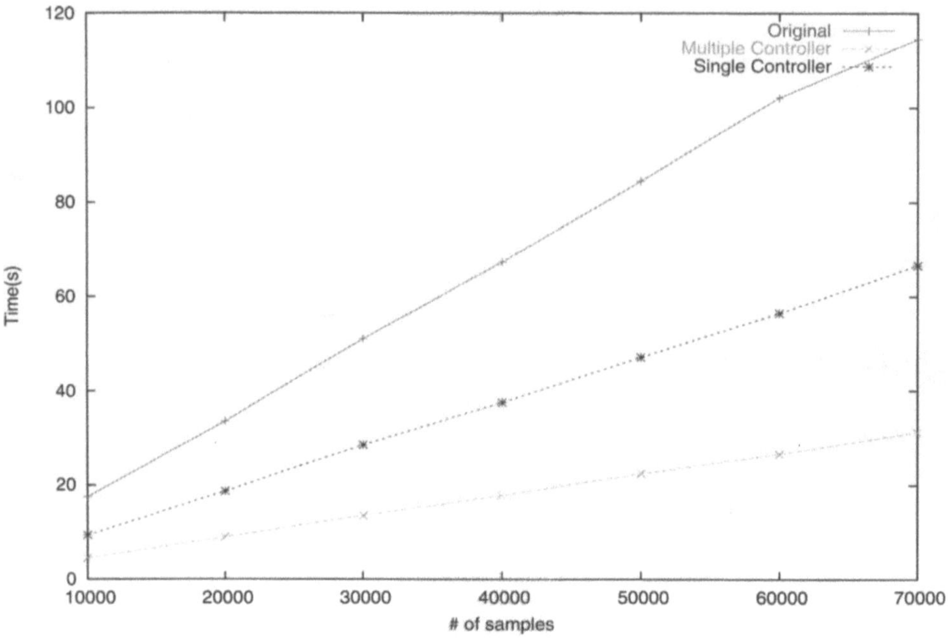


*Figure 12.3.* FFT Transfer Rate

This transformer will take in as input a configuration file that contains a list of modules of the model(one does not have to specify the threads or methods involved - The tool will detect them by itself). The output of this system will be the set of modules transformed into methods. The salient features of this tool include the ability to extract information on whether a thread exists and consequently, whether a module needs to be subject to transformation. It also detects suspension points in branches. At the present moment, it handles a linear flow of suspension points and suspension points inside *if-else* constructs. Support for transformation of suspension points inside *switch* and *while* loops are under development at the present time.

## 5. Conclusion and Future work

In this work, we have developed an algorithm to transform system level threaded models into their non-threaded counterparts in order to improve simulation efficiency. The reasoning behind this transformation was, as shown, in reducing the interaction of the kernel with the thread and hence improve simulation efficiency. This algorithm is applied to four hardware systems, namely, FFT, SOBEL, DMA and RISC CPU systems and the corresponding improvement in Simulation efficiencies have been shown. This algorithm is further implemented in the form of a tool, EPT, that takes in a threaded model as an

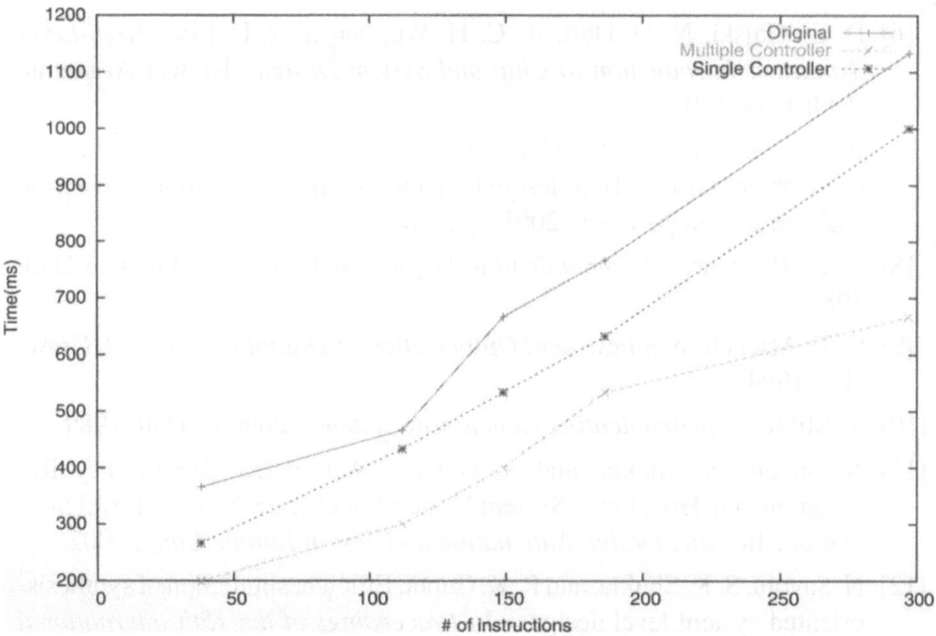


*Figure 12.4.* RISC CPU Operation Times

input and gives a an equivalent non-threaded model as its output. Current focus in this area is on improving EPT to be compatible to any sort of threaded system, including multiple-levels of nesting and different numbers of suspension points in branches. This work is a precursor to a set of important research objective to be achieved in the form of formal verification of System Level models in SystemC, transformation of systems to Multi-threaded graphs and developing design patterns for simulation efficiency.

## References

[1] Accellera. SystemVerilog 3.0 Language Reference Manual. http:www.accellera.org3.0_LRM.pdf.

[2] DMA read and write transfer schematics: http://www.mip.sdu.dk/~fonseca/bachelor_project/html_sections/se-isa-bus-dma-transfer.html.

[3] S. Edwards. An Esterel Compiler for Large Control Dominated Systems. *IEEE Transactions on Computer Aided Design*, Feb 2002.

[4] Fast Fourier Transforms: http://www.daqarta.com/0tffft4.htm.

[5] D. D. Gajski, R. Domer, and J. Zhu. IP-Centric Methodology and Design with the SpecC Language. In *System-Level Synthesis*. Kluwer Academic

Publishers, 1999.

[6] D. D. Gajski, N. D. Dutt, A. C.-H. Wu, and S. Y.-L. Lin. *High-Level Synthesis: Introduction to Chip and System Design*. Kluwer Academic Publishers, 1992.

[7] P. Garg, S. Shukla, and R. Gupta. Efficient Usage of Concurrency Models in an Object-Oriented Co-design Framework. In *Proc. Design Automation and Test in Europe Conf.*, 2001.

[8] C. A. R. Hoare. *Communicating Sequential Processes*. Prentice Hall, 1985.

[9] G. D. Micheli. *Synthesis and Optimization of Digital Circuits*. McGraw-Hill, 1994.

[10] R. Milner. *Communicating Concurrent Systems*. Prentice Hall, 1989.

[11] N. Savoiu, S. Shukla, and R. Gupta. Automated Concurrency Re-assignment in High Level System Models For efficient System-Level Simulation. In *Proc. Design Automation and Test in Europe Conf.*, 2002.

[12] N. Savoiu, S. K. Shukla, and R. K. Gupta. Efficient simulation of synthesis-oriented system level designs. In *Proceedings of the 15th international symposium on System Synthesis*, pages 168–173. ACM Press, 2002.

[13] M. Sgroi, L. Lavagno, Y. Watanabe, and A. Sanglovanni-Vincentelli. Quasi-static scheduling of embedded software using equal conflict nets. In *Proc. of ICATPN'99*, volume 1630, pages 208–227. Springer-Verlag, June 1999.

[14] Sobel Operator Algorithm: www.visc.vt.edu/armstrong/ee5514/as4_00.pdf.

[15] SystemC. OSCI. Home page: http://www.systemc.org.

Chapter 13

# XFM: EXTREME FORMAL METHOD FOR CAPTURING FORMAL SPECIFICATION INTO ABSTRACT MODELS

David Berner,[1] Syed Suhaib,[2] Sandeep Kumar Shukla,[2] and Jean-Pierre Talpin,[1]

[1] *INRIA, project Espresso*
*IRISA, Campus de Beaulieu*
*F-35042 Rennes Cedex, France*

[2] *Department of Electrical and Computer Engineering*
*Virginia Tech*
*Blacksburg, Virginia, USA*

**Abstract** In this chapter we introduce an agile formal method (named XFM) based on extreme programming concepts to construct abstract models from a natural language specification of a complex system. Building formal models for verification purposes is being used in the industry for two different usage modes: (i) Descriptive Formal Models (DFM) are used to capture an implementation into an abstract model to submit to analysis by model checking tools, (ii) Prescriptive Formal Models (PFM) are used to capture natural language specifications into a formal model to analyze consistency of the specification and also as a reference model to compare a DFM against it. We propose XFM as a methodology to incrementally build a correct PFM from a natural language specification. We illustrate the the benefits of the proposed methodology with the help of two examples: a control intensive traffic light controller, and the DLX pipeline. Our experiments show that this methodology not only constructs abstract models in sufficiently shorter time than the time taken in constructing ad hoc abstract models from implementation or specification, but also provides models that are constructively correct, closer to the intended specification and better structured.



*R. Gupta et al. (eds.), Formal Methods and Models for System Design,* 331-349.

# 1. Introduction

Due to the increasing recognition of the need for formal verification and formal methods in hardware and software industry, there has been a surge in activities in the industry to build formal models and apply formal model checking, or use the models for test generation (e.g. [15]). From our experience, building formal models for verification purposes is being used in the industry for two different usage modes: (i) Descriptive Formal Models (DFM) [2] are used to capture an implementation into an abstract model to submit to analysis by model checking tools, (ii) Prescriptive Formal Models (PFM) [3, 16] are used to capture natural language specifications into a formal model to analyze the consistency of the specification and also as a reference model to compare a DFM against it.

We propose Extreme Formal Models (XFM) as a methodology to incrementally build a correct PFM from a natural language specification. Our approach is necessitated by the absence of a prescription on how to go about building these models from natural language documents in the literature. For example, in [15, 16] SMV specifications of Bus Protocols are developed as PFM but the goal of these PFM is to check consistency or test generation and it does not prescribe any incremental methodology based on regression. In [3] the specification language ESL is described in which all properties are specified together and then an automaton is synthesized from the complete ESL specification. This wholesome approach often has the problem that (i) the inconsistency in the properties or mistakes in capturing the intended property are found late, (ii) the synthesis of automata may explode in size when everything is considered together. We have tried our hands on some of the $\omega$-automata synthesis tools [14] and usually when the number of properties are sizable, such synthesis tools do not work well. It might be better and more feasible to construct the model by hand incrementally as is explained later in this chapter and regressively model check it to ensure constructively correct models.

Extreme programming (XP) has been popularized in the object oriented software community in recent years. It introduces novel guidelines and concepts of an agile methodology that seem to increase programming productivity significantly while producing higher quality error free code [21] [20]. Some of the features of extreme programming are usage of 'user stories', 'test-first' development, 'refactoring', 'continuous regression' etc., which we find very useful in creating more dependable software if applied properly. In the spirit of the successful use of extreme programming in software development, we decided to experiment with a parallel methodology in formal model construction. As it turns out, these agile methods that got a wide acceptance with the XP boom, are particularly suited for formal methods. They lessen some of the aforementioned problems such as inconsistency and state space explosion, while at the same time reducing the barrier for design engineers to use formal models in the design process.

We show the flow and demonstrate the benefits of this XFM based methodology with two illustrative examples; a control intensive traffic light controller, and the DLX pipeline. Our experiments show that this methodology not only constructs abstract models with sufficiently shorter time than the time taken in constructing ad hoc abstract models from implementation or specification, but also provides models that are constructively correct and closer to the intended specification.

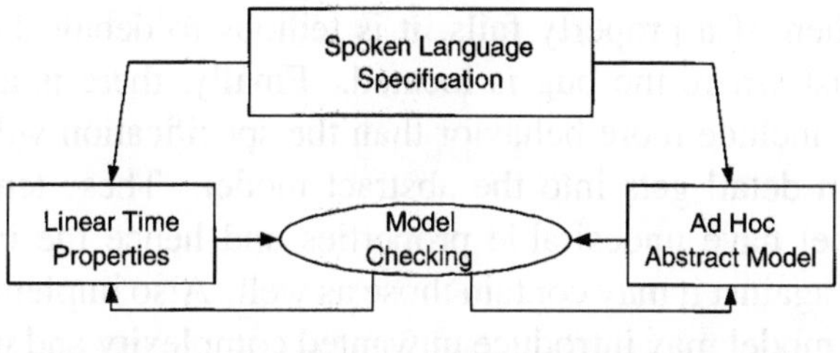


*Figure 13.1.* State of the art to capture a formal model

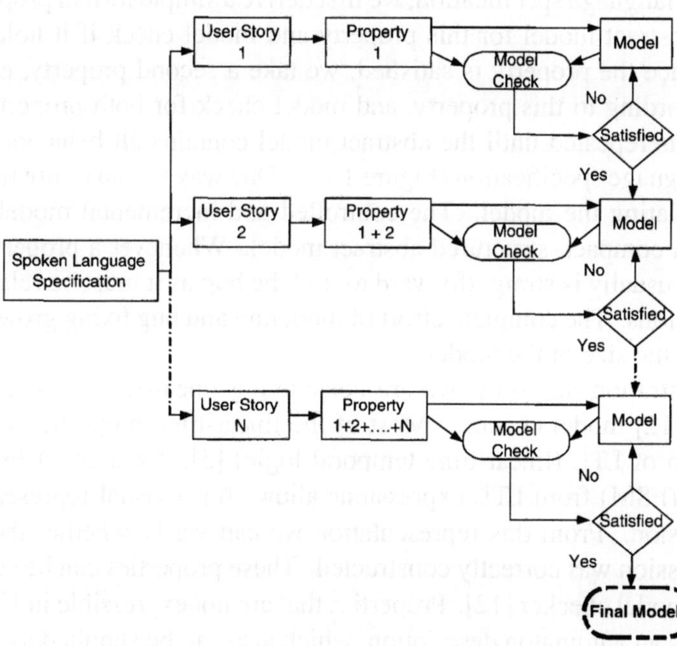


*Figure 13.2.* Capturing a formal model with XFM

Figure 13.1 and 13.2 show a comparison of the current state of the art in capturing formal specifications against our XFM approach. Figure 13.1

displays that an ad hoc abstract model is usually built from a spoken language specification and checked against formal properties with a model-checker. At times, the ad hoc abstraction is built from an implementation itself, which is then checked against the implementation for conformance. In our view that defeats the purpose of formal verification, except that the abstraction step might uncover bugs unknown to the implementers. There are several drawbacks in this approach. First of all, the ad hoc building of both the model and the properties is error prone and the effort of model building and debugging grows exponentially along with the size of the model. Next, as there is no way to control the inclusion of all properties, some may be overlooked, thus reducing the significance of the model. Then, if a property fails, it is tedious to debug the model. Few indications exist where the bug is located. Finally, there is a tendency that the model will include more behavior than the specification will allow. Often implementation detail gets into the abstract model. These tendencies might make the model have undesirable properties and hence the implementation being checked against it may contain those as well. Also implementation detail in the abstract model may introduce unwanted complexity and may later cause problems in a conformance check.

Figure 13.2 presents XFM's incremental approach to formal modeling. From the spoken language specification, we first derive a simple formal property, then build an abstract model for this property and model check if it holds for the model. Once the property is satisfied, we take a second property, extend the model according to this property, and model check for both properties. This procedure is repeated until the abstract model contains all behavior from the spoken language specification (Figure 13.2). One way to make sure that it does is by simulating the model. The controlled and incremental model building results in a compact, structured abstract model. Whenever a property fails to validate, it usually is straightforward to find the bug as it must be related to the latest additions. The complete effort of modeling and bug fixing grows linearly along with the size of the model.

For illustration purposes, we are considering linear-time properties using the SPIN [12] model checker. Most of the linear-time properties we use are in the form of LTL (linear-time temporal logic) [5]. Generating Finite State Machines (FSM) from LTL expressions allows for a visual representation of the expression. From this representation we can verify whether the original LTL expression was correctly constructed. These properties can be supplied to the SPIN model checker [12]. Properties that are not expressible in LTL can be modeled as an automaton description, which also can be supplied to SPIN as a property.

The remainder of this chapter is organized as follows. Section 2 describes related work in the field. Section 3 points out the major contributions. In

Section 4 we discuss in detail the methodology of XFM. In Section 5 we present two examples for XFM and Section 6 concludes this chapter.

## 2. Related Work

General information about XP and agile techniques can be found in [1, 6, 20, 21]. There are some projects that also use agile methods in the domain of modeling and verification. Herranz and Moreno-Navarro from TU Madrid describe in [10, 11] the integration of some XP practices to formal methods using the SLAM software tool [7]. Their environment generates sequential programs from formal expressions using an assertion based JAVA development framework. While our work involves the use of XP to model complex concurrent hardware systems their approach is directed towards sequential software programs. Henzinger et.al. [9] show how to reuse previous model checking results to incrementally model check the modification of a model. While this a completely different approach, these technique can be used in XFM in order to speed up the single model checking steps. Besides the typical method to capture formal models, there is other research going on to speed up this process, such as the automatic synthesis of models from temporal constraints [18], but high quality models are still created manually by qualified engineers.

## 3. Major Contributions

Our preliminary model building exercises show that an XFM based methodology features several advantages compared the traditional way of capturing formal specification. One of the key benefits is that XFM involves an iterative technique. The evolving model facilitates debugging whenever a property is found unsatisfied. After each stage, we make sure that the model concurs to the specification through model checking of all previously specified properties. If a property fails, the error can be located in the parts of the model that have been changed in the last iteration, which considerably restricts the amount of code to search for.

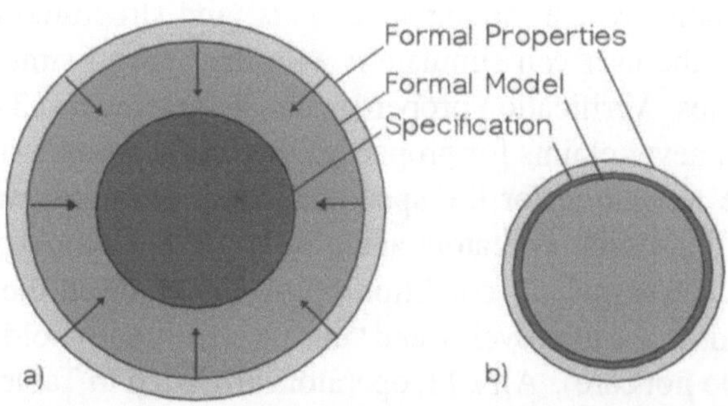


*Figure 13.3.* Behavior during the modeling process (a) and for the modeling result (b)

Another contribution is that our model is built using formal properties. Hence, it inherently includes less unintentional details. Figure 13.3 illustrates how the amount of behavior for the properties and the abstract model develop during the capturing process. At any point, the behavior of the formal properties is more general than that of the abstract model. At the beginning, however, they are both much more general than the behavior of the specification. In each iteration step their behavior is confined by adding more properties and details to the model. Obviously the behavior of the specification does not vary during this process since the specification is not altered. Ideally, in the end all three behaviors are identical with the specified behavior, but in practice there will always be a small gap. However this gap will be much smaller for the XFM methodology than for the traditional approach. The fact that the behavior of the model is closely linked to the properties, entails a close to complete set of properties once the model is complete; simulation of the model will help reveal missing functionality. In the conventional approach, however, the model tends to contain much more functionality than specified, but less properties than needed as there is no mechanism that guarantees the exposure of all properties of the specification. The overall time to build and validate the model is substantially less, especially for large systems. This is mainly due to the iterative aspect. Since the model is checked for each property after each iteration, the time needed to debug is less. This is in contrast to debugging the entire model at once for satisfying all properties for the traditional approach.

# 4. Modeling Approach

## 4.1 Tools

We chose the SPIN model checker as the model checking environment since it is a widely used model checkers for concurrent systems. However, any model checker could be used in order to obtain the same compelling results. In SPIN, models are specified with the PROcess MEta LAnguage (PROMELA), a system description language. Its basic building blocks are asynchronous processes, message channels, synchronizing statements, and structured data. Once the model is built, the user can simulate it with the built-in simulator and verify formal properties. Verification properties can be entered as LTL or in the form of PROMELA never claims for properties that are not expressible in LTL.

LTL is one technique for the specification of temporal rules. It extends propositional logic with operators auch as "always" (condition holds always in the future), "eventually" (condition holds sometime in the future), "next" (condition holds in the next cycle), and "until" (condition A holds until condition B, afterward do not care). All LTL operators are listed in Table 13.1. As to the expressiveness of LTL, it is complete with respect to first order logic [8].

Small changes in an LTL property, such as the misplacement of a parenthesis, can change the meaning of the property completely. For example [](a→<> (b U c)) represents a simple 2-state automaton, if we move one parenthesis before the <>, it is a 7 state automaton. Even if these kinds of mistakes are hard to detect, it is especially important that properties are correctly defined before checking the model for it. There is a tool called LTL 2 BA [14] (LTL to Büchi automata) that generates a Büchi automaton representing any LTL expression. This visualization can facilitate the verification that the expression matches the specification. LTL 2 BA also generates PROMELA code from an LTL expression. Therefore it could - in theory - be used to obtain the abstract model directly and automatically from the LTL properties. However, in practice this does not work since it is neither possible to describe concurrent processes in LTL nor to describe implementation details such as initial states or changing state variables.

*Table 13.1.* LTL operators

| | |
|---|---|
| !A | negation |
| A→B | implication |
| A ↔ B | equivalence |
| A && B | and |
| A \|\| B | or |
| []A | Always |
| A U B | until |
| <> A | eventually |
| X A | next |

## 4.2 Extreme Programming Techniques

As stated earlier, XFM works on the lines of XP. Many of the XP rules can be applied directly and successfully in XFM. For instance one of the main XP rules is to write tests before the actual code. In XFM this rule maps to specifying the LTL property before writing the abstract model. Another important XP technique is to add functionality as late as possible, keeping the model simple for as long as possible. Iterations are small steps in the development process. At the start of each iteration the goals are identified and written down in the form of "user stories" - individual cards that point out specific implementation details and requirements. These user stories act as a detailed guideline for the programmer. To refactor problems as much as possible, to update tests after a bug is found, and to work in pairs are other basic XP principles that are as beneficial to the capturing of formal methods as they are for common programming projects. The benefit of further XP techniques such as a stand up meeting in the mornings, collective code ownership, and to moving people

*Table 13.2.* Formal steps of the XFM approach

{For a given system, we have the behavior in terms of a natural language specification that is converted into user stories}

Let $US\{us_1, us_2, ..., us_n\}$ be the set of all user stories for the system
Let $\Pi(us_i) = \{\pi_j, \pi_{j+1}, ..., \pi_k\}$ be the set of properties for a user story
Let $\Pi = \cup_i \Pi(us_i) = \{\pi_1, \pi_1, ..., \pi_m\}$ be the complete set of properties
Let $\Pi_i = \{\pi_1, \pi_1, ..., \pi_i\} \subseteq \Pi$, so $\Pi_i$ represents the first $i$ properties.
Let $X(\Pi_i)$ be the model that satisfies all properties in $\Pi_i$

`Initial` $X = \emptyset$, $i = 0$,
`Step1:` $i := i + 1$
`Step2: Build Abstract Model` $X(\Pi_i)$`; the model is built to satisfy all the properties in` $\Pi_i$ `simultaneously.`

  $X := X(\Pi_i)$
  *ModelCheck*$(X, \Pi_i)$ /* This is the regression step */
  **if** *ModelCheck* `fails for a` $\pi_k$
    `go to Step2 /* to change the model so the failing property can pass */`
  **else if** $i = m$
    $X$ is the required model
  **else**
    go to Step1

around depends on the type of the project, the size of the team, and on personal and corporate preferences.

## 4.3 Modeling Steps

Figure 13.2 illustrates the flow of the presented modeling approach. First user stories have to be defined from the spoken language specification. These user stories define specific behaviors of the system. Then these user stories are ordered and the corresponding LTL properties are defined. The ordering is based on the ease of modeling. It is important and the quality of the model as well as the ease of iteration process can depend on it. Here we will not go into the details of ordering optimizations, however [19] discusses in detail several algorithms for various ordering approaches.

We first construct a model that satisfies the properties of the first user story. This model is then model checked for the corresponding properties and possibly modified until it passes the check. Now it is extended in order to satisfy the properties of the first and the second user story, and so on, until it satisfies all properties. Table 13.2 features a formal description of this proceeding. In an external loop, user stories are added, and in an internal loop the model is extended, and checked. The systematic steps break down the building of the big formal model to small steps that are comparably easy to follow.

## 4.4 Difficulties and Details

The XFM steps are not always as simple as they might seem. Since most problems encountered are known from the traditional capturing of formal models, they can be dealt with in the same way. The first challenge is to get a correct and complete natural language specification. Then, the order in which the user stories are specified can make a difference in the complexity of the model. Attempts to determine good orders are presented in [19], but even if an initial choice was not optimal, the model can be simplified in a subsequent refactoring step.

The building of formal properties is not always straightforward. However, it can be assisted with methods and tools such as specification patterns [4] and property elucidation [17], and their sanity can be checked for with tools such as LTL 2 BA [14], which transforms an LTL property into a finite state machine. The building of the formal models in XP requires the engineers to only model the functionality specified in the property. Refactoring is done in order to keep the model concise. Throughout the modeling procedure we make sure that the model contains all specified functionality from the previous modeled properties. Again there is no general solution for this, but simulation is one way to get more confidence about this. Contradicting properties may be discovered when during an iteration some properties can not be satisfied any longer.

Putting this new methodology to work does not mean ignoring all previous results in overcoming known obstacles, but contrarily it encourages the use of all these techniques for the single steps to obtain an even better result. As always, results significantly improve and can be obtained faster once the engineer gets more experienced and familiar with the process.

## 4.5 Advantages

Following an XFM based methodology, brings about a number of advantages. One of them is speed. The small debugging steps between the iterations can be done faster than debugging the complete model in the end. Besides speed, the quality of the resulting formal model is more precise with respect to the actual specification. Figure 13.3a illustrates how the amount of behavior for the properties and the abstract model develop during the capturing process. At any point, the behavior of the formal properties is more general than that of the abstract model. Both are more general than the behavior of the specification. In each iteration step their behavior is confined by adding additional properties and details to the model. Since the specification is not altered, the behavior of the specification does not vary during this process. After the model is completely built, ideally all three behaviors are identical, but in practice there are always small gaps (Figure 13.3b). The gap between the formal properties and the specification indicates how much system functionality is not expressed in the formal properties, and the gap between the formal model and the specification marks the amount of functionality the model contains that is unaccounted for in the specification. The latter stays small in this approach since we only add functionality from simple user stories. The former gap is small as the process makes sure that the gap between properties and model never gets big in the first place.

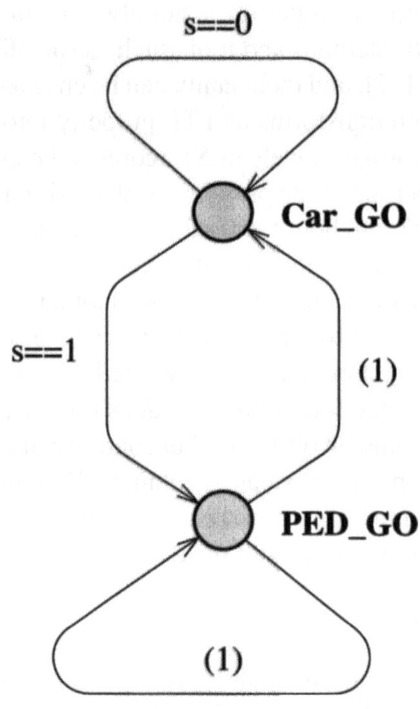


*Figure 13.4.* Sketch of the pedestrian traffic light

## 5. Examples and Results

In order to demonstrate the power of XFM we present two examples from different domains and of different complexity. A simple traffic light will illustrate the main steps, tools, and techniques involved. The design of a DLX [13] microprocessor pipeline will show how this works for a bigger model, and how the model evolves with the incremental approach.

*Table 13.3.* LTL properties of the traffic light (c = cars stop, p = ped stop, sw = button is pressed)

| | |
|---|---|
| never both have a green signal | $[]!(!c\&\&!p)$ |
| If cars cannot go, they will go eventually | $[](c \rightarrow <>!c)$ |
| No button pressed, cars keep going | $[]((!c\&\&!sw) \rightarrow X!c)$ |
| No button pressed the pedestrians cannot walk | $[]((p\&\&!sw) \rightarrow Xp)$ |
| When the switch is pressed while the cars go, pedestrians will go before the switch is turned off. $[](sw\&\&p \rightarrow swU(!p\&\&sw)\&\&(!p\&\&sw \rightarrow !pU(!p\&\&!sw)))$ | |

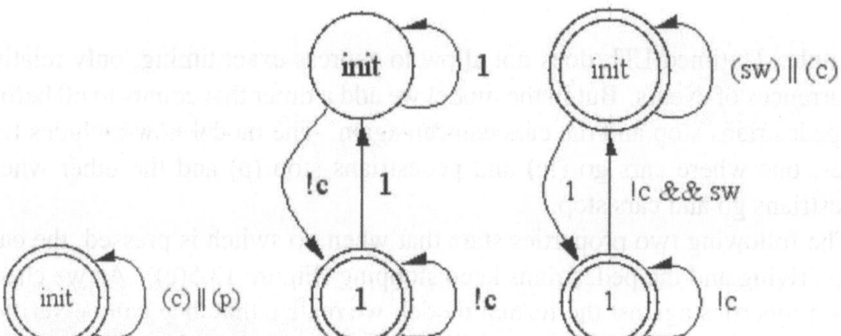


*Figure 13.5.* FSMs of traffic properties 1 (a), 2 (b), 3 (b)

## 5.1 Traffic Light Model

This is a very simple example of a pedestrian crossing with a traffic light (Figure 13.4). Whenever a pedestrian pushes a button, the lights turn red, and the pedestrians can walk. After thirty units the pedestrians get a red light and the car's red light is turned off. This description is the spoken language specification. Now, we construct LTL properties describing this system. We start with the most important property stating that both pedestrian and car, can never get the go signal at the same time: []!(!p && !c). We verify that the property concurs with the specification by means of LTL 2 BA. Figure 13.5(a) shows the automaton corresponding to this property. The corresponding model is just as simple, just one state.

The next property we come up with states that whenever the cars stop, they will eventually go (Figure 13.5(b)). Table 13.3 lists all LTL properties for this

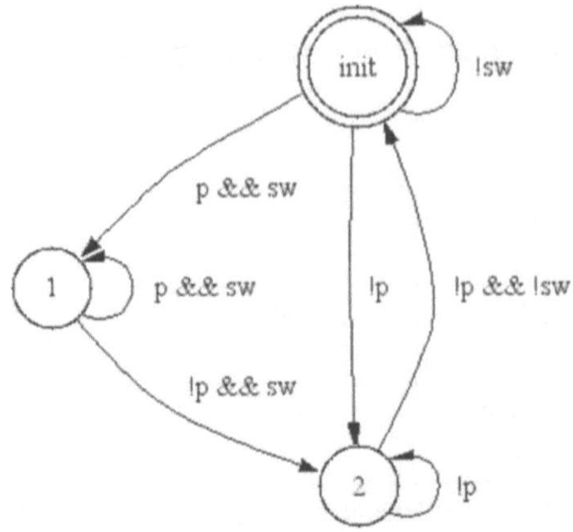


*Figure 13.6.* Graph for traffic property 5

example. Untimed LTL does not allow to express exact timing, only relative occurrences of events. But in the model we add a timer that counts to 60 before the pedestrians stop and the cars can run again. The model now includes two states, one where cars go (!c) and pedestrians stop (p) and the other where pedestrians go and cars stop.

The following two properties state that when no switch is pressed, the cars keep driving and the pedestrians keep stopping (Figure 13.5(c)). As we check these properties against the formal model, we realize that they can be verified without making any modifications to the system, and a closer look at the properties shows that their behavior is already satisfied by property 1 (Figure 13.5(a)). One functionality that is still missing is the inclusion of the switch. When cars go and the switch is pressed, eventually pedestrians should be allowed to walk before the switch turns off. This property is a bit longer than the others, and without LTL 2 BA it is not easy to figure out if it is correct (Figure 13.6). After implementing the functionality of these properties into the model, simulation shows that it works as specified, so we have found all properties. Figure 13.7 shows the PROMELA code for the complete traffic light model.

## 5.2 Model of a DLX pipeline control

The actual power of the XFM approach develops when working on large systems. The pipeline control of the DLX RISC processor model [13] is a well known and reasonably large example to show the use of XFM. The DLX has a 5-stage pipeline, which means up to five instructions can run concurrently. The cycles for the instructions are instruction fetch (IF), instruction decode (ID), execute (EX), memory access (MEM), and write back (WB). However, not all

```
bool sw, c, p;
int time;
active [0] proctype signal()
{
cargo:
  p=1; c=0;
  if
    :: (1) -> sw =1; time=30; goto pedgo
    :: (1) -> goto cargo
  fi;
pedgo:
  c=1; p=0; sw =0;
  time = time -1;
  if
    :: (time > 0) -> goto pedgo
    :: (time == 0) -> goto cargo
    fi
}
init {
  p = 1; c = 0; sw = 0;
  run signal();
}
```

*Figure 13.7.* Promela code for pedestrian crossing

instruction types use the same cycles in the same order. Table 13.4 shows the cycle usage for the different instruction types.

*Table 13.4.* Cycles for different instruction types

| | *IF* | *ID* | *EX* | *MEM* | *WB* |
|---|---|---|---|---|---|
| *Arithmetic* | X | X | X | | X |
| *Load* | X | X | X | X | X |
| *Store* | X | X | X | X | |
| *Branch* | X | X | X | X | |

Starting from this system description, we identify the first user story. One of the most basic behaviors states that each instruction executes in a certain order. So, generally speaking, instructions execute in the order $IF \rightarrow ID \rightarrow EX \rightarrow MEM \rightarrow WB$. In LTL this can be expressed as $[](if \rightarrow Xid)$, always ID after IF and then the same for ID and EX, EX and MEM, MEM and WB, and finally WB and IF. The automaton generated with LTL 2 BA (Figure 13.9(a)) shows that the LTL expression is sound. These five properties can be represented with a circular automaton that satisfies our first user story.

The second user story is the fact that this order of execution still has to hold when we consider five concurrent instructions in the pipeline. In order to keep the model small we decide to use five concurrent processes each of which handles one instruction (Figure 13.11). Since the processes run independently, the first property does not hold any more. It is not guaranteed that directly after the first instruction is in the fetch stage it advances to the decode stage, since in the meantime other processes may get execution time. What we can guarantee however, is that we will never go directly into any of the other stages. Now this has to be expressed for each cycle in each instruction, which means we get 25 LTL properties such as cat1 in Table 13.5.

```
proctype instruction1() {
    inst_if:
      if
      :: st1=fet; goto inst_id fi;
    inst_id:
      if
      :: st1=dec; goto inst_ex fi;
    inst_ex:
      if
      :: st1=ex;  goto inst_mem fi;
    inst_mem:
      if
      :: st1=mem; goto inst_wb fi;
    inst_wb:
      if
      :: st1=wb;  goto inst_if fi;  }
```

*Figure 13.8.* PROMELA code for one single instruction

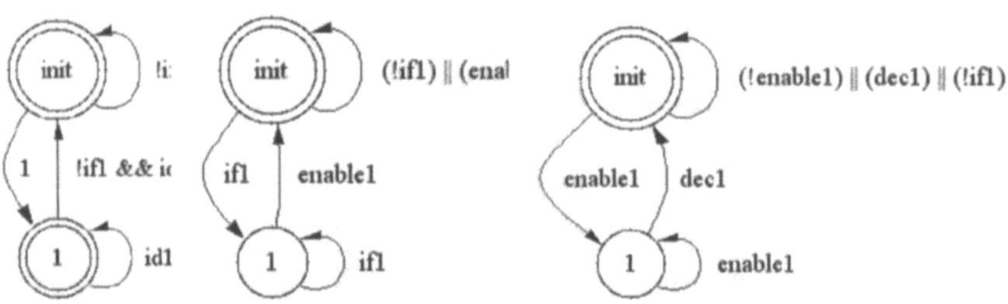


*Figure 13.9.* Graphs of pipeline properties 1 (a), 3 (b), and 4 (c)

In the next iteration we introduce the possibility to control the instructions from outside. This is done by "enable signals", one for each instruction. The LTL expression will say that an instruction will not advance unless the enable signal is given (Figure 13.9(b)). Again we obtain 25 properties in the style of cat2 in Table 13.5. The changes in the model for these properties are small, so all of them can be verified without problems.

The following iteration is adding some synchronization. Our user story says that the control enables each instruction in each cycle. Once the instruction advances, it is setting its enable signal to zero, thus signaling the control that

*Table 13.5.* LTL properties for pipeline (examples)

| | |
|---|---|
| cat1 | $[](if1 \rightarrow !(Xex1 \| \| Xmem1 \| \| Xwb1))$ |
| cat1b | $[]((ex1\&\&(load1 \| \| store1 \| \| branch1)) \rightarrow !(Xif1 \| \| Xwb1 \| \| Xdec1 \| \| Xwait1))$ |
| cat2 | $[]((if1\&\&!enable1) \rightarrow (if1 U enable1))$ |
| cat2b | $[]((wait1\&\&!enable1) \rightarrow (wait1 U enable1))$ |
| cat3 | $[](if1 \rightarrow ((enable1 U dec1) \| \| !enable1))$ |
| cat3b | $[]((ex1\&\&(load1 \| \| store1 \| \| branch1)) \rightarrow ((enable1 U mem1) \| \| !enable1))$ |
| cat4 | $[]((if1\&\&enable1) \rightarrow ((!(if2\&\&enable2)) \| \| (!(if3\&\&enable3)) \| \| (!(if4\&\&enable4)) \| \| (!(if5\&\&enable5)))U!enable1))$ |

it is ready for the next cycle. This category of properties is somewhat more complex, but with the help of the LTL 2 BA tool we finally find cat3 in Table 13.5. It reads that whenever a stage is true, it will change to the next stage before the enable sign goes down, unless the enable sign is already low (Figure 13.9(c)). Again we get one of these properties for each stage that is 25. Once the correct properties are found, the changes in the model are small, and after all properties verified (including the previous ones), we can check the correct behavior with the built in SPIN simulator.

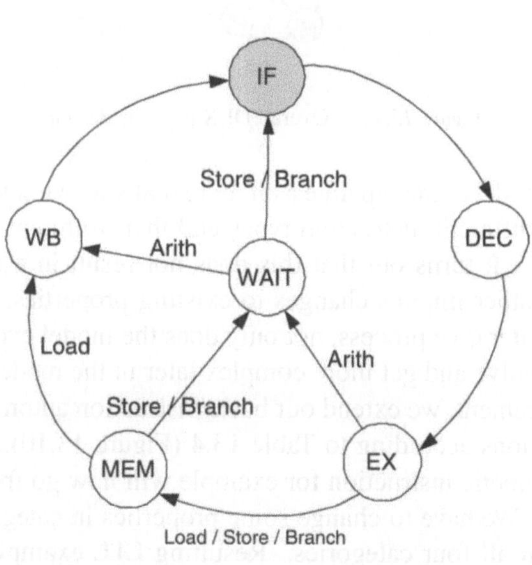


*Figure 13.10.* Automaton for one instruction

Another important behavior of a pipeline is to prohibit the multiple usage of resources. If at no time the fetch, decode, execute, address bus, and data bus units are used by more than one instruction there are no resource conflicts. Cat4 in Table 13.5 expresses this in LTL for the fetch cycle of the first instruction. Again the category will consist of 25 properties, one for each cycle. In order to satisfy this property in the model we are introducing a control process that in an initialization phase will start each instruction successively, and later makes sure that the every instruction advances in each cycle. Again the verification of all properties and simulation finishes up this iteration step. With only 4 categories of properties the basic functionality of the pipeline is now verified and working.

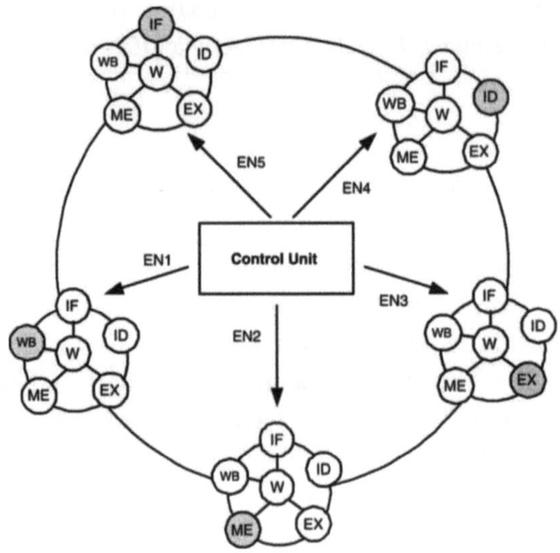


*Figure 13.11.* Overall DLX pipeline design

To make the model of the pipeline a bit more realistic, we select the user story that defines the different instruction types and their different cycle sequences from Table 13.4. It turns out that this does not result in a new category of properties, but rather implies changes to existing properties. This step illustrates that in the iterative process, not only does the model evolve, but also the properties can evolve and get more complex later in the modeling process. To satisfy the requirement, we extend our basic instruction automaton with a wait stage and transitions according to Table 13.4 (Figure 13.10). This will make sure that an arithmetic instruction for example will now go from EX to WAIT and then to WB. We have to change some properties in category 1 and 3, and add properties in all four categories. Resulting LTL examples are shown in cat 1b and 3b in Table 13.5. Changes in the abstract model to reflect this are limited to update the FSM description for each instruction to the automaton of Figure 13.10 that means introducing the notion of an instruction type, and adding the transitions to and from the wait stage. For the control these changes are transparent since after the changes still each instruction takes 5 cycles to finish, therefore preventing the occurrence of structural hazards. Figure 13.11 shows the complete structure of the pipeline model. Of course there would still be many more details that could be added to the pipeline, such as data dependencies and forwarding, but the steps will always be the same, so we will not continue this for this chapter.

## 6. Conclusion

In this chapter we present and demonstrate the usage of agile methods for the formal modeling of hardware systems. The presented methodology focuses on the concept of incremental formal modeling based on properties from a natural language specification. Each property represents a specific behavior based on which the abstract model is constructed. The incremental approach used in the chapter helps that the constructed abstract model satisfies all properties, and it ensures that the model contains only little unwanted behavior that might cause the failure of a later conformance check against an implementation.

Since XFM involves an iterative technique, the evolving abstract model facilitates debugging whenever a property is found unsatisfied. In each iteration step the behavior is confined by adding additional properties and details to the abstract model. The fact that the behavior of the abstract model is closely linked to the properties entails a close to complete set of properties once the abstract model is complete. In the conventional approach, however, the abstract model tends to contain much more functionality than specified, but less properties than needed as there is no mechanism that provides for the exposure of all properties contained in the specification.

## References

[1] Kent Beck, *Extreme programming explained: Embrace change*, Addison Wesley, 2000.

[2] Bob Bentley, *Validating the intel pentium 4 microprocessor*, Design Automation Conference, 2001, pp. 244–248.

[3] Edmund M. Clarke, Steven M. German, Yuan Lu, Helmut Veith, and Dong Wang, *Executable protocol specification in ESL*, Formal Methods in Computer-Aided Design, 2000, pp. 197–216.

[4] Matthew B. Dwyer, George S. Avrunin, and James C. Corbett, *Property specification patterns for finite-state verification*, Proc. 2nd Workshop on Formal Methods in Software Practice (FMSP-98) (New York), ACM Press, 1998, pp. 7–15.

[5] E. A. Emerson, *Temporal and modal logic*, Handbook of Theoretical Computer Science (J. van Leeuwen, ed.), vol. B, Elsevier Science Publishers, Amsterdam, 1990.

[6] Michigan eXtreme Programming Enthusiasts (MXPE), *Extreme programming: A gentle introduction.*, http://www.extremeprogramming.org/.

[7] Babel group at TU Madrid, *The SLAM website*, http://lml.ls.fi.upm.es/slam.

[8] Jesper G. Henriksen, *Logics and automata for verification - expressiveness and decidability issues*, Ph.D. thesis, Basic Research in Computer Science (BRICS), University of Aarhus, Denmark, June 2000.

[9] Thomas A. Henzinger, Ranjit Jhala, Rupak Majumdar, and Marco A.A. Sanvido, *Extreme model checking*, Proceedings of the Int. Symposium on Verification: Theory and Practice, Lecture Notes in Computer Science, Springer-Verlag, 2003.

[10] A. Herranz and J.J. Moreno-Navarro, *Formal extreme (and extremely formal) programming*, 4th International Conference on Extreme Programming and Agile Processes in Software Engineering, XP 2003 (Genova, Italy) (Michele Marchesi and Giancarlo Succi, eds.), LNCS, no. 2675, May 2003, pp. 88–96.

[11] ______, *Rapid prototyping and incremental evolution using slam*, 14th IEEE International Workshop on Rapid System Prototyping, RSP 2003) (San Diego, California, USA), June 2003.

[12] Gerhard J. Holzmann, *The spin model checker: Primer and reference manual*, Addison Wesley, Boston, MA, September 2003.

[13] David Goldberg John L. Hennessy, David A. Patterson, *Computer architecture: A quantitative approach*, 3rd ed., Morgan Kaufmann, San Mateo, CA, may 2002.

[14] Dennis Oddoux, *LTL 2 BA : fast algorithm from LTL to buchi automata*, http://www.liafa.jussieu.fr/~oddoux/ltl2ba/.

[15] K. Shimizu and D. Dill, *Deriving a simulation input generator and a coverage metric from a formal specification*, 2002.

[16] Kanna Shimizu, David L. Dill, and Alan J. Hu, *Monitor-based formal specification of PCI*, Formal Methods in Computer-Aided Design, 2000, pp. 335–353.

[17] Rachel L. Smith, George S. Avrunin, Lori A. Clarke, and Leon J. Osterweil, *Propel: an approach supporting property elucidation*, Proceedings of the 24th international conference on Software engineering, ACM Press, 2002, pp. 11–21.

[18] B. Steffen, T. Margaria, and M. von der Beeck, *Automatic synthesis of linear process models from temporal constraints: An incremental approach*, 1997.

[19] Syed Suhaib, D. Matai, and Sandeep Shukla, *Property Ordering in Extreme Formal Modeling: ISA and Pentium Pro Bus: Case Studies.*, Technical Report no. 2004-02, Virginia Tech, FERMAT Lab, Blacksburg, VA, February 2004.

[20] Laurie Williams, *The XP programmer - the few minutes programmer*, IEEE Software **20** (2003), no. 3, 16–20.

[21] William A. Wood and William L. Kleb, *Exploring XP for scientific research*, IEEE Software **20** (2003), no. 3, 30–36.

V

# POST-PRODUCTION FORMAL METHODS

# Chapter 14

# ENGINEERING CHANGES IN FIELD MODIFIABLE ARCHITECTURES

Satoshi Komatsu, Hiroshi Saito, Kenshu Seto, Yoshihisa Kojima, and Masahiro Fujita

*The University of Tokyo, Japan*

**Abstract** Because engineering changes to fix design errors and satisfy design constraints are required even after chip fabrication, design flexibility and debuggability are extremely important for system-level designs to provide reliable and flexible designs and shorten time-to-market. In this paper, we present a new VLSI architecture called field modifiable architecture (FMA) and its design method which enables hardware/software co-design with considerable flexibility. Because of the nature of programmability in both hardware and software of FMAs, all of the engineering changes in designs are resolved at specification-level in terms of instruction extension and code re-compilation. As a result, engineering changes even after chip fabrication can be realized easily without wasting design time. The efficiency of our proposed method is proved by applying it to state-of-the-art design examples.

**Keywords:** Hardware/Software Co-Design, Field Modifiable, Engineering Change

## Introduction

Because of the great advance in VLSI process and device technologies, systems which consist of general purpose processors, custom hardware, memories, and other peripheral components can be realized on a chip (i.e., system-on-a-chip). To efficiently enjoy the properties of systems such as performance, power dissipation, cost, and so on, system designs based on hardware/software co-design are getting to be indispensable[1]. In addition to hardware/software co-design, system-level verification is also getting to be indispensable.



*R. Gupta et al. (eds.), Formal Methods and Models for System Design,* 353-372.

This is because most of design time is spent for verification and debugging at the current system design flow. This situation is becoming worse and worse because the size of integration will increase in future shrunk technology. As a result, it is difficult to expect that verification and debugging processes complete before the tape-out of designs. Since engineering changes (ECs) caused by these processes are required even after chip fabrication, design flexibility and debuggability are extremely important for system-level designs to provide reliable designs and shorten time-to-market.

In this paper, we present a new architecture called field modifiable architecture (FMA) and its design method which enables hardware/software co-design with considerable flexibility. The basic organization of FMA is as follows. The software part of FMA consists of a general purpose processor and instruction/data memories. On the other hand, the hardware part consists of functional units, local memories, register files, a controller for the hardware part, and a HW-SW interface. These hardware components are controlled by a micro-controller by means of application specific instructions. We assume that the hard-macro library has all of the hardware components with their information such as physical layout, physical area, performance, power dissipation, and so on. Because of such design method, high flexibility and high debuggability are realized in FMAs.

The rest of this paper is organized as follows. In section 2, we describe the basic concept of FMAs with several related researches. In section 3, we describe the FMA design method. In section 4, we show how to realize ECs on FMAs. In section 5, we present implementation results by using an encryption algorithm on our proposed design method. In section 6, we show case studies of ECs. Finally, in section 7, we give concluding remarks.

## 1. Field Modifiable Architectures

In this section, we present an overview of Field Modifiable Architecture. First we describe the basic concept of FMAs and how FMAs accomplish flexibility and debuggability. Then we present several related works which propose hardware/software co-design methods with various approaches.

### Basic Concept

Figure 14.1 shows the basic concept of traditional hardware/software co-design approaches, as is proposed in [2]. In such approaches, computationally critical parts in the initial specification are implemented in customized hardware and the rests of the specification are implemented in software. After the chip fabrication, the customized hardware in these approaches is completely fixed because it is synthesized with behavioral and logic synthesis tools and the

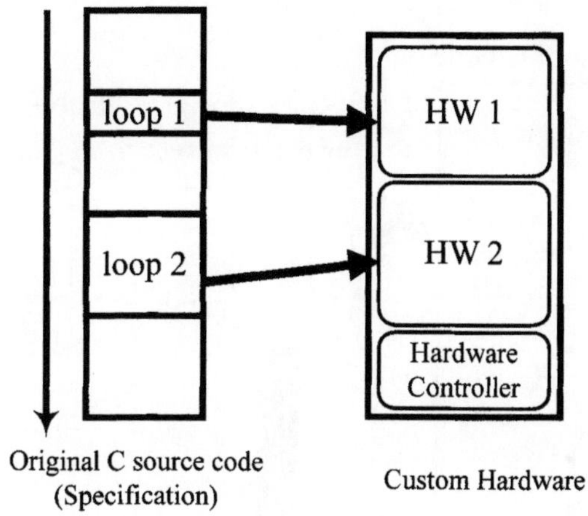


*Figure 14.1.* A hardware/software co-design based on traditional approaches

physical layout is also fixed. Therefore, it takes long time for ECs because of less design flexibility.

Figure 14.2 shows a simple example of FMA designs. An FMA is composed of a main processor and instruction/data memories for its software part, a micro-controller, a register file, hard-macros, and local memories for its hardware part. We define the hardware part as "specialized hardware." Like other recent works [5, 6], computationally critical parts in the specification are implemented on the specialized hardware. In FMA, the specialized hardware is composed of only hard-macros which the design library has.

In our proposed design method, a system description is written in C language as the initial specification. By profiling the specification using the instruction set simulator of the main processor, computationally non-critical parts in the specification are implemented on the software part, while critical parts are implemented on the hardware part. Then, we generate a set of application specific instructions for critical parts, which can execute several operations in one cycle to improve performance. For this purpose, we develop an instruction generator based on multi-terminal binary decision diagram[3]. In addition to the instruction generator, we develop a code generator based on code generation finite state machine (CGFSM)[4], which compiles optimum instruction codes of critical parts by using the generated instruction set.

As shown in the following, there are two main advantages in FMA and its design method.

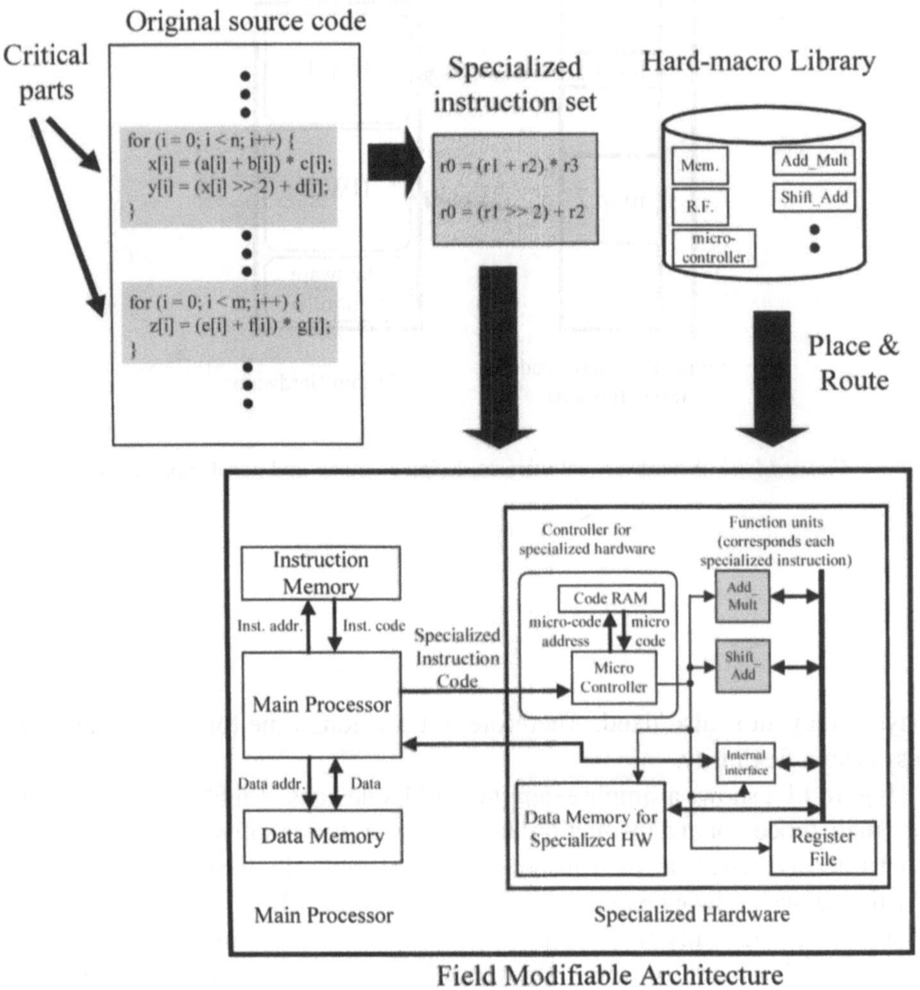


*Figure 14.2.* A simple example of FMA designs

- **Simplified design.** Since the hardware part of FMAs consists of a set of hard-macros [1], only placement and routing operations are required to complete the design. The remaining processes are 1) generating specialized instructions to control the hard-macros via micro-controller, 2) generating optimum codes for the micro-controller, and 3) generating a

[1] In this work, hard-macros are assumed to basic ones such as adder, multiplier, and so on, which are well used in design experiments. In addition, processors, micro-controllers, and memories which are well distributed are also regarded as a hard-macro.

physical layout for these macros. From such a view point, the logic synthesis process is not required in FMA design method. It means that the design time is shortened because the logic synthesis and the verification of customized circuits can be omitted.

Different from ASIC designs, since logic synthesis/verification phases are not concerned [2], the design method may significantly reduce design time. Compared to FPGA designs, FMAs have high performance especially for the arithmetic functions. Of course FPGAs can realize such functions, but the transistor level circuit optimization is required to improve their performance.

- **High flexibility and debuggability.** From the nature that both software and hardware parts are controlled by instruction codes, all of the ECs in designs are resolved at specification-level in terms of instruction extension and code re-compilation. It means that ECs in the software part are realized just by re-compiling the modified specification by using the compiler of the general purpose processor. On the other hand, ECs in the hardware part are realized just by re-compiling the modified specification by using our code generator with generated application specific instructions. However, if new operations which are caused by the ECs cannot be executed by the above re-compilation, our instruction generator generates new instructions so that all of the operations in the modified specification are executed by generated application specific instructions. As a result, ECs even after chip fabrication are easily realized without wasting design time.

  In case of ASIC designs, they have almost no debuggability after chip fabrication. In case of FPGAs, though they have feasible debuggability, the ECs require many processes (logic synthesis, verification, placement and routing, and so on) which take a long time. Compared with them, FMAs are much more flexible and debuggable with short design/debug time.

In section 2, we present the design method of FMA.

## Related Work

Recently, processor-based system LSIs have become indispensable for many commercial electric products. In addition, shortening the design time is one of the key requirements in the current system designs. From such view points,

[2] Logic synthesis and verification are carried out only when someone prepare the hard-macros. However, in this work, we assume that a hard-macro library already has a enough number of macros.

hardware/software co-design methods are indispensable for designing high performance system LSIs with short design time and many hardware/software co-design methods have been proposed.

COSYMA system[2] from Technical University of Braunschweig is a study of hardware/software co-design. Based on the profiling of a given initial specification, timing critical parts in the specification are identified and implemented into a specialized hardware. Different from our approach where specialized hardware is controlled by micro-codes of specialized instructions, the specialized hardware is implemented as an ASIC. As mentioned in the previous, since ASIC designs are lacking flexibility, it is difficult to realize ECs at hardware parts.

Proceler[5] is developing a tool that generates an application specific processor which consists of a main processor and so-called soft instruction set architecture (ISA) processor implemented in FPGA. Compared to FMAs, although it may have much flexibility because of FPGA implementation, the performance, power, and cost of it may be worse than ours because we use hard-macros which are well customized rather than FPGA.

Xtensa[6] from Tensilica is a RISC processor which is configurable and extensible. Configuration is carried out by using additional multiply-accumulation (MAC) units, by changing cache size, or by using different number of interrupt vectors. Besides the configurability, user defined instructions can be added to Xtensa by using so-called Tensilica Instruction Extension (TIE) language. However, the main drawback is that the compiler does not utilize the user defined instructions automatically. In our FMA designs, extensions of instructions are also realized as specialized instructions which are used to control hard-macros in specialized hardware. Although they are restricted to available hard-macros, we support both specialized instruction generator and code generator for micro-controller.

PICO (program in, chip out)[7] from HP Labs is a template base hardware/software co-design architecture and its design methodology. A PICO system design contains a VLIW processor and a nonprogrammable accelerator (NPA) subsystem consisting of one or more NPAs. Each NPA is customized to execute a computationally critical loops to improve the system performance. Though the basic idea of PICO is similar to FMA, there are manifest difference between them in terms of the flexibility and the debuggability since once the hardware is fixed the functionality cannot be changed in PICO system. In other words, FMA has considerable flexibility at the expense of the performance.

AccelChip[8] from AccellChip, Inc. is a straightforward MATLAB-to-RTL solutions for FPGA based digital signal processing systems. It synthesizes from MATLAB specifications to RTL designs from the view point of high-level synthesis. Though it realizes high performance DSP chips and short design time, it requires logic synthesis and P&R flow to implement the system.

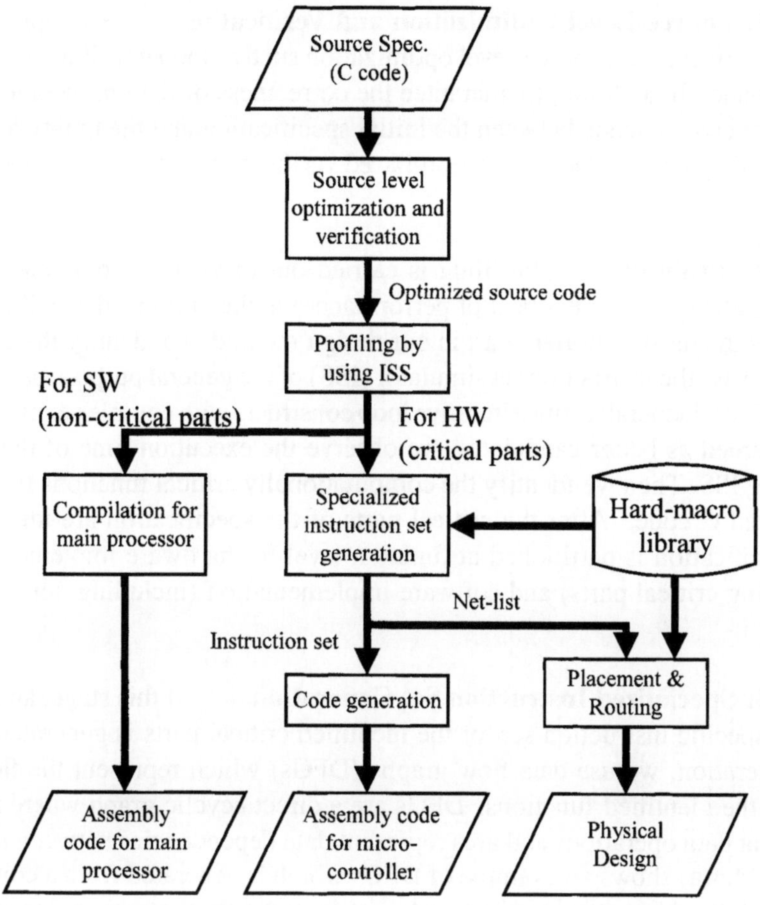


*Figure 14.3.* Design flow of Field Modifiable Architecture.

Our method has almost no logic synthesis flow, which means that it realizes much shorter design and debug time.

## 2. FMA design method

Figure 14.3 shows the flow of FMA design method. In the design flow, a system description written in C language is accepted as the initial specification. By using the design method, finally, the assembly codes for the general purpose processor and the micro-controller are automatically (or manually) generated. In addition, the physical design which is composed of hard-macros is also generated. In the following, we describe the overview of each procedure of FMA design method.

**Stage 1: Source-Level Optimization and Verification.** To improve the overall performance, source-level optimization such as loop unrolling is carried out by hand. In addition, to guarantee the correctness of the optimization, the functional equivalence between the initial specification and the optimized one is verified by using either simulation-based verification or formal method such as [9].

**Stage 2: Profiling.** Profiling is carried out to identify computationally critical parts (i.e., in the aspect of performance) in the optimized specification, which is the most important part in our design method. To identify the critical parts, we use the instruction set simulator (ISS) of the general purpose processor as a profiler. Generally, since iterative loop constructs with complex expressions are regarded as better candidates, we observe the execution time of the parts by using ISS. Then we identify the computationally critical functions from the optimized C code. After the critical parts of the specification are identified, the specification is partitioned at function-level for hardware implementation (including critical parts) and software implementation (including non-critical parts).

**Stage 3: Specialized Instruction Set Generation.** At this stage, an application specific instruction set for the identified critical parts is generated. For the generation, we use data flow graphs (DFGs) which represent the flow of data in the identified functions. DFGs are a direct acyclic graph where nodes represent data operations and arcs represent data dependencies between nodes. Figure 14.4(a) shows an example of DFGs which is generated from a compiler such as [10]. Note that terminal nodes such as A and X represent inputs and outputs respectively. We assume that each specialized instruction consists of one or more operation nodes in the DFG. In addition, each specialized instruction must correspond to the function unit in Fig. 14.2 because of the nature of FMA.

The generation of application specific instructions is considered as a covering problem of DFGs by using matches, where the match is the combination of DFG nodes. Figure 14.4(a) shows an example of DFG and Fig. 14.4(b) shows the generated matches for the DFG, i.e., one match corresponds to one instruction. To improve performance, not only matches which come from single DFG node but also matches which come from several DFG nodes. By considering the cost of generated instructions (i.e., how many functional units is used), the covering problem is solved by our instruction generator based on multi-terminal binary decision diagram (MTBDD)[3].

**Stage 4: Code Generation.** Based on the generated specialized instruction set, an optimum instruction code for micro-controller is generated by using our

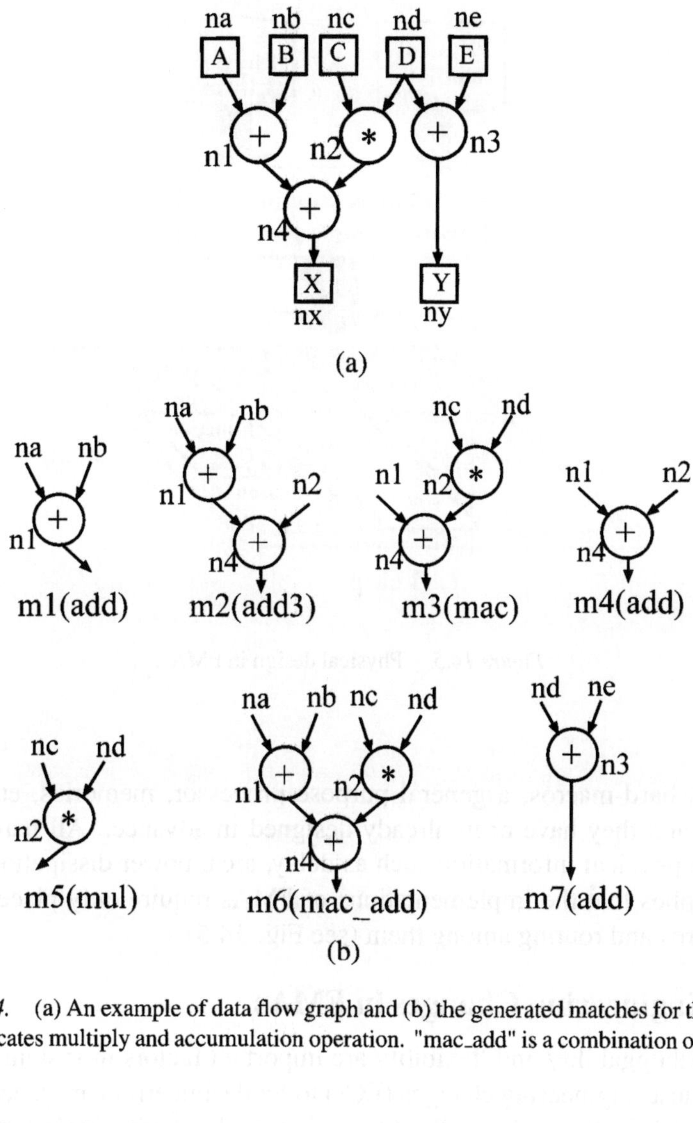


*Figure 14.4.* (a) An example of data flow graph and (b) the generated matches for the example. "mac" indicates multiply and accumulation operation. "mac_add" is a combination of "mac" and "add".

code generator based on code generation finite state machine (CGFSM)[11]. The code generator accepts the DFG and the generated instructions for the function which includes critical parts. Then, the optimum instruction code is generated so that the number of execution cycles of each function is minimized. Instructions which have no data dependency are executed in parallel as in very large instruction word (VLIW) architecture. Finally, the generated instruction code is stored into the instruction memory of the hardware implementation.

**Stage 5: Layout.** The implementations of FMAs consist of the configuration of hard-macros. The hard-macro library has all components (a micro-

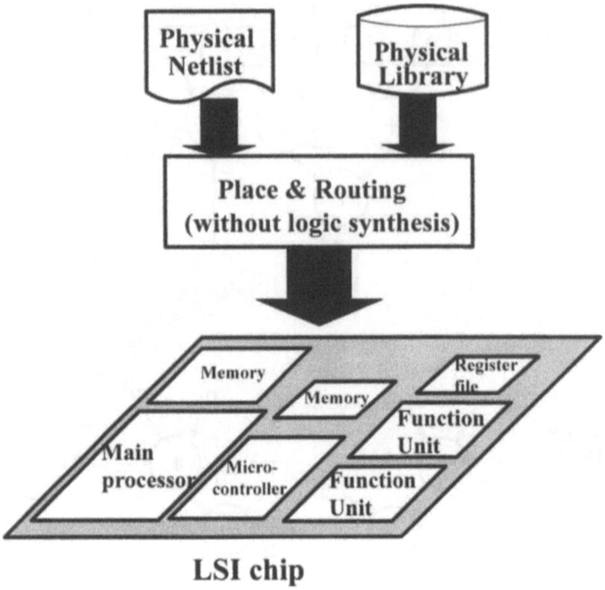


*Figure 14.5.* Physical design in FMAs.

processor, hard-macros, a general purpose processor, memories, etc.) used in FMAs and they have been already designed in advance. All hard-macros have their physical information such as delay, area, power dissipation, and so on. It implies that the implementations of FMAs require only placements of hard-macros and routing among them (see Fig. 14.5).

## 3. Engineering Changes in FMAs

High debuggability and flexibility are important factors in system-level designs because engineering changes (ECs) to fix design errors are required even after chip fabrication. Generally, ECs are carried out after logic synthesis or physical designs. However, ECs at the lower level of designs are getting to be more and more difficult because the size of integration is rapidly increased in the future shrunk technology. From this reason, a simple EC strategy at higher levels is indispensable.

In FMAs, ECs are carried out at the specification-level with a simple strategy, i.e., instruction extension and code re-compilation. This is because in FMAs the computation on both software part and hardware part is realized by instruction codes generated from the compiler of the general purpose processor and our code generator of the micro-controller. In the following, we explain how to realize ECs on FMAs by using an illustration shown in Fig. 14.6.

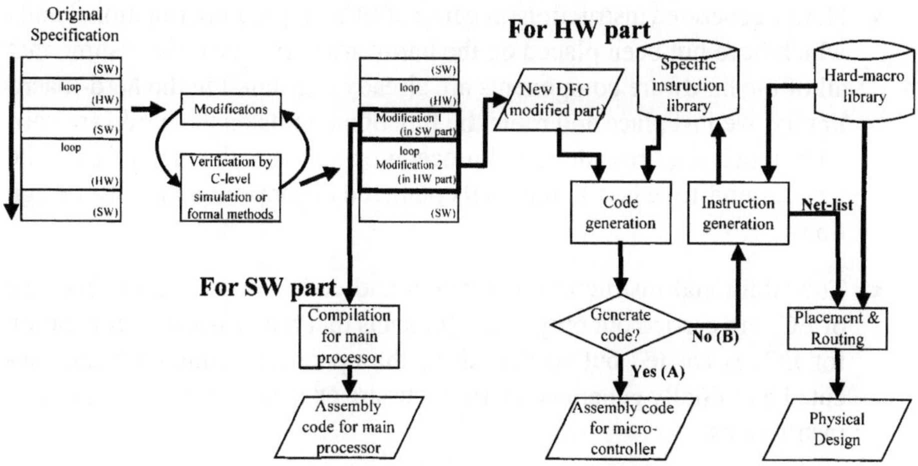


*Figure 14.6.* Flow of engineering changes.

First of all, we need to identify which functions of the specification are modified. In Fig. 14.6, there are two modifications, modification 1 and modification 2, in the initial specification. Note that the functional equivalence between the initial specification and the modified one is checked by C-level simulation or formal methods.

If the modifications are carried out at the software part (the modification 1 in Fig. 14.6), the modified parts are re-compiled by using the compiler of the general purpose processor. On the other hand, if the modifications are carried out at the hardware part (the modification 2 in Fig. 14.6), the DFGs of the modified functions are extracted. If all of the operations in the DFGs can be executed by the initially generated instruction set, the instruction code of the modified part is generated via our code generator (see branch (A) in Fig. 14.6). However, if there are operations which cannot be executed by the initial instruction set, our instruction generator generates new instructions for such operations (see branch (B) in Fig. 14.6). As a result, all of the operations in the modified part are executed by the updated application specific instructions. Finally, our code generator generates the instruction code for the micro-controller.

There are a few important notes for the ECs of the hardware part.

- If the new DFGs are covered by the initially generated instruction set completely, the ECs can be carried out just by re-compiling the source code for the micro-controller. It is the same way as ECs of the software part.

- Newly generated instructions because of ECs may require functional units which have not been placed on the hardware part. Since we assume that all of the hardware components are already registered in the hard-macro library, we just place and route the functional units on the hardware part. Although it requires additional procedures for the realization of ECs, the turn-around-time is still marginal compared to conventional ASIC design flows.

- The additional instruction generation and code re-compilation because of ECs are carried out only once. It means that the instruction generation for ECs is carried out so that all of the operations which are not executed by initially generated instructions are executed by newly generated instructions.

Although there is a trade-off between design flexibility and design cost, FMA designs support a simple mechanism for ECs. This is one of the most important characteristics in the FMA design method.

# 4. Implementation of the Rijndael Encryption Algorithm

In this section, we present the implementation results of FMAs by using Rijndael encryption algorithm.

The Rijndael encryption algorithm[12] accepts a data block of plain text and transforms it into the blocked cipher text using the cipher key, where a variable block length and a variable key length are allowed (i.e., 128, 192, or 256 bit).

## Organization of the target FMA

To present the implementation examples of FMA design flow in this section, we define the target FMA as follows:

- P0, a microprocessor core from Pacific Design Inc.[13], is used as the micro-controller.

- The instruction set is generated under the constraint that the maximum number of basic nodes in each instruction is limited to 3.

- The parallelism of the specialized hardware is limited to 4.

- The instruction set generation and the code generation are carried out ignoring the delay time of each function unit since our current implementation cannot consider it. We only optimize the total number of clock cycles.

To realize above hardware architecture, we use VUPU architecture from Pacific Design Inc.[13]. VUPU is composed of a small processor (P0) and a

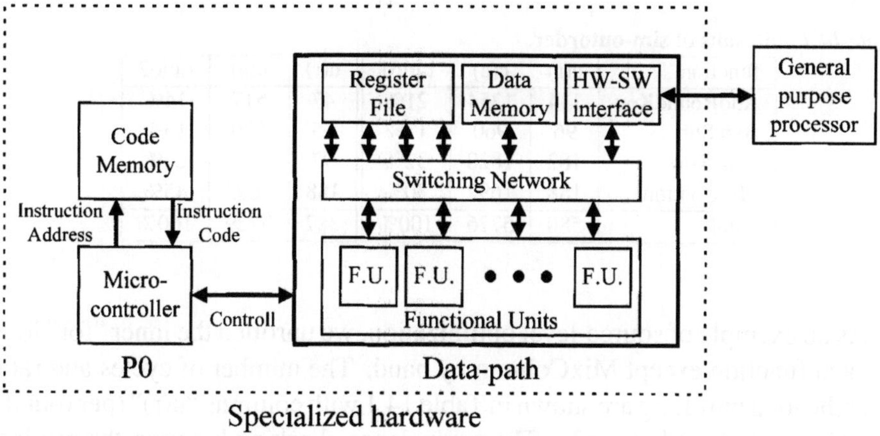


*Figure 14.7.* Specialized Hardware Architecture of the FMA design using P0.

user-defined co-processor, which is applicable for the specialized hardware of FMAs with small modification. Figure 14.7 shows the specialized hardware architecture of the FMA design. P0 controls the specialized hardware according to the VLIW-like instruction code. The data-path of the specialized hardware has a register file, a data memory, functional units, and a switching network which realizes the parallelism of the specialized hardware (up to 4 parallel instructions).

In the following subsections, we show the detailed design example at each design step of the FMA design method.

## Profiling Result

For profiling, we used SimpleScalar version 3.0[14] as the instruction set simulator. In particular, **sim-outorder**, one of the simulators of SimpleScalar, was used to simulate the Rijndael encryption algorithm. **sim-outorder** simulates the target instruction code compiled from the source C code with a pipelined out-of-order execution. As our target instruction set, PISA was used which is a default instruction set of SimpleScalar. Through profiling flow, we focused on four functions in the Rijndael encryption algorithm (AddRoundKey, ByteSub, ShiftRow, and MixColumn) because all of them had nested loop configurations with a set of logical and arithmetic expressions.

The number of cycles profiled by **sim-outorder** are shown in Table 14.1 with columns "c(r)" and "c(a)", where "c(r)" shows the number of cycles for each function per a round (round means one instance of loop body) and "c(a)" shows the total number of cycles for each function. Column "ratio1" represents the ratio for each function with respect to total cycles.

*Table 14.1.* Results of **sim-outorder**.

| function | c(r) | c(a) | ratio1 | u(r) | u(a) | ratio2 |
|---|---|---|---|---|---|---|
| AddRoundKey | 114 | 1254 | 21% | 47 | 517 | 14% |
| ByteSub | 96 | 960 | 17% | 55 | 550 | 15% |
| ShiftRow | 182 | 1820 | 32% | 97 | 970 | 26% |
| MixColumn | 188 | 1692 | 30% | 188 | 1692 | 45% |
| Total | 580 | 5726 | 100% | 387 | 3729 | 100% |

As an example of source-level optimization, we unrolled the inner "for" loop of each function except MixColumn by hand. The number of cycles and ratio after the loop unrolling are shown in Table 14.1 with columns "u(r)" (per round), "u(a)" (in total), and "ratio2". The equivalence checking between the original source code and the optimized one was also carried out by C-level functional simulation.

## Specialized Instruction Generation and Optimum Instruction Code Generation

In this experiment, we generated a set of specialized instructions for all of four functions. The optimum instruction code for each function was generated via our code generator. To show the effectiveness of FMA designs, we performed and compared two experiments, the first one was the code generation for an FMA with specialized instructions and the second one was the code generation for a VLIW (Very Long Instruction Word) architecture without using any specialized instruction. For the specialized instruction generation in the first experiment, up to sequentially ordered 3 DFG nodes can be configured as an instruction (one instruction takes one cycle for its execution). For the code generation of a micro-controller, our code generator allows parallel executions of up to 4 instructions if there is no dependency. On the other hand, we assume that there is no limitation for parallelism in the case of the VLIW architecture (i.e. there is no resource constraints and no restriction for parallelism.).

In the first experiment, 29 specialized instructions were generated based on the hard-macro library. Table 14.2 shows the results of the first experiment. The columns "sw" and "FMA" show the number of cycles in SimpleScalar and our FMA design, respectively. For the number in the column "FMA", a communication overhead because of hardware/software partitioning is slightly reflected. Similarly, the columns "swu" and "FMAu" means the number of cycles in SimpleScalar and our FMA design after the loop unrolling. From the result, we know that our FMA design is 1.86 times (for non-loop unrolling) faster than the case of SimpleScalar per round (1.85 times for loop unrolling).

Columns "vliw" and "vliwu" in Table 14.2 show the result of the second experiment. From this, we can see that our FMA design is 1.24 times (for

*Table 14.2.* Results of code generation.

| | non-loop unrolling | | | loop unrolling | | |
|---|---|---|---|---|---|---|
| function | sw | FMA | vliw | swu | FMAu | vliwu |
| AddRoundKey | 114 | 72 | 72 | 47 | 38 | 34 |
| ByteSub | 96 | 52 | 72 | 55 | 27 | 39 |
| ShiftRow | 182 | 98 | 134 | 97 | 51 | 47 |
| MixColumn | 188 | 90 | 110 | 188 | 93 | 117 |
| Total | 580 | 312 | 388 | 387 | 209 | 237 |

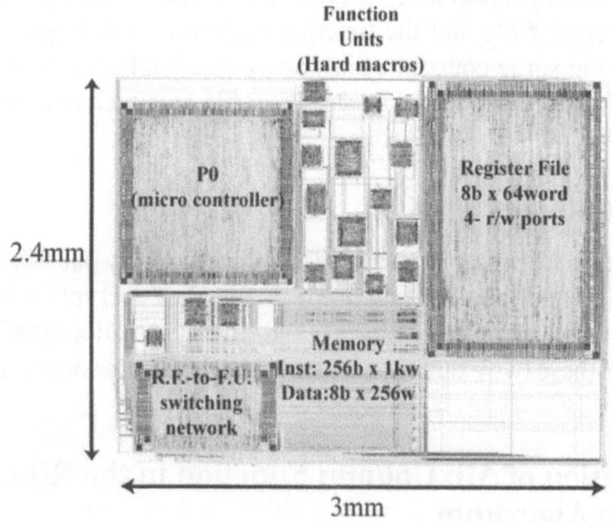


*Figure 14.8.* Layout result for specialized hardware.

non-loop unrolling) faster than the VLIW architecture (1.13 times for loop unrolling).

## Layout Result

To roughly estimate the hardware implementation results, we designed the above example into the physical layout by using commercial EDA tools. The net-list of the hardware part was generated considering the instruction set for Rijndael encryption algorithm. In addition, the size of memory was decided from the results of C-level functional simulation. We used Apollo[3] for the physical design. Figure 14.8 shows the layout result of the Rijndael encryption algorithm. In this experiment, we used 5-metal layer 0.18$\mu$m CMOS technology for the target technology and the clock frequencies of all hard-macros ( a

[3]Apollo is P&R tool from Synopsys Inc.

*Table 14.3.* Results of an EC in MixColumn.

| function | FMAu | | vliwu | |
|---|---|---|---|---|
| | before EC | after EC | before EC | after EC |
| MixColumn | 93 | 77 | 117 | 113 |

micro-controller, hard-macros, and so on) were constrained to 200 MHz. Since the parallelism of the specialized hardware is 4, the switching network circuit between the register file and the function units was very large. It will be resolved by setting some constraints on the connection between the register file and the function units. A partitioned register file architecture is also one of the solutions. They are our future work.

## 5. Case studies of ECs

In this section, we present two case studies how to carry out ECs effectively in the FMA design method. The first one uses Rijndael encryption algorithm and the second one uses an inverse discrete cosine transformation (IDCT) algorithm [15]. Through these two examples, we show the flexibility and the debuggability of FMAs.

### A Modification of MixColumn Function in the Rijndael Encryption Algorithm

First, we demonstrate an EC at the function MixColumn in the Rijndael encryption algorithm. Since MixColumn is one of the most computationally critical parts, we replaced shift operations of the original MixColumn function (*xtime* as in Fig. 14.9(a)) to table accesses (*xtimetbl* as in Fig. 14.9(b)) to improve performance. We modified the loop unrolled source code which was derived in the previous section. The functional equivalence between the original MixColumn function and the modified one was checked by C-level functional simulation.

As shown in the previous section, since MixColumn was implemented at specialized hardware, the target of this EC was a re-compilation of the modified MixColumn by using our code generator with the specialized instruction set. This process corresponds to modification 2 in Fig. 14.6. During the code generation, although the functionalities at C source level were changed, new DFG which had modified parts was covered by the initial specialized instruction set successfully. Therefore, the re-compiled code was re-mapped onto the local memory located at the specialized hardware and utilized by the micro-controller.

Table 14.3 shows the numbers of cycles of MixColumn before/after the EC process. According to this EC, the performance is improved 1.20 times faster

```
word8 xtime (word8 tm){
 if((tm>>7)==1){
  tm <<=1;
  return tm^0xb;
 }else{
  retrun tm<<1;
 }
}

void mixcolumn(word8 State[4][4]){
 word8 tm, tmp, s;
 int i;

 for(i = 0; i < 4; i++) {
  s = State[0][i];
  tmp = State[0][i]^State[1][i]
  ^State[2][i]^State[3][i];
  tm = State[0][i]^State[1][i];
  tm = xtime(tm);
  State[0][i] ^= tm^tmp;
        :
 }
}
```

(a)

```
word8 xtimetbl[] = {
 0x00, 0x02, 0x04 .....
        :
}

void mixcolumn(word8 State[4][4]){
 word8 tm, tmp, s;
 int i;

 for(i = 0; i < 4; i++) {
  s = State[0][i];
  tmp = State[0][i]^State[1][i]
  ^State[2][i]^State[3][i];
  tm = State[0][i]^State[1][i];
  tm^=xtimetbl[tm] ^ tmp;
  State[0][i] = tm;
        :
 }
}
```

(b)

*Figure 14.9.* An EC in MixColumn.

than the original (see the column "FMAu" in Table 14.3). The column "vliwu" shows the effect of this EC at the VLIW architecture.

## A Modification of Invert Discrete Cosine Transform

Another example of ECs is considered on an inverse discrete cosine transform (IDCT) used in MPEG encoder/decoder systems. In IDCT algorithm, for the transformation of 8 * 8 block element data (an element has 8 bits), function IdctRow is used for the transformations on the rows and function IdctColumn is used for the transformations on the columns respectively.

Initially, IdctRow was implemented at the specialized hardware of FMA. At the initial implementation, our instruction generator generated 23 specialized instructions. Note that one instruction could execute up to two parallel operations in the DFG of IdctRow in this experiment. The number of total cycles for IdctRow was 144 (18 cycles per function call * 8 rows).

To carry out the EC, we modified function IdctRow so that the total number of arithmetic operations is reduced. Figure 14.10 (a) and 14.10 (b) show the original C source code of function IdctRow and the modified one (bold characters represent the modifications). Unfortunately, the code re-compilation of the modified C description by using the initial 23 instructions was failed. Since it implies that in the DFG of the modified C description there are some operations

```
static void IdctRow(short b0, ...)
{
    int x0,x1,x2,x3,x4,x5,x6,x7,x8;

        :
        :

    x8 = W7 * (x4 + x5);
    x4 = x8 + (W1 - W7) * x4;
    x5 = x8 - (W1 + W7) * x5;
    x8 = W3 * (x6 + x7);
    x6 = x8 - (W3 - W5) * x6;
    x7 = x8 - (W3 + W5) * x7;

    x8 = x0 + x1;
    x0 -= x1;
    x1 = W6 * (x3 + x2);
    x2 = x1 - (W2 + W6) * x2;
    x3 = x1 + (W2 - W6) * x3;
    x1 = x4 + x6;
    x4 -= x6;
    x6 = x5 + x7;
    x5 -= x7;

        :
```

(a)

```
static void IdctRow(short b0, ...)
{
    int x0,x1,x2,x3,x4,x5,x6,x7,x8,tmp;

        :
        :

    tmp = x4;
    x4  = W7 * x5  + W1 * tmp;
    x5  = W7 * tmp - W1 * x5;
    tmp = x6;
    x6  = W3 * x7  + W5 * tmp;
    x7  = W3 + tmp - W5 * x7;

    x8  = x0 + x1;
    x0  -= x1;
    tmp = x2;
    x2  = W6 * x3  - W2 * tmp;
    x3  = W6 * tmp + W2 * x3;
    x1  = x4 + x6;
    x4  -= x6;
    x6  = x5 + x7;
    x5  -= x7;

        :
```

(b)

*Figure 14.10.* An EC in IdctRow.

which cannot be realized by the initial specialized instructions, the modification is realized only on a main processor as the software part. By using SimpleScalar simulation, we found out that it took 472 cycles (59 cycles per function call * 8 rows) if IdctRow was realized at software part. This means that the modification leads to considerable performance degradation.

To explore further flexibility and debuggability, we added primitive instructions (ex. add, mult, and so on) which are included in the original DFG to the initial instruction set. Actually, we added a primitive multiply instruction to the initial instruction set and we re-compiled the modified source code with the updated instruction set. As a result, the re-compilation was carried out successfully and the number of cycles was 88 (11 cycles per function call * 8 rows). This experiment indicates that the flexibility and the debuggability can be improved in FMAs if the initial instruction set has the optimized instructions and the primitive instructions which the target DFG has. Though this strategy increases the redundancy in the specialized hardware, the redundancy improve the flexibility and the debuggability of FMAs.

Even if these ECs fail, the code generator can indicate which parts are not covered by the initial instruction set. Then the instruction generator can generate extra instructions which can cover the above uncovered DFG nodes (see the flow (B) in Fig. 14.6). It can realize fast ECs by modifying the specialized hardware since the modification is very small in many cases.

From such results, we assume that FMAs provide high flexibility and debuggability though the flexibility and the debuggability extremely depend on the degree of the modification.

## 6. Conclusion and Future Work

In this paper, we have proposed a new architecture (FMA) and its design method which enable designers to modify and debug their chips even after chip fabrications. In FMAs, since not only software part with a main processor but also hardware part with a micro-controller and hard-macros are controlled by instruction codes, all of the ECs can be resolved at specification-level just by re-compiling a modified specification. For this purpose, we have developed an instruction generator and an optimal code generator for the micro-controller.

To evaluate the applicability of the method, the state-of-the-art encryption algorithm was analyzed and successfully implemented onto an FMA. In addition, to evaluate the flexibility of FMAs, we demonstrated two case studies of ECs by using realistic design examples. Because of the flexibility of FMAs, the EC was carried out successfully in Rijndael encryption algorithm example. In addition, we presented the simple strategy that the instruction set had all primitive instructions (though they are redundant in the initial instruction set) as well as the optimized instructions improved the flexibility of FMAs in IDCT example.

As part of ongoing and future work, we are focusing on the following extensions:

1 comprehensive experiments in terms of EC possibilities and refinement of FMA architecture based on them.

2 implementing the instruction set generation tool which can treat physical parameters such as delay time (both of function units and wires), power dissipation, and so on.

3 constructing a sophisticated tool which can fully support FMA design method.

Thus, we believe that our framework will help in developing flexible and debuggable design methodology for future embedded systems.

## Acknowledgements

This work is supported by VLSI Design and Education Center (VDEC), the University of Tokyo in collaboration with Hitachi Ltd., Synopsys, Inc., and Cadence Design Systems, Inc.

## References


[1] W. H. Wolf, "Hardware-Software Co-design of Embedded Systems," *Proc. Ieee*, vol. 82, pp. 967–989, Jul. 1992.

[2] A. Osterling *et al.*, "The COSYMA System, in Hardware/Software Co-Design: Principles and Practice," *Kluwer Academic Publishers*, 1997.

[3] Yoshihisa Kojima *et al.*, "Field Modifiable Architecture and Its Design Methodology — System Design Without Logic Synthesis —," *IEEE/ACM 11th International Workshop on Logic & Synthesis*, pp. 103–108, New Orleans, Jun. 2002.

[4] Kenshu Seto, *et al.*, "Field Modifiable Architecture and its Design Method," Student Research Forum (Poster Session), ACM SIGPLAN 2002 Conference on Programming Language Design and Implementation (PLDI), Berlin, Germany, 2002.

[5] Proceler web site: http://www.proceler.com/

[6] Tensilica web site: http://www.tensilica.com/

[7] V. Kathail *et al.*, "PICO: Automatically Designing Custom Computers," *IEEE Computer*, pp. 39–47, Vol. 35, No. 9, Sep. 2002.

[8] AccelChip web site: http://www.accelchip.com/

[9] H. Saito *et al.*, "An Equivalence Checking Methodology for Hardware Oriented C-based Specifications" *IEEE Int. Workshop on HLDVT*, pp. 139–144, Oct. 2002.

[10] ACE web site, http://www.ace.nl/

[11] K. Seto *et al.*, "Co-Design of Custom VLIW-DSP Type Data-path Architecture and its Parallel Program for Loops based on Formal Verification Technique," *SCOPES*, Mar. 2001.

[12] J. Daemen *et al.*, "AES Proposal: Rijndael, Document Version 2," Sep. 1999.

[13] Pacific Design Inc. web site, http://www.pdi.co.jp/

[14] Simple Scalar web site, http://www.simplescalar.com/

[15] Z. Wang, "Fast Algorithms for the Discrete W Transform and for the Discrete Fourier Transform", *IEEE Trans. on Acoustics, Speech, and Signal Processing* vol. ASSP-32, no. 4, pp.803–816, 1984.

MIX
Papier aus verantwortungsvollen Quellen
Paper from responsible sources
FSC® C105338

If you have any concerns about our products,
you can contact us on
**ProductSafety@springernature.com**

In case Publisher is established outside the EU,
the EU authorized representative is:
**Springer Nature Customer Service Center GmbH**
**Europaplatz 3, 69115 Heidelberg, Germany**

Printed by Libri Plureos GmbH
in Hamburg, Germany

